AF353008

Marine Nearshore Biodiversity

Marine Nearshore Biodiversity

Editor

Thomas J. Trott

Basel • Beijing • Wuhan • Barcelona • Belgrade • Novi Sad • Cluj • Manchester

Editor
Thomas J. Trott
Suffolk University
Boston, MA, USA

Editorial Office
MDPI
St. Alban-Anlage 66
4052 Basel, Switzerland

This is a reprint of articles from the Special Issue published online in the open access journal *Diversity* (ISSN 1424-2818) (available at: https://www.mdpi.com/journal/diversity/special_issues/nearshore_biodiversity).

For citation purposes, cite each article independently as indicated on the article page online and as indicated below:

Lastname, A.A.; Lastname, B.B. Article Title. *Journal Name* **Year**, *Volume Number*, Page Range.

ISBN 978-3-0365-8682-3 (Hbk)
ISBN 978-3-0365-8683-0 (PDF)
doi.org/10.3390/books978-3-0365-8683-0

Cover image courtesy of Thomas J. Trott

Contents

About the Editor

Thomas J. Trott

Thomas J. Trott received his Ph.D. from the Boston University Marine Program in 1989. He is currently Emeritus Professor of Biology at Suffolk University, Boston, where he taught for 19 years before assuming the position of benthic ecologist for the Maine Coastal Program. He is an associate editor of two peer-review journals and a reviewer for many others. He has served on various regional committees focused on coastal management and monitoring, has played key roles in developing and executing young citizen science programs, and presented at many national and international conferences and public venues. His early career interests focused on chemoreception by marine organisms using electrophysiology and behavior, followed by field studies aimed at determining small- and large-scale patterns of distribution among nearshore populations of macroinvertebrates. This work contributed to the Census of Marine Life through the support from the History of Marine Animal Populations, NaGISA, and Gulf of Maine Census Programs. He has been fortunate to work across a variety of habitats in different countries and benefit from rewarding collaborative research. He is particularly interested in assessing, documenting, and unravelling the influences of coastal processes on marine nearshore biodiversity, and how changes in coastal ecosystems evolve to affect the quality of life for people living in coastal communities.

Preface

Nearshore ecosystems contain most of the ocean's highly productive waters and varied habitats that support a range of phyla more diverse than that which terrestrial ecosystems hold. This band of water extends seaward from the intertidal zone out through the subtidal and envelopes most of the marine biodiversity hotspots. Among the marine ecosystems, nearshore biodiversity has the deepest history of exploration, exploitation, and benefits to society. Yet, due to societal influence on ocean conditions, the populations of the nearshore benthos have been reduced, restructured, and replaced. These consequences are revealed by altered patterns in the abundance, genetic diversity, range of species, and ecosystem functions. The rapid advances in our knowledge and methodologies in recent years show the pervasiveness of these trends. Nevertheless, understanding the patterns and processes that lie at the heart of the problem is frustrated in many instances by a scarcity of baseline information as simple as knowing what species occupy different nearshore habitats.

This book highlights recent advances in research concerning marine nearshore biodiversity. In summary, this collection aims to present, in a broad sense, a global comparison of nearshore biodiversity and the drivers of change. Continued studies of marine nearshore biodiversity will benefit from the wide range of topics and methods presented. Covering many aspects of marine nearshore biodiversity, from habitat monitoring and biogeography through to species discovery and extinction, as well as delivering tools, techniques and backgrounds for practical applications, this book provides a resource for many different researchers, practitioners and contexts.

I kindly thank all the authors for their participation and for the contribution of their valuable work to this book, in addition to all the reviewers whose efforts have helped to improve the quality of this publication. Finally, I express my sincere appreciation for the excellent support from the editorial staff of MDPI and the journal *Diversity*, which was essential to the production of this volume.

May our collective efforts advance the understanding of marine nearshore biodiversity and inspire research that outpaces the rapid onslaught of global change.

Thomas J. Trott
Editor

Editorial

Marine Nearshore Biodiversity: Introduction to the Special Issue

Thomas J. Trott [1,2]

1   Biology Department, Suffolk University, Boston, MA 02108, USA; ttrott@suffolk.edu
2   Maine Coastal Program, Department of Marine Resources, West Boothbay Harbor, ME 04575, USA

Citation: Trott, T.J. Marine Nearshore Biodiversity: Introduction to the Special Issue. *Diversity* **2023**, *15*, 838. https://doi.org/10.3390/d15070838

Received: 29 June 2023
Accepted: 30 June 2023
Published: 7 July 2023

Millions are nourished, economies are fueled, and culture is inspired—these are just a few of the extraordinary benefits stemming from the coastal waters adjoining the shorelines of the world. These nearshore waters serve as the first step to understanding the marine world, a springboard for scientific curiosity, and a gateway to less accessible oceanic spaces. The importance of the nearshore is clear by positing its benefits to society. What economies could possibly be sustained without the services provided by the nearshore? And yet with its deep history of importance and exploration, the nearshore continues to yield new insights to understanding global patterns in the distributions of algae and animals and the influence of societal interactions with them.

This Special Issue assembles pioneering research on nearshore ecosystems distributed among all world oceans, from waters spanning the Artic and Antarctic (Figure 1). These works are organized according to the major themes of biodiversity, biogeography, and species distribution, bridged by the subjects of species discovery, species loss, and habitat change. Imbedded topics relevant to the present day direct and indirect societal impacts on marine nearshore biodiversity include fishery management, coral reef biodiversity, and extinction. The use of innovative molecular tools to address taxonomic questions and biodiversity distinguishes this collection from any published before the relatively recent development of the field of metagenomics. Likewise, the advantages of global databases founded within the last 15 years, like the citizen science platform iNaturalist, are demonstrated with their application in reporting species richness in difficult environments that challenge observations. In summary, this Special Issue presents an informed overview of current and pressing topics in marine nearshore biodiversity.

Of great concern is the loss of nearshore biodiversity, and bottom-trawl fisheries using destructive catch methods have indiscriminate effects on non-target species and benthic habitats [1]. The effectiveness of fishery management in minimizing effects is examined by Fondo et al. [2] for a shallow-water bottom-trawl prawn fishery in Kenya. Nine years of catch data and four years of catch composition data following the enactment of regulations indicate their effectiveness in restraining declines in the status of the stock and integrity of the bays examined based on diversity and trophic indices. The authors highlight the benefits of technologies which reduce effects on non-target species and recommend more by-catch be retained and its economic value maximized in local markets and elsewhere.

Conceptually, marine protected areas reduce the biodiversity loss resulting from resource harvesting, whether they be commercially valuable species, minerals, or hydrocarbons [3]. Baselines of species incidence and richness are key for evaluating the effectiveness of conservation protections, and Ginsburg and Huang [4] provide an updated one for Santa Catalina Island, California. Their survey illustrates the high biodiversity of the region and identifies a number of species that are either introduced or are range shifters, and others that are vulnerable and endangered species deserving protection.

Globally abundant in nearshore coastal areas, ubiquitous, and ecologically diverse, *Roseobacter*, a marine bacterium, plays important roles in biochemical cycles and climate change [5]. Using a highly conserved gene transfer agent, the g5 gene, Zeng et al. [6]

extend the current knowledge of the biogeography of roseobacters in polar marine waters. Interestingly, bipolar distributions exist and with others endemic to either the Antarctic or Arctic. Since GTA-related gene transfer is widely considered a mechanism for maintaining metabolic flexibility in changing conditions, these discoveries may relate to the adaptation of *Roseobacter* g5 clades to polar environments.

The importance of local oceanographic features in shaping marine nearshore biodiversity cannot be overstated. In the Gulf of Maine, Trott [7] shows that the similarity of rocky intertidal species assemblages is correlated with latitude and is distinguishable into two groups that correspond with the two principal branches of the Gulf of Maine Coastal Current. Thermogeography of the nearshore is largely influenced by these hydrographic features, and the dissimilarity of the two Gulf regions is significantly related to temperature. Consequences of the rapid warming of the Gulf of Maine [8] on rocky intertidal community patterns are forecast as species range shifts and non-native species introductions disrupt assemblage composition and community dynamics.

Marine nearshore biodiversity can be difficult to estimate, particularly in subtidal habitats located in environments that challenge costs for sampling associated with accessibility, time, and expense. Adapting the rover diver method for non-destructive sampling of benthic taxa, Bravo et al. [9] successfully demonstrate the effectiveness of this sampling procedure when paired with photography in kelp forests at the sub-Antarctic Bécasses Island, located in the Beagle Channel, Argentina. Their innovative use of the citizen science platform iNaturalist to archive photographs, thus creating records of species occurrence, permits transparency in taxonomic curation and facilitates data sharing. Long-term monitoring of subtidal benthos like that associated with kelp forests, globally threatened by climate change [10], can use this approach for cost-effective surveys and reporting.

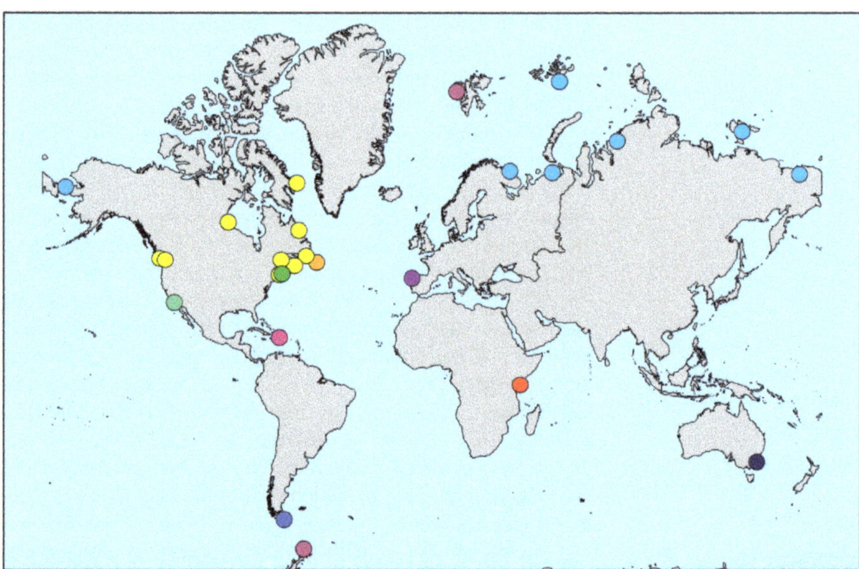

Figure 1. Nearshore study areas investigated by authors published in this Special Issue. Symbol and reference: ● [2], ● [4], ● [6], ● [7], ● [9], ● [11], ● [12], ● [13], ● [14], ● [15], ● [16].

Regional-scale management strategies to mitigate the degradation of nearshore habitats and diversity rely on similarly scaled observations and not ones from only one or a few locations. Steneck and Torres [11] present differences in trends among Caribbean coral reefs monitored for health in six regions within three sectors of the Dominican Republic coastline for 7 years. Country-wide declines in coral cover and reef fish are shown, most steep for reefs once among the Caribbean's best. However, the degree of negative trends is not the same among all sectors, a result that can steer management and continued monitoring. The abundant and increasing macroalgal cover that seriously interferes with reef recovery from disturbances could be mitigated by beneficial gains from improved fishery management.

Knowledge gaps from understudied habitats and taxa compromise the assessment of biodiversity, the detection of change, and extinction in extreme cases. Worldwide, marine flatworms, i.e., polyclads, present a prime example of this situation, where the dearth of basic ecological knowledge for this taxonomically challenging group, like habitat preferences, seriously hinders an accurate evaluation of species occurrence and richness. Tosetto et al. [12] tackle this problem in surveys of intertidal boulder beaches in southeastern Australia and report distribution patterns related to beach exposure, boulder size, and latitude. Their work constitutes one of the few studies of this kind for marine polyclads and will stimulate more investigations of these understudied predators.

There are few pan-Arctic studies focused on nearshore biodiversity. These are of particular need considering the rapid environmental alterations to polar seas resulting from climate change. With so few works to date, the scarce knowledge of this region has fueled some disagreement about nearshore community structure. Denisenko and Denisenko [13] settle a long-standing debate about the degree that bryozoans contribute to benthic biomass in coastal regions of the Arctic by evaluating samples spanning 43 years of collection throughout the Eurasian seas prior to the onset of rapid warming. They reveal biogeographic patterns in the distribution of dominant, key-biomass species related to oceanography and bottom type. The intensive coastal erosion of permafrost and consequential increase in turbidity in some regions may influence colony growth in shallow depths by interfering with suspension feeding.

Many monitoring programs aim to assess nearshore biodiversity and changes due to societal impacts, but surveys can be costly, and the taxonomic identification of retrieved organisms is time intensive. Since some habitats like hard bottom communities are difficult to sample, this problem is approached by deploying artificial substrates and monitoring their colonization. Using a cost-effective and innovative molecular approach, Leite et al. [14] compare hard-bottom macrozoobenthic species colonization of different standardized structures. They report that shape and structural complexity strongly affect colonization, with some taxa exclusively colonizing more dimensionally rich simulated seaweed. Monitoring programs using artificial structures can better assess biodiversity when habitat complexity is modelled by more than one kind of artificial substrate at a time.

Marine organisms with life histories characterized by alternation between generations with stages that are strikingly different in appearance pose challenges to ecologist and taxonomists alike, particularly when only one form is known, or each stage has been described as a different species. These situations can lead to a mismeasure of biodiversity and misrepresent biogeography. Focusing on Pacific and Atlantic Canada shores, Saunders and Brodie [15] use taxon-targeted metabarcoding to explore these domains for red algae in the order Bangiales, for which only the cryptic sporophyte (Conchocelis) stage is known. Their work extends the vertical (depth), host, and biogeographical ranges of an asexual Conchocelis-only species and uncovers known and possibly new species among their samples. Taxon-targeted metabarcoding is forecast to bring significant gains in understanding bangialean ecology and reveal its dark contribution to nearshore biodiversity.

Habitat-forming species, ecosystem engineers, enhance species colonization and increase biodiversity. Kelps, corals, and mussels are a few examples. Rhodoliths, free-living nodules of coralline red algae, can aggregate under favorable conditions to form rhodolith beds, dimensionally complex benthic habitats supporting highly diverse communities in otherwise somewhat featureless bottoms. In a sub-Arctic rhodolith bed, Bélanger and Gagnon [16] track the variability in structural complexity and macrofaunal diversity for nearly a year. In addition to relating macrofaunal diversity to rhodolith complexity, the unprecedented fine taxonomic resolution of their study supports the notion that rhodolith beds are biodiversity hotspots. Changes in macrofauna abundance are due to seasonality, but a disturbance from sporadic intensive physical forcing from storms, for example, can rework beds. The predicted intensification of wind and wave storms may pose challenges to the resilience of these biodiverse habitats.

Global biodiversity is facing an extinction crisis, the Sixth Mass Extinction [17]. But if comparisons of numbers of publications on topics make sound evaluations of importance, the wealth of papers devoted to estimating the number of species in the world oceans published in the past few decades (for examples [18,19]) assigns the topic of marine extinctions second place at best. This observation defies a common sense of importance and urgency for knowledge. Both topics confront the difficulty of observing (or not) organisms that are "hidden" beneath the ocean waves, so what makes the study of marine extinctions so different? Carlton [20] answers the hidden complexity of this otherwise simple question by providing reasons for the resistance to declare marine invertebrate species extinct. His call for inventories of globally missing marine invertebrates provides practical guidelines to sway the current state of affairs largely driven by global authority definitions of extinction and absence of evidence.

Investigations of marine nearshore biodiversity are needed now more than ever, the situation fueled by the predicted changes in ocean climates driven by societal impacts. The collection of papers in this Special Issue address many of the most vital topics related to this invitation. They provide a source of inspiration for further research to help understand and guide decisions about global changes in marine nearshore biodiversity.

Conflicts of Interest: The author declares no conflict of interest.

References

1. Watling, L. The global destruction of bottom habitats by mobile fishing gear. In *Marine Conservation Biology: The Science of Maintaining the Sea's Biodiversity*; Elliot, N.E.A., Crowder, L.B., Eds.; Island Press: Washington, DC, USA, 2005; pp. 198–210.
2. Fondo, E.N.; Omukoto, J.O.; Wambiji, N.; Okemwa, G.M.; Thoya, P.; Maina, G.M.; Kimani, E.N. Diversity of shallow-water species in prawn trawling: A case study of Malindi–Ungwana Bay, Kenya. *Diversity* **2022**, *14*, 199. [CrossRef]
3. Halpern, B.S. Making marine protected areas work. *Nature* **2014**, *506*, 167–168. [CrossRef] [PubMed]
4. Ginsburg, D.; Huang, A. Over, under, sideways and down: Patterns of marine species richness in nearshore habitats off Santa Catalina Island, California. *Diversity* **2022**, *14*, 366. [CrossRef]
5. Luo, H.; Moran, M.A. Evolutionary ecology of the marine *Roseobacter* clade. *Microbiol. Mol. Biol. Rev.* **2014**, *78*, 573–587. [CrossRef] [PubMed]
6. Zeng, Y.; Li, H.; Luo, W. Gene transfer agent g5 gene reveals bipolar and endemic distribution of *Roseobacter* clade members in polar coastal seawater. *Diversity* **2022**, *14*, 392. [CrossRef]
7. Trott, T. Mesoscale spatial patterns of Gulf of Maine rocky intertidal communities. *Diversity* **2022**, *14*, 557. [CrossRef]
8. Saba, V.S.; Griffies, S.M.; Anderson, W.G.; Winton, M.; Alexander, M.A.; Delworth, T.L.; Hare, J.A.; Harrison, M.J.; Riosati, A.; Vecchi, G.A.; et al. Enhanced warming of the Northwest Atlantic Ocean under climate change. *J. Geophys. Res. Oceans* **2016**, *121*, 118–132. [CrossRef]
9. Bravo, G.; Kaminsky, J.; Bagur, M.; Alonso, C.; Rodríguez, M.; Fraysse, C.; Lovrich, G.; Bigatti, G. Roving diver survey as a rapid and cost-effective methodology to register species richness in sub-Antarctic kelp forests. *Diversity* **2023**, *15*, 354. [CrossRef]
10. Wernberg, T.; Krumhansl, K.; Filbee-Dexter, K.; Pedersen, M.F. Status and Trends for the World's Kelp Forests. In *World Seas: An Environmental Evaluation*, 2nd ed.; Sheppard, C., Ed.; Academic Press: Cambridge, MA, USA, 2019. [CrossRef]
11. Steneck, R.; Torres, R. Trends in Dominican Republic coral reef biodiversity. *Diversity* **2023**, *15*, 389. [CrossRef]
12. Tosetto, L.; McNab, J.; Hutchings, P.; Rodríguez, J.; Williamson, J. Fantastic flatworms and where to find them: Insights into intertidal polyclad flatworm distribution in southeastern Australian boulder beaches. *Diversity* **2023**, *15*, 393. [CrossRef]
13. Denisenko, N.; Denisenko, S. Large-scale variation in diversity of biomass-dominating key bryozoan species in the seas of the Eurasian sector of the Arctic. *Diversity* **2023**, *15*, 604. [CrossRef]
14. Leite, B.; Duarte, S.; Troncoso, J.; Costa, F. Artificial seaweed substrates complement ARMS in DNA metabarcoding-based monitoring of temperate coastal macrozoobenthos. *Diversity* **2023**, *15*, 657. [CrossRef]
15. Saunders, G.; Brooks, C. Metabarcoding extends the distribution of *Porphyra corallicola* (Bangiales) into the Arctic while revealing novel species and patterns for Conchocelis stages in the Canadian flora. *Diversity* **2023**, *15*, 677. [CrossRef]
16. Bélanger, D.; Gagnon, P. Spatiotemporal variability in subarctic *Lithothamnion glaciale* rhodolith bed structural complexity and macrofaunal diversity. *Diversity* **2023**, *15*, 774. [CrossRef]
17. Cowie, R.H.; Bouchet, P.; Fontaine, B. The sixth extinction: Fact, fiction, or speculation? *Biol. Rev.* **2022**, *97*, 640–663. [CrossRef] [PubMed]
18. Costello, M.J.; Wilson, S.; Houlding, B. Predicting total global species richness using rates of species description and estimates of taxonomic effort. *Syst. Biol.* **2012**, *61*, 871–883. [CrossRef] [PubMed]

19. Mora, C.; Tittensor, D.P.; Adl, S.; Simpson, A.G.B.; Worm, B. How many species are there on earth and in the ocean? *PLoS Biol.* **2011**, *9*, e1001127. [CrossRef] [PubMed]
20. Carlton, J. Marine invertebrate neoextinctions: An update and call for inventories of globally missing species. *Diversity* **2023**, *15*, 782. [CrossRef]

 *diversity*

Article

Diversity of Shallow-Water Species in Prawn Trawling: A Case Study of Malindi–Ungwana Bay, Kenya

Esther N. Fondo [1,*], Johnstone O. Omukoto [1,2], Nina Wambiji [1], Gladys M. Okemwa [1], Pascal Thoya [1], George W. Maina [3] and Edward N. Kimani [1,*]

[1]   Kenya Marine and Fisheries Research Institute, Mombasa P.O. Box 81651-80100, Kenya; jomukoto@gmail.com (J.O.O.); nwambiji@gmail.com (N.W.); gladysokemwa@gmail.com (G.M.O.); pascalthoya@gmail.com (P.T.)
[2]   Lancaster Environment Centre, Lancaster University, Lancaster LA1 4YQ, UK
[3]   The Nature Conservancy, Africa Regional Office, Nairobi P.O. Box 19738-00100, Kenya; gwmaina@tnc.org
*   Correspondence: efondo@yahoo.com (E.N.F.); edwardndirui@yahoo.com (E.N.K.)

Abstract: Bottom trawling is a common fishing method that targets bottom-dwelling fisheries resources. It is non-selective and large amounts of by-catch are discarded, raising serious sustainability and ecosystem conservation concerns. In this study, a shallow-water bottom-trawl fishery was evaluated using logbook catch data between 2011 and 2019 and the species composition data collected by fisheries observers between 2016 and 2019. The logbook data showed a twenty-fold increase in the annual catches with a ten-fold increase in fishing effort and an increase in the proportion of retained catch from 2011 to 2019. The observer data showed that for prawn, the by-catch ratio ranged from 1:3 to 1:9 during the four years. Multivariate analysis revealed significant differences between the compositions of retained and discarded catches mainly attributed to *Pellona ditchela*, *Nematopalaemon tenuipes*, and *Secutor insidiator*. There was no significant decline in species diversity and the trophic level of the catches over the 4-year observer period indicating no marked impact of trawling on the stock at the current level of fishing effort. This study provides baseline information on the prawn trawl fishery against which the performance of the management regulations may be evaluated towards the Ecosystem Approach to Fisheries management.

Keywords: shell-fish; fish; by-catch; discards; species composition

Citation: Fondo, E.N.; Omukoto, J.O.; Wambiji, N.; Okemwa, G.M.; Thoya, P.; Maina, G.W.; Kimani, E.N. Diversity of Shallow-Water Species in Prawn Trawling: A Case Study of Malindi–Ungwana Bay, Kenya. *Diversity* **2022**, *14*, 199. https://doi.org/10.3390/d14030199

Academic Editors: Thomas J. Trott and Bert W. Hoeksema

Received: 4 February 2022
Accepted: 6 March 2022
Published: 9 March 2022

Publisher's Note: MDPI stays neutral with regard to jurisdictional claims in published maps and institutional affiliations.

1. Introduction

Bottom trawling and dredging contribute a significant part to capturing finfish and shellfish worldwide. Global evaluation of the contribution of bottom trawling and dredging to capture fisheries indicate approximately 28% [1], while long-term FAO data show that trawling has contributed about 25% of capture fisheries between 1950 and 2018 [2]. However, evaluation shows that bottom trawling generates the most waste in fisheries, accounting for nearly 60% of the fish dumped back into the ocean [1]. The sustainability and conservation concerns of bottom trawl fisheries have attracted attention in the past [1,3–5]. Trawling is considered a wasteful and destructive fishing method associated with large amounts of discarded by-catch leading to changes in trophic structure and loss of fishery resources [6,7].

Changes in the trophic structure and function of benthic communities have important implications on primary production and the wider functioning of the marine ecosystem [8,9]. Ecological studies on bottom trawling have focused on ecosystem impacts through widespread physical disturbance of the bottom substrate, excessive removal of juveniles, and the potential of modifying ecosystems' trophic structure [10,11]. There are claims that any trawl fishery is unsustainable due to its environmental and ecosystem impacts [12,13], and there have been suggestions for bans on some types of trawling [3,14,15]. However, bottom trawling continues to be one of the most common fishing methods and contribute a significant part of demersal fish and shallow and deep-water crustaceans in many parts of the world's oceans [1].

Bottom trawling within the Western Indian Ocean contributes significantly to industrial shallow-water prawn fisheries in South Africa, Mozambique, Tanzania, Kenya, and Madagascar. The few recent reports on bottom trawling in the region indicate sustainability and conservation challenges. Prawn trawling around Bagamoyo/Sadani and the Rufiji Delta in southern Tanzania, landed between 400 and 1000 tons by 16–26 vessels annually [16–18]. In Mozambique, trawling at the Sofala Bank region yielded 6000 to 9000 tons annually, landed by between 50 and 90 vessels between 1980 and 2014 [19]. There was a marked decline in the catches to 1800 tons after 2012 [20]. A relatively small prawn trawl fishery operates in the Thukela Bank in South Africa, landing a total of 350 tons [21,22]. Prawn trawling in the northern and west coast of Madagascar undertaken by a maximum of 77 vessels in 1996 declined to 53 in 2007, whereas landings varied between 2600 and 4000 tons annually [23]. In Kenya, between 5 and 20 trawlers operated annually within the Malindi–Ungwana Bay, landing between 334 and 640 tons of shallow-water prawns annually during the last few decades [21,23–25]. These fisheries have continued to contribute to coastal economic activities, but with little scientific information on the ecosystem impacts to support their sustainable management.

Conflicts between prawn trawlers and small-scale fishers, as well as environmental concerns in the Malindi–Ungwana Bay (Kenya), resulted in the suspension of the trawl fishery in 2006 by the government pending the development of a management plan to address the social as well as the environmental issues in the fishery [26]. The management plan for the fishery was developed through extensive stakeholder consultations, and regulations for the fishery were enacted in 2010 [27]. The key regulations prescribed in the management plan included restricting the number of vessels to a maximum of four, the mandatory use of turtle excluder devices, regulation on mesh sizes, zoning of fishing area, seasonal bans, restricting trawling time, and submission of a business plan as part of the application for a trawl fishing license. To address environmental concerns, the plan required details for full use of by-catch to be part of the business plan. The plan also introduced a fisheries observer program to collect scientific data and information to evaluate the status of the fishery to support reviews of the regulations in the plan. The fisheries observer program began by deploying observers on four trawlers in 2016 providing an opportunity to evaluate the impacts of the fishery on the ecosystem.

In this study, the catch and effort data (2011–2019) from the fishery and the retained and discarded by-catch data between 2016 and 2019 reported by observers, was evaluated to determine changes in the catch and species composition over time, and the impacts of fishing on the trophic structure of the fish stocks. The results provide information to support the management and planning of fishery to guide sustainable use of the resource.

2. Materials and Methods

2.1. Study Area

The industrial prawn trawl fisheries operate within the Malindi–Ungwana Bay between latitudes 3°30′ S and 2°30′ S and longitudes 40°00′ N and 41°00′ N, covering the Malindi and Ungwana Bay Complex (Figure 1). The bay is shallow with a wide continental shelf, extending between 15 and 60 km offshore. It is one of the areas suitable for trawling along the Kenyan coast due to the wide continental shelf and absence of coral reefs [28,29]. The benthic habitats are muddy and sandy, some with seagrasses and seaweeds and some rocky areas. The mean depth at high spring tide is 12 m at 1.5 nm and 18.0 m at 6.0 nm. The depth increases rapidly to 100 m after 7 nm and generally decreases northwards. The sub-stratum of the whole of Malindi–Ungwana Bay is mainly composed of siliciclastic sediments [30]. The area has one of the most productive marine fisheries in Kenya as a result of the mangroves forests surrounding the bay, topography of the continental shelf in the bay, and the runoff from the two rivers Sabaki and Tana that drain from a large part of the central and eastern regions of Kenya [31,32] (Figure 1).

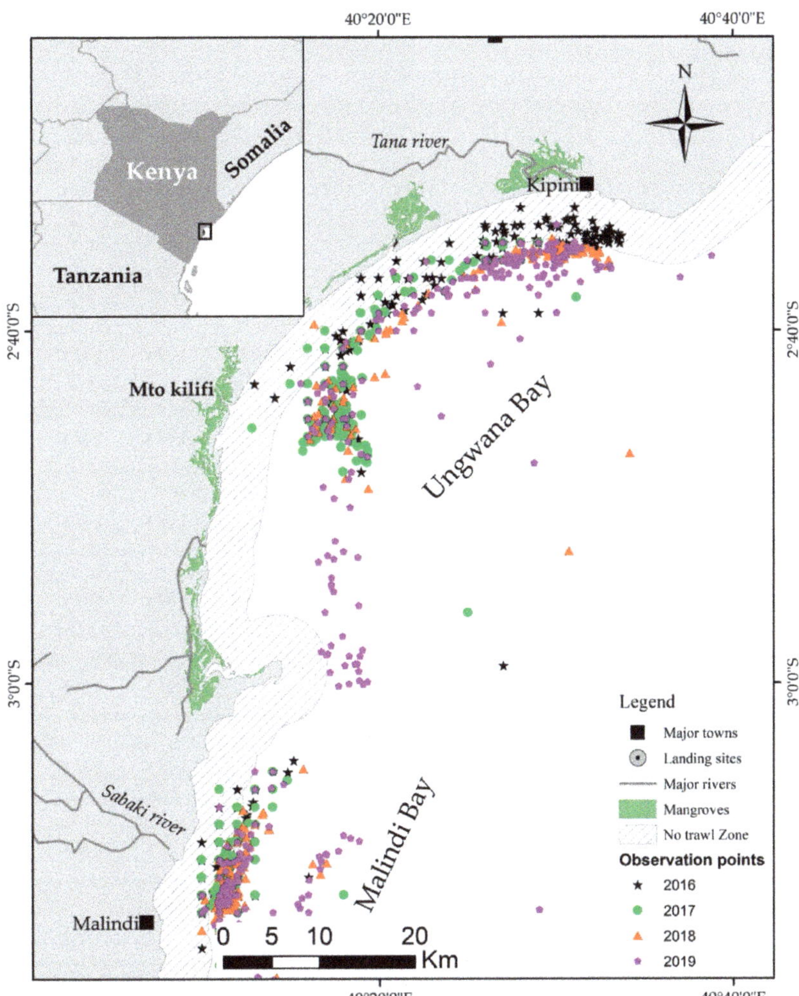

Figure 1. Map of Kenya (inset) and the Kenyan coastline showing the location of Malindi and Ungwana Bays, the Sabaki and Tana rivers, and the trawling observations in Malindi–Ungwana Bay from 2016–2019.

The bay is influenced by two dominant offshore current regimes: the Northeast monsoon (NEM) and the Southeast monsoon (SEM). During the SEM, which occurs between April and October, the current circulation is dominated by the northward flow of the East African Coastal Current. During this season, the bay also receives the heaviest river discharge from the Tana and Sabaki Rivers [32]. During the NEM, between November and March, the northward-flowing East African Coastal Current meets the southward-flowing Somali Current to form the Equatorial Counter Current, which flows east into the Indian Ocean [33].

2.2. Data Collection

Fisheries and observer data collected on four industrial trawlers licensed to fish in the Malindi–Ungwana Bay during the prawn fishing season between 1 April and 31 October every year from 2016–2019 were used in this study. The trawlers are all Kenyan flagged, with one smaller trawler being 22 m long, 9 m wide, and with a gross registered tonnage of 117. The three others were 25 m long, 9.3 m wide, and with a gross registered tonnage of 140. All the trawlers had an engine capacity of 300 HP and were fitted with double rigged trawl nets with a mesh size of 55–60 mm and 40–45 mm at the funnel and cod end, respectively. The trawl type was steel beam, with square doors opening 6–15 m, footrope length of 38–200 m, with a net mouth vertical opening of 3–6 m and horizontal opening of 6–20 m, and bottom-line armor of chain. During the shallow-water trawl fishing operations, trawling duration ranged between 2 to 3 h. The captain used GPS and a fish finder installed in the vessel to locate the fishing ground. The observer was positioned on the upper deck of the vessel where he could observe and record the catch and discards following the observer protocol. The catch from each haul was emptied onto a steel sorting table on the deck, large live animals mainly sharks, rays, and turtles were quickly returned to the sea to optimize their chances of survival. These large animals were recorded by the observer. The prawns were collected, graded, cleaned, treated, and packed into 2-kg cartons and blast frozen. The fish were sorted into species to be retained and discarded. The retained fish were put in 25-kg plastic crates, cleaned using pressurized seawater, packed into labeled 5-kg cartons, and blast frozen. As the retained fish were separated from the catch, the remaining unwanted catch was discarded into the sea. The catch that was discarded included small-sized and low-value species. The vessel captains kept a record of the fishing operation and the catch for each haul and sent a weekly report to the Kenya Fisheries Service (KeFS).

Logbook catch data were used to evaluate the catches and fishing effort, between 2011 and 2019. The data consisted of details on each fishing event, including the start and end times and the GPS positions of each haul, catch weights for prawns, finfish, octopus, squids, lobsters, crabs, and others. The catch composition data were collected by scientific observers from the Kenya Marine and Fisheries Research Institute (KMFRI), following sampling protocols adopted from Athayde [34]. At the beginning of each observer trip, the vessel and fishing gear information was recorded. For each fishing event, the start and end fishing positions and times were recorded. Large individuals, including fish, sharks, and rays were first removed from the catch, identified, and recorded. The prawns were sorted from the catch and weighed (kg) on a top-loading balance. The remaining catch was then separated into retained and discards. The retained catch and discards for each haul were weighed and sub-samples for identification were collected. The individuals in the samples were separated into species following standard species identification guides for the region [35–38], counted, and weighed. The catch composition was obtained by multiplying the sample data with the raising factor (i.e., number of portions of which main catch was divided), using the catch composition estimation method. The total catch weight was obtained, by adding the total weight of non-random samples (large fish put aside) and the scaled-up weights obtained from the samples. The data were recorded in a standard data sheet developed for the observer program.

On disembarking from a completed observer trip, an observer coordinator verified the data during a debriefing session with the observer to correct any mistakes before the data were entered into a spreadsheet. The data were cleaned by making sure that all names of species were correct, and the dates, times, GPS positions, and weights were correctly entered in standard units.

2.3. Coverage of Fishery Observers

Thirty-seven observers were deployed between 2016 and 2019 and recorded a total of 1371 hauls. Observations were undertaken in all the months of the shallow-water prawn fishing season in 2017, while a few months of the fishing season were not observed in 2016, 2018, and 2019 (Table 1). Overall, between 11% and 19% of the fishery was observed every year. The trawling observations taken from 2016 to 2019 are shown in Figure 1.

Table 1. Number of observers, deployments, and hauls during the study period.

Year	Months Observed	Number of Observers	Number of Deployments	Total Number of Trawls	Units of Trawls Recorded	% Observed Trawls
2016	Jun, July, Aug, Sept, Oct	7	9	1843	318	17.3
2017	Apr, May, Jun, Jul, Aug, Sep, Oct	6	10	1963	376	19.2
2018	May, Jun, Jul, Aug, Oct	6	7	2400	281	11.7
2019	Aug, May, Jun, Jul, Aug, Sep, Oct	9	11	2325	396	17.0

2.4. Data Analysis

The variation in the nominal total catch, discarded, and retained catch was evaluated using time-series graphs. Nominal catches were used to allow comparisons with previous studies, which also used nominal catches. The species composition of the retained and discarded catches was described using two metrics: species diversity (Shannon) and mean trophic level. The trophic level for each species was obtained from FISHBASE [39]. The mean trophic level was calculated as:

$$\mathrm{TL_L} = \sum_{i=1}^{n} Y_i \cdot \mathrm{TL}_i / Y_\mathrm{L} \tag{1}$$

where Y_i is the catch of species i, TL is the trophic level of species i.

The Mann–Whitney U-test was then used to compare the differences in species diversity and the mean trophic level of the total catch in the two bays for the four years (2016–2019).

The species composition of the catches was investigated using nonmetric multidimensional scaling (nMDS) ordination on standardized and square root transformed data to compare differences between the months, years, and retained vs. discarded species. A hierarchical group-average clustering based on a Bray–Curtis similarity matrix was overlaid to elucidate similarities between seasons, depth, and sites [40]. The relative distance of the data points provided a measure of similarity. A posterior analysis of similarity (ANOSIM) test was applied to check for significant differences in the species composition between years, retained and discarded species, followed by a similarity of percentage (SIMPER) analysis, which identified species that contributed most to dissimilarities between the years, retained and discarded species. The statistical analyses were conducted using STATISTICA (StatSoft, Inc., Tulsa, OK, USA) and PRIMER [41].

3. Results

3.1. Catch and Effort

The annual trends in fishing effort and total catches obtained from the fishery data between 2011 and 2019 showed increasing catches with an increase in fishing effort over time (Figure 2). The fishing effort increased tenfold from 437 h in 2011 to 5102 h in 2019, with the steepest increase from 2013 to 2016. The by-catch increased more than 20 times within the nine years, from 20 tons in 2011 to 450 tons in 2019. Prawn catches also increased more than 20 times with increasing effort, but the increase was gradual, from 6 tons in 2011 to 133 tons in 2019.

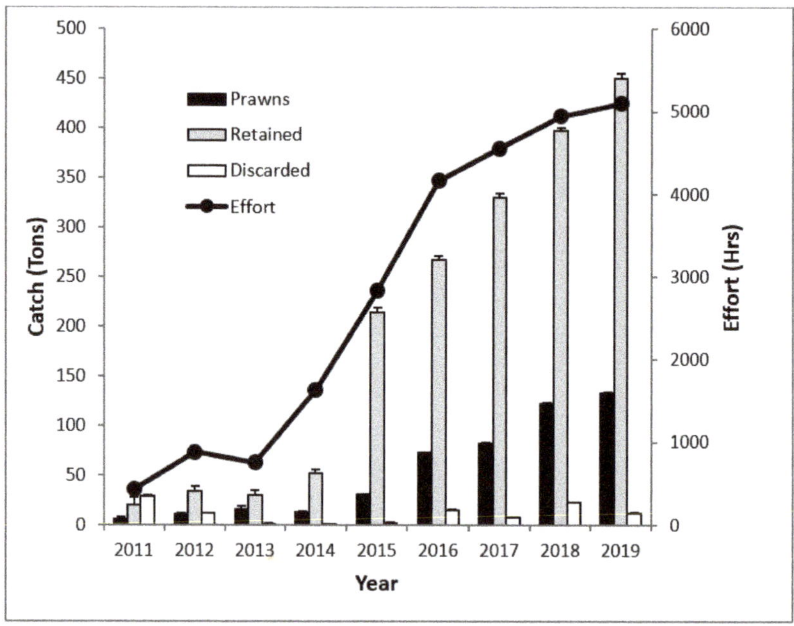

Figure 2. Annual trends in trawl catch and effort (±SE) in Malindi–Ungwana Bay in the 2011–2019 period.

3.2. Spatial and Seasonal Variation in the Catch

The Bray–Curtis similarity analysis of species composition between the seasons, depth, and site (Figure 3) showed a significant difference in the depth, with the depth of >60 m being different from the other depths (0–20, 21–40, and 41–60). There were no significant differences in species composition between seasons and sites.

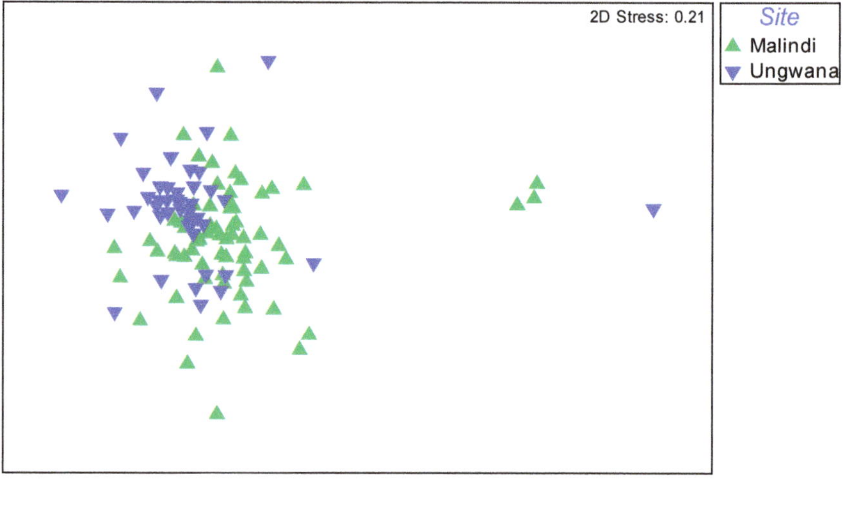

(a)

Figure 3. *Cont.*

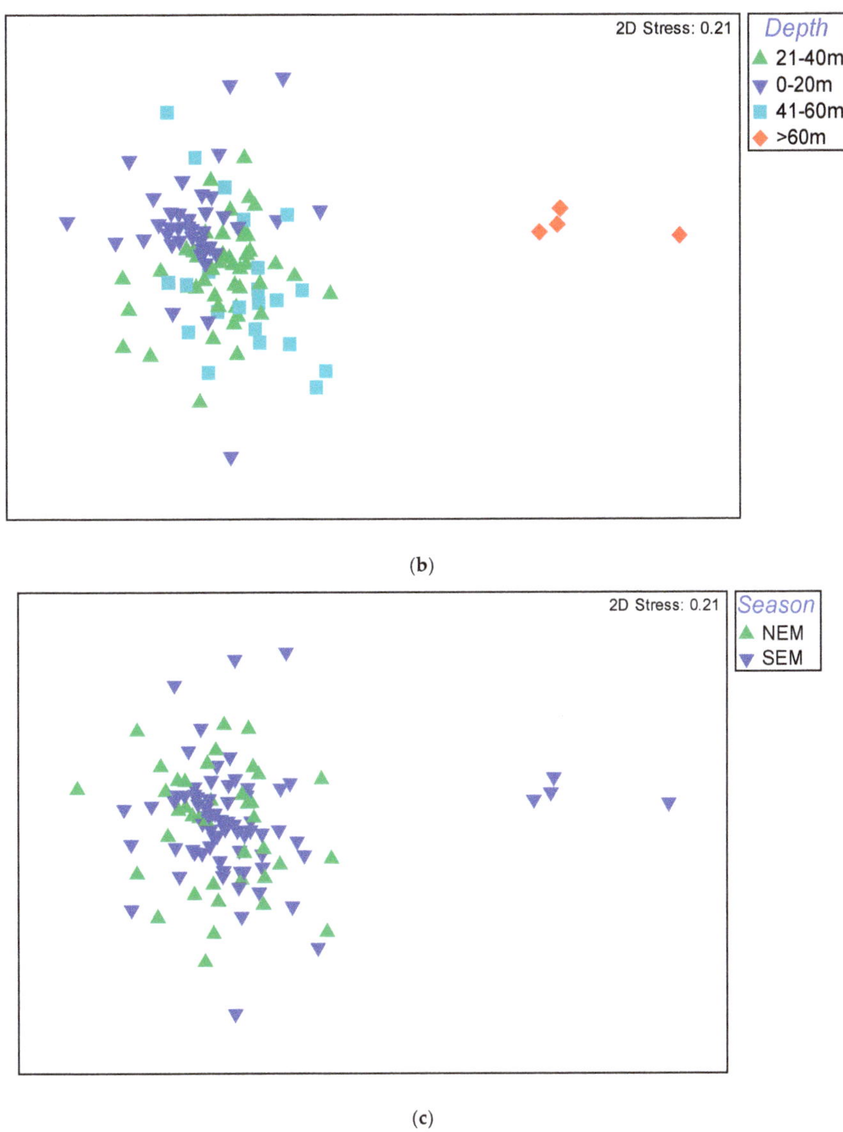

(b)

(c)

Figure 3. Bray–Curtis similarity plots of species composition between the (**a**) sites, (**b**) depth, and (**c**) seasons.

3.3. Variation in Diversity and Trophic Levels

There were no significant differences in the species diversity between the two bays and the years (Mann–Whitney U-test, U = 8, P = 1.00). The annual Shannon diversity index for both bays together ranged from 2.6 (±0.10 SE) in 2018 to 3.0 (±0.1 SE) in 2017. Malindi Bay had the highest species diversity of 3.1 (±0.15 SE) in 2017 and the lowest species diversity of 2.6 (±0.12 SE) in 2018 (Figure 4). Ungwana Bay had the highest species diversity (2.9 ± 0.13 SE) in 2017 and 2019.

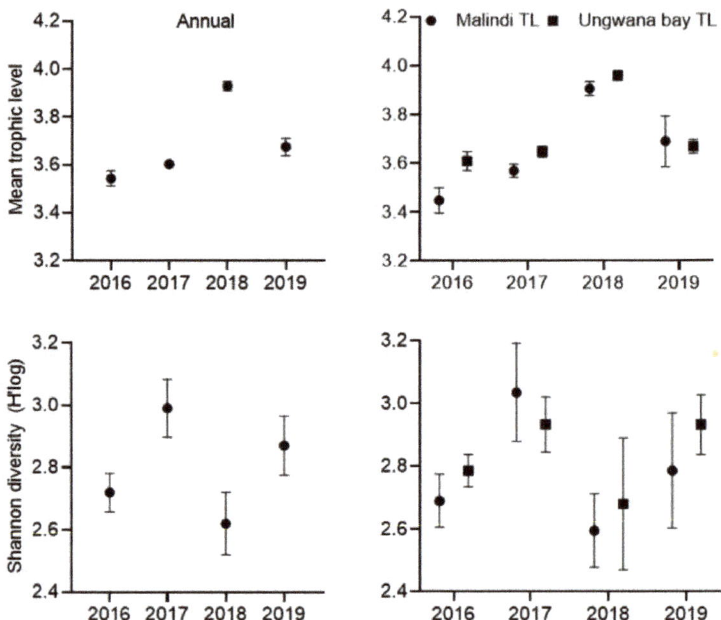

Figure 4. Annual mean trophic level and mean species diversity: general and on each bay. Error bars: SE.

The results of the Mann–Whitney U-test on the mean trophic levels showed no significant differences between the two bays and the years (Mann–Whitney U-test, U = 6, P = 0.739). The annual mean trophic level for both bays combined from 2016 to 2019 ranged from 3.5 ($\pm$0.03 SE) to 3.9 ($\pm$0.02 SE). The highest mean trophic level was in 2018. The mean trophic levels of the two bays were similar from 2016 to 2019 with Ungwana Bay having slightly higher mean trophic levels than Malindi Bay. Both bays had a high mean trophic level in 2018 (Figure 4).

3.4. Retained and Discarded Catch

The relative amounts of catch recorded by observers showed that retained catch increased between 2016 and 2019 (Figure 5). The discarded catch was highest in 2017 but gradually decreased in 2018 and 2019. The overall target prawn: total by-catch ratio was 1:9 from 2017 to 2019 compared to 1:3 during 2016. The prawn: discarded catch ranged from 1:1.7 (2016) to 1:3.3 (2017). The trends in the target catch, retained by-catch, and discarded by-catch indicated a relative reduction in target catch (Figure 5). On average the total catch comprised 16% target, 59% retained, and 25% discards. The proportion of the target prawns was highest in 2016 but declined through the other years (Figure 5).

3.5. Composition of Retained and Discarded Catch

Overall, 475 species were recorded by observers during the study period with the highest number of species (275) recorded in 2019. Among the top 10 retained and discarded species in terms of weight, *Otolithes ruber* and *Pomadasys maculatus* were captured in all years, with *O. ruber* comprising 10 to 20% of the retained catch (Table 2). None of the targeted prawn species were in the list of top-ten retained species. Three species *Pellona ditchela*, *Galeichthys feliceps*, and *Secutor insidiator* were captured in the top 10 of the discarded species in all years, with *P. ditchela* being the most discarded species (comprising 14% of discards). Some species retained also appeared in the list of discards, e.g., *P. ditchela*, *S. insidiator*, *G. feliceps*, *Trichiurus lepturus*, and *Leiognathus equulus*.

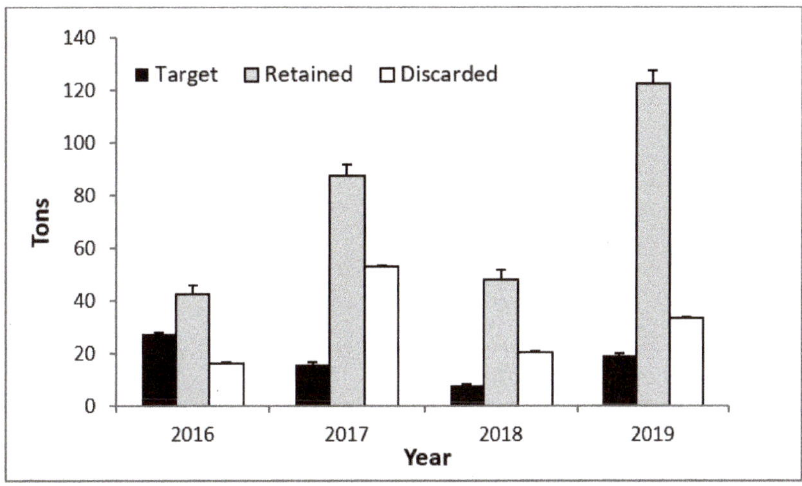

Figure 5. Trawl-fishery target retained and discarded catch (±SE) recorded by observers between 2016 and 2019 in Malindi–Ungwana Bay.

Table 2. Top-ten species by weight of retained and discarded species recorded from 2016 to 2019. Weight in kg (percentage).

	2016	2017	2018	2019
Retained				
Otolithes ruber	8581.48 (20.37)	9183.81 (10.5)	6505.43 (13.35)	13,649.81 (11.12)
Pomadasys maculatus	3063.06 (7.27)	3687.24 (4.22)	2711.51 (5.56)	5777.21 (4.71)
Galeichthys feliceps	2770.25 (6.58)	2760.77 (3.16)		
Johnius dussumieri	2579.50 (6.12)			
Lobotes surinamensis	2167.08 (5.14)			
Upeneus sulphureus	1965.95 (4.67)	5326.79 (6.09)		
Sphyraena jello	1695.13 (4.02)	2667.26 (3.05)	1835.36 (3.82)	
Leiognathus equulus	1531.83 (3.64	3360.49 (3.84)		
Trichiurus lepturus	1381.24 (3.28)			
Pomadasys multimaculatus	1335.00 (3.17)			
Discards				
Pellona ditchela	2393.18 (14.6)	7605.41 (14.33)	2695.23 (13.1)	4133.17 (12.35)
Trichiurus lepturus	1729.85 (10.55)	4465.77 (8.41)	1119.30 (5.44)	
Galeichthys feliceps	1531.19 (9.34)	2634.68 (4.96)	1198.54 (5.83)	1570.19 (4.69)
Secutor insidiator	1394.63 (8.51)	5668.83 (10.68)	2206.81 (10.73)	2226.35 (6.65)
Photopectoralis bindus	1177.52 (7.18)	1418.39 (2.67)		
Nematopalaemon tenuipes	1078.18 (6.58)		1347.20 (6.55)	
Thryssa vitrirostris	982.82 (6)		1257.57 (6.11)	1136.33 (3.4)
Thryssa malabarica	567.22 (3.46)			
Leiognathus equulus	407.09 (2.48)		2543.74 (12.37)	1850.31 (5.53)
Johnius amblycephalus	371.22 (2.27)	1527.96 (2.88)		
Secutor ruconius	2554.29 (4.81)			

The result of the nMDS ordination analysis of the catch data recorded by observers showed a clear difference between the compositions of retained and discarded species, an indication that the selection is not random (Figure 6). ANOSIM revealed a strong dissimilarity between the composition of retained and discarded species (R = 0.709, *p* = 0.001). Three species were most responsible for 86.38 of the average dissimilarity between retained and discarded species: *P. ditchela, Nematopalaemon tenuipes,* and *S. insidiator* (Table 3).

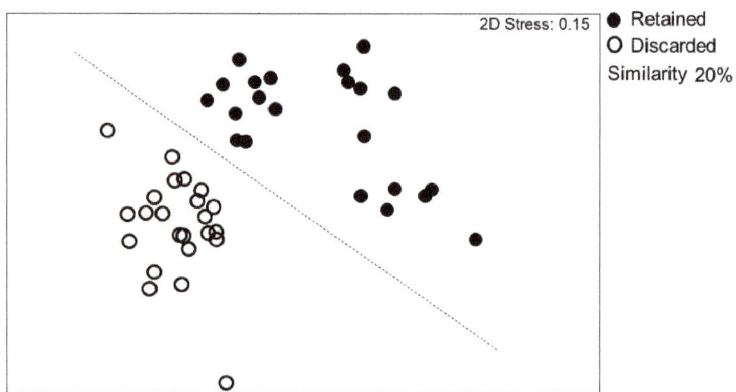

Figure 6. Non-metric multidimensional scaling ordination plots of retained vs. discarded species.

Table 3. Summary of SIMPER analysis showing the average dissimilarity in the species composition between retained and discarded catch and three species that contributed most to the overall dissimilarities.

Species	Average Abundance (%)		Dissimilarity	Contrib%	Cum.%
	Retained	Discarded	Av.Diss = 86.38		
Pellona ditchela	1.11	3.5	2.83	3.28	3.28
Nematopalaemon tenuipes	0.01	2.3	2.58	2.99	6.27
Secutor insidiator	0.39	2.66	2.49	2.88	9.15

The nMDS ordination of the catches between years, grouped 2016 and 2017 as more similar in species composition (ANOSIM, R = 0.23) influenced by the composition of retained species (Figure 7), while all derived pairwise comparisons with 2018 and 2019 catches were strongly dissimilar (ANOSIM R values of 0.99). However, the observed dissimilarities between years were not statistically significant ($p > 0.05$). The species composition of discarded species did not have a clear pattern as that of the retained catch. The discarded species in 2016 and July and October 2018 were dissimilar from those of 2017, some months of 2018, and 2019 (Figure 8). Discarded species in September 2017 and May 2019 were similar, while those in August 2017 and July 2019 were dissimilar to all the other observed months (Figure 8).

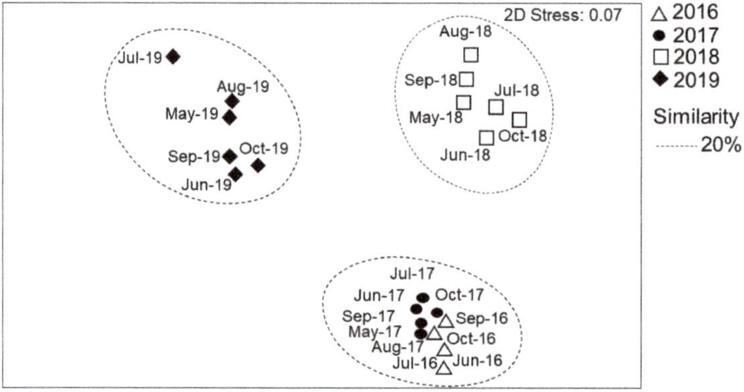

Figure 7. Non-metric multidimensional scaling ordination plots for retained catches in the months of different years.

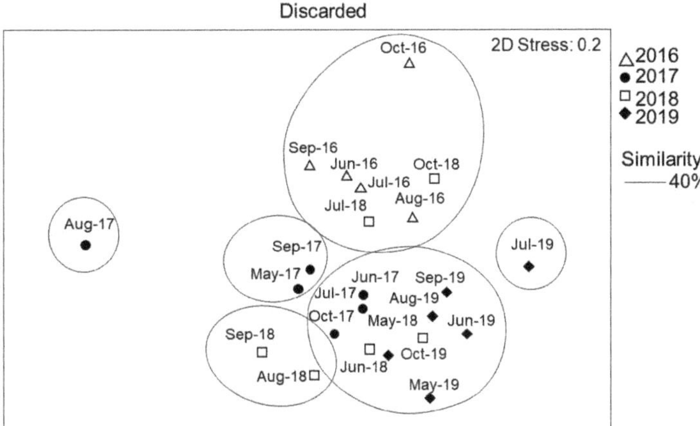

Figure 8. Non-metric multidimensional scaling ordination plots for discarded catches in the months of different years.

Result of SIMPER for the retained species identified *O. ruber*, *Panulirus homarus*, *Penaeus indicus*, and *Penaeus japonicus* as most responsible for the dissimilarity between 2016 and the other years (2017, 2018, and 2019), while *P. homarus* was most responsible for the dissimilarity between 2019 and the other years (2016, 2017, and 2018), contributing between 4.4 and 4.8% (Table 4). For discarded species, *N. tenuipes* contributed most to the dissimilarity between 2016 and the other years, contributing between 4 to 11% (Table 5).

Table 4. Summary of SIMPER analysis showing the average dissimilarity in the composition of retained species across the years studied and three species that contributed most to the overall dissimilarities.

Species	Average Abundance (%)		Dissimilarity	Contrib%	Cum.%
	2016	2017	Av.Diss = 50.63		
Upeneus vittatus	0.6	1.68	1.51	2.98	2.98
Pellona ditchela	1.89	1.92	1.42	2.81	5.79
Plesionika martia	1.26	0.37	1.36	2.69	8.48
	2016	2018	Av.Diss = 81.98		
Otolithes ruber	3.62	0.06	3.49	4.25	4.25
Penaeus japonicus	0.68	3.9	3.2	3.91	8.16
Galeichthys feliceps	2.72	0.02	2.64	3.23	11.38
	2016	2019	Av.Diss = 86.74		
Panulirus homarus	0	3.89	3.78	4.36	4.36
Penaeus indicus	3.64	0	3.43	3.95	8.31
Otolithes ruber	3.62	0	3.41	3.93	12.24
	2017	2018	Av.Diss = 79.88		
Penaeus japonicus	0.87	3.9	2.95	3.69	3.69
Metapenaeus stebbingi	0.26	3	2.62	3.28	6.97
Penaeus indicus	3.64	1.14	2.54	3.18	10.15
	2017	2019	Av.Diss = 83.23		
Panulirus homarus	0	3.89	3.73	4.49	4.49
Penaeus indicus	3.64	0	3.42	4.11	8.6
Uroconger lepturus	0	2.51	2.44	2.93	11.53
	2018	2019	Av.Diss = 82.48		
Panulirus homarus	0	3.89	3.96	4.8	4.8
Penaeus japonicus	3.9	1	2.99	3.63	8.43
Metapenaeus stebbingi	3	0.23	2.72	3.3	11.72

Table 5. Summary results of SIMPER showing the average dissimilarity in the discarded species composition across the years studied and three species that contributed most to the overall dissimilarities.

Species	Average Abundance (%)		Dissimilarity	Contrib%	Cum.%
	2016	2017	Av.Diss = 68.29		
Nematopalaemon tenuipes	6.81	0.73	7.48	10.95	10.95
Thryssa setirostris	0.67	2.7	2.55	3.73	14.69
Trichiurus lepturus	2.2	3.03	2.52	3.69	18.38
	2016	2018	Av.Diss = 64.54	Contrib%	Cum.%
Nematopalaemon tenuipes	6.81	2.25	6.87	10.65	10.65
Leiognathus equulus	0.82	1.98	2.35	3.65	14.3
Secutor insidiator	1.74	3.24	2.34	3.63	17.93
	2016	2019	Av.Diss = 69.51	Contrib%	Cum.%
Nematopalaemon tenuipes	6.81	0.17	8.09	11.63	11.63
Pellona ditchela	2.6	4.29	2.44	3.5	15.14
Thryssa setirostris	0.67	2.38	2.17	3.12	18.25
	2017	2018	Av.Diss = 65.07	Contrib%	Cum.%
Nematopalaemon tenuipes	0.73	2.25	3.07	4.72	4.72
Trichiurus lepturus	3.03	2.08	2.36	3.63	8.34
Antennarius pictus	2.05	0	2.34	3.59	11.93
	2017	2019	Av.Diss = 66.28	Contrib%	Cum.%
Trichiurus lepturus	3.03	1.34	2.4	3.62	3.62
Pellona ditchela	2.94	4.29	2.34	3.53	7.15
Antennarius pictus	2.05	0	2.14	3.22	10.38
	2018	2019	Av.Diss = 61.76	Contrib%	Cum.%
Nematopalaemon tenuipes	2.25	0.17	2.86	4.63	4.63
Gazza minuta	0.16	1.94	2.2	3.56	8.19
Leiognathus equulus	1.98	2.28	2.19	3.54	11.73

4. Discussion

4.1. Catch and Effort

This study analyzed the catch trends using fisheries logbook data and observer data with the aim of determining the impact of fishing on the ecosystem. The catch and effort increased during the nine years of operation with the prawn and fish catch ranging from 6 to 133 tons and 20 to 450, respectively, from 2011 to 2019. Previously reported average annual landings in the bay were higher between 257 and 400, and 315 and 602 tons of prawns and fish, respectively, between 2001 and 2006 [42]. The average prawn landings in the Malindi–Ungwana Bay were 400 tons per year in the 1970s, 1980s, and 1990s [43]. Thus, the prawn catch is yet to reach these earlier reported catches. The lower total annual catches could be attributed to the restricted fishing effort, with six months annual fishing season and only during the daytime from 6 a.m. to 6 p.m. required by the management plan regulations. In addition, the regulations in the management plan zoned < 3 nm offshore as no trawling to reduce the interaction of trawlers with small-scale fishing gears. Stock surveys have shown that prawn stocks are higher close to the shore and the estuaries [42,44]. The zoning in the management regulations moved the trawlers from the centers of prawn stocks concentration resulting in the lower catch.

4.2. Spatial and Seasonal Variation

Overall, in this study, no significant differences in the species composition between the seasons and the two bays (Malindi and Ungwana) were found. However, distinct abundance and species composition patterns have been reported in the same bay for prawns driven by the bottom type and depth [45]. However, depth showed a significant difference in the species composition at depths > 60 m. These findings indicate that there were no changes in the species composition over the four years of observations, and the dominant species in the catches have remained the same. However, the species composition in the bay is influenced by the depth.

4.3. Diversity and Trophic Levels

Species diversity in the catch did not vary significantly and was dominated by a few species and families, similar to other trawl fisheries in the tropics [46,47]. A previous study reported *G. feliceps, O. ruber, Johnius amblycephalus, Johnius dussumieri, Lobotes surinamensis, L. equulus, P. maculatus* as the dominant species in the trawl catches of Malindi–Ungwana Bay [48]. These species were also reported as dominant in this study.

Results of the nMDS analysis indicated that species composition from prawn trawling differed over the four years. The species contributing to differences in the retained catch in the four years were *P. indicus, O. ruber,* and *P. homarus* with *Otolithes ruber* being most abundant. In particular, *P. homarus* contribute to the dissimilarity of the catches in 2019 indicating a shift in the distribution of fishing effort to deeper water. Based on numbers, *N. tenuipes* and *P. ditchella* were responsible for the difference in species composition of the discarded catch. *Nematopalaemon tenuipes* is discarded because it is considered to be too small and of low economic value [49]. *Pellona ditchela* also appeared in the top 10 species in terms of weight. In a previous study, *G. feliceps* and *P. ditchela* were the most dominant species contributing to the highest spatial dissimilarity in the inshore areas of the bay [50]. Overall, in this study, no significant differences in the species composition between the seasons and the two bays (Malindi and Ungwana) were found. However, depth showed a significant difference in the species composition at depths > 60 m. These findings indicate that there were no changes in the species composition over the four years of observations, and the dominant species in the catches have remained the same. However, the species composition in the bay is influenced by the depth.

Mean trophic levels indicate the status of resource exploitation and is an indicator of fishery-induced impacts at the food web level [51,52]. The trophic levels in this study (3.45–3.96) were comparable with the values between 3.2–4 recorded in an earlier study [48], an indication of no marked variation and a relatively stable ecosystem, meaning that the impacts of fishing at the current level has not surpassed the self-regulatory capacity of the bay [51,53].

4.4. Retained and Discarded Catch

Globally, the trawl fishery is characterized by a high by-catch rate, with prawn trawling reporting prawn to by-catch ratios of between 1:3 to 1:15 [54,55]. In this study, the prawn to by-catch ratio decreased through the four years and the prawn to discarded catch ratio of 1:1.7 to 1:3.3 is comparable to the prawn to discarded by-catch ratio of 1:1.5 obtained in 2012 [42]. The by-catch in this study made about 84% of the catch of which 59% comprised retained by-catch and 25% of discarded by-catch. Other studies estimated the total by-catch from prawn trawling in the Malindi–Ungwana Bay to be 70–80% by weight [43]. In Mozambique, the by-catch comprised about 80% of the total catches [21]. These estimates of by-catch are comparable with those found in this study and are an indication of high amounts of by-catch resulting from prawn trawling.

The reasons for discarding by-catch are attributed to lack of storage space and the low value of small fish [56]. The discarded by-catch reported for the Malindi–Ungwana Bay comprised different families, mainly the Leiognathidae, Clupeidae, Dasyatidae, and Carcharhinidae (This study, [43]). Non-commercial fishes contributed more than 43% of the discards, whereas juveniles of some commercially important species, such as *O. ruber* and *Johnius* sp. (Sciaenidae), and *Pomadasys* sp. (Haemulidae), made up 25% of discards [43,57]. In comparison to our study, the families that were discarded were mainly Pristigasteridae, Trichiuridae, and Ariidae while the species retained included *O. ruber, P. maculatus, G. feliceps* among others. Thus, the commercially important species that were previously being discarded were now being retained. Species that were previously discarded may have gained acceptance and value in the local market making retaining them cost-effective. The fishing industry has been exploring ways of maximizing the use of by-catch.

4.5. Composition of Retained and Discarded Catch

Overall, the number of all species reported in this study increased with fishing effort over the years, with the highest 275 in 2019. The authors of [48] reported 223 species, while in a bottom trawl survey in the bay in 2012, 66 fish species in 43 families were found with the highest biomass in the shallow areas (<50 m) [58]. In an earlier survey, the number of species collected was 160 species belonging to 61 families [44]. This shows that the number of species reported varied from the different studies, with the highest reported in this study attributed to the length of time over which the fishery observations were made.

Fish species contributed higher biomass compared to the target prawns, which were ranked below the 10 top species. This is the general observation in most prawn trawl fisheries of the world where large amounts of non-target species are caught [59,60]. In Malindi–Ungwana Bay, of the five target penaeid prawns, *P. indicus* is the most dominant [61]. Though seasonal variations influence prawn catches, *P indicus* along with *P. monodon, P. monocerous, P. semisulcatus,* and *P. japonicus* are common prawn species in the bay [25,49,50,62].

Concerns have been raised regarding the overfishing by trawlers of species important in small-scale fisheries in the Western Indian Ocean (WIO) region, including *O. ruber* [63]. The species is a common trawl fishery by-catch along the East African coast and is usually retained for its high commercial value. Reduced abundance of *O. ruber* along with other common species associated with prawn trawling could result in ecological changes (e.g., altered predator–prey relationships) and impact the artisanal fishers' catches [63]. Predator–prey relationships between finfish and prawns may contribute to the resulting high abundance of finfish by-catch [64]. Most of the finfish abundant in the by-catch, such as *O. ruber, P. maculatus, P. ditchela, Thryssa vitrirostris,* and *L. equulus, Terapon jarbua,* are predators of penaeid prawns [64,65].

4.6. Reduction and Use of By-Catch

Discarded catch from prawn trawling has been a concern for a long time, with pressure to reduce the capture of non-commercial species increasing [5,60]. The by-catch in prawn trawling can be reduced but cannot be eliminated, and it is estimated that present selectivity technology and management can reduce by-catch by 30 percent at most [21]. By-catch reducing devices are increasingly being used in prawn trawling to reduce the amounts of by-catch, and in some areas, they have been successful and beneficial to prawn fisheries [60,66–68].

Besides the efforts to reduce the by-catch, the complete utilization of catch is also considered an important way of increasing the benefits from the fisheries [69]. In China, the by-catch was used for the aquaculture industry [70]. In Madagascar, by-catch is normally sold for human consumption [69]. In Kenya, the retained by-catch is offloaded in urban centers, mainly Malindi and Mombasa, and sold to women in the fish retail business (locally known as *"Mama karanga"*) [71]. Increasing the amount of by-catch that reaches the market would support more livelihoods through trade and support food security, particularly in the urban centers within the coast region. The utilization of by-catch as food reduces the ethical argument against the un-selective fishing of trawl fisheries.

5. Conclusions

The Malindi–Ungwana Bay fishery is a good example of the competing interests of fisheries resource use and ecosystem conservation needs between resource users; including industrial fishing, small-scale fishing, as well as recreational use of the ecosystem, in which industrial fishing has been criticized for environmental degradation and large wastage in form of discards. Consequently, several studies have been undertaken to assess the Malindi–Ungwana Bay fishery addressing the status of the fishery [49,61], ecological indicators affecting the fishery [72], and resource use and distribution [42,50]. This study evaluated the state of the shallow-water prawn trawl fishery of the Malindi–Ungwana Bay based on some catch and ecosystem-based indices, after the implementation of the management plan in 2010. The results showed an increase in fishing effort and catch over the four years, 2016 to 2019. The levels of by-catch remained high, while the proportion of retained by-catch increased over the years. The species composition of the trawl catches in the two bays was similar and the dominating species in the Malindi–Ungwana Bay remained the same over the years. There was a distinct difference between the retained and discarded species, and differences in species composition of retained catch over time. However, the evaluation showed no marked decline in the status of the stock in the bay based on the diversity and tropic indices. We recommend that more of the discarded species are retained to ensure that the fishery is less wasteful.

Author Contributions: Conceptualization, E.N.F. and E.N.K.; Data curation, E.N.F., J.O.O., N.W., P.T. and E.N.K.; Formal analysis, E.N.F., J.O.O., G.M.O. and E.N.K.; Methodology, E.N.F., J.O.O., G.M.O., P.T. and G.W.M.; Validation, E.N.F., J.O.O., N.W., E.N.K., G.M.O. and G.W.M.; Writing—original draft, E.N.F.; Writing—review and editing, E.N.F., J.O.O., N.W., G.M.O., P.T., G.W.M. and E.N.K. All authors have read and agreed to the published version of the manuscript.

Funding: This research received no external funding.

Institutional Review Board Statement: Not applicable.

Data Availability Statement: The data presented in this study are available on request from the corresponding authors. The data are not publicly available due to confidential information included in the data.

Acknowledgments: We are grateful to the Kenya Marine and Fisheries Research Institute (KMFRI), Kenya Fisheries Service (KeFS), and the fishing industry (East African Sea Foods Ltd. and Ittica Ltd., Mombasa Kenya) that provided the logistical support of all the shallow-water prawn trawling observer program surveys. We appreciate the cooperation and support from the captains of the fishing vessels and the entire fishing crew who provided the scientific observers with the necessary support to collect data and samples. The support, cooperation, and commitment of KMFRI fishery observers (Boaz Orembo, Masudi Juma Zamu, Rashid Anam, James Gonda, Jibril Olunga, Ben Ogola, Sammy Kadhengi, Benard Kimathi, Geoffrey Odhiambo Otieno, Justus Andati, Nimrod Ishmael, Hafidh Ishmael) during the field and laboratory work are appreciated. We are grateful to our respective institutions (KMFRI and The Nature Conservancy) for supporting authors' time in this study. We also acknowledge the reviewers who gave constructive changes and recommendations to improve the manuscript.

Conflicts of Interest: The authors declare no conflict of interest.

References

1. Cashion, T.; Al-Abdulrazzak, D.; Belhabib, D.; Derrick, B.; Divovich, E.; Moutopoulos, D.K.; Noël, S.L.; Palomares, M.L.D.; Teh, L.C.L.; Zeller, D.; et al. Reconstructing global marine fishing gear use: Catches and landed values by gear type and sector. *Fish. Res.* **2018**, *206*, 57–64. [CrossRef]
2. Pauly, D.; Zeller, D.; Palomares, M.L.D. (Eds.) *Sea around Us Concepts, Design and Data*; Oxford University Press: Oxford, UK, 2003; Available online: https://www.seaaroundus.org/ (accessed on 21 November 2020).
3. Roberts, C. *Ocean of Life: How Our Seas Are Changing*; Penguin: London, UK, 2012; p. 400.
4. FAO. *Report of the Technical Consultation on Reduction of Wastage in Fisheries. 28 October–1 November 1996 Tokyo, Japan*; FAO Fisheries Report No. 547; FAO: Rome, Italy, 1996; p. 27.
5. Kelleher, K. *Discards in the World's Marine Fisheries. An Update*; FAO Fisheries Technical Paper. No. 470; FAO: Rome, Italy, 2005; p. 131.
6. Jennings, S.; Kaiser, M.J. The effects of fishing on marine ecosystems. *Adv. Mar. Biol.* **1998**, *34*, 201–212.
7. Zeller, D.; Cashion, T.; Palomares, M.; Pauly, D. Global marine fisheries discards: A synthesis of reconstructed data. *Fish Fish.* **2018**, *19*, 30–39. [CrossRef]
8. Jennings, S.; Pinnegar, J.K.; Polunin, N.V.C.; Warr, K.J. Impacts of trawling disturbance on the trophic structure of benthic invertebrate communities. *Mar. Ecol. Prog. Ser.* **2001**, *213*, 127–142. [CrossRef]
9. Hiddink, J.G.; Jennings, S.; Sciberras, M.; Szostek, C.L.; Hughes, K.M.; Ellis, N.; Rijnsdorp, A.D.; McConnaughey, R.A.; Mazor, T.; Hilborn, R.; et al. Effects of bottom trawling on seabed. *Proc. Natl. Acad. Sci. USA* **2017**, *114*, 8301–8306. [CrossRef]

10. Newsome, T.M.; Ballard, G.-A.; Fleming, P.J.S.; van de Ven, R.; Story, G.L.; Dickman, C.R. Human-resource subsidies alter the dietary preferences of a mammalian top predator. *Oecologia* **2014**, *175*, 139–150. [CrossRef]
11. Fondo, E.N.; Chaloupka, M.; Heymans, J.J.; Skilleter, G.A. Banning Fisheries Discards Abruptly Has a Negative Impact on the Population Dynamics of Charismatic Marine Megafauna. *PLoS ONE* **2015**, *10*, e0144543. [CrossRef]
12. Pusceddua, A.; Bianchelli, S.; Martín, J.; Puig, P.; Palanques, A.; Masquéd, P.; Danovaroa, R. Chronic and intensive bottom trawling impairs deep-sea biodiversity and ecosystem functioning. *Proc. Natl. Acad. Sci. USA* **2014**, *111*, 8861–8866. [CrossRef]
13. Jacquet, J.; Pauly, D.; Ainley, D.; Holt, S.; Dayton, P.; Jackson, J. Seafood stewardship in crisis. *Nature* **2010**, *467*, 28–29. [CrossRef]
14. Morton, B. At last, a trawling ban for Hong Kong's inshore waters. *Mar. Pollut. Bull.* **2011**, *62*, 1153–1154. [CrossRef]
15. Watling, L. Deep-sea trawling must be banned. *Nature* **2013**, *501*, 7. [CrossRef]
16. Jiddawi, N.; Ohman, M.C. Marine Fisheries in Tanzania. *Ambio* **2002**, *31*, 518–527. [CrossRef] [PubMed]
17. Abdallah, A.M. *Management of the Commercial Prawn Fishery in Tanzania*; Final Project Report; University of Iceland: Reykjavík, Iceland, 2004; p. 42.
18. Haule, W.V. Reducing the impact of tropical shrimp trawling fisheries on living marine resources through the adoption of environmentally friendly techniques and practices in Tanzania. In *Tropical Shrimp Fisheries and Their Impact on Living Resources*; FAO Fisheries Technical Paper; FAO: Rome, Italy, 2001; Volume 974, pp. 216–233.
19. de Sousa, L.P.; Abdula, S.; Palha de Sousa, B.; Penn, J.W. Assessment of the shallow water shrimp fishery of Sofala Bank. *Mozamb. Fish. Res. Inst. Rep.* **2014**, *33*, 76.
20. WWF. *A Sustainable Shrimp Fishery for Mozambique*; WWF Fact Sheet: Morges, Switzerland, 2017; p. 4.
21. Fennessy, S.T.; Mwatha, G.K.; Thiele, W. (Eds.) *Report of the Regional Workshop on Approaches to Reducing Shrimp Trawl By-Catch in the Western Indian Ocean. Mombasa, Kenya, 13–15 April 2003*; FAO Fisheries Report. No. 734; FAO: Rome, Italy, 2004; p. 49.
22. Ayers, M.J.; Scharler, U.M.; Fennessy, S.T. Modelling ecosystem effects of reduced prawn recruitment on the Thukela Bank trawling grounds, South Africa, following nursery loss. *Mar. Ecol. Prog. Ser.* **2013**, *479*, 143–161. [CrossRef]
23. Razafindrainibe, H. *Baseline Study of the Shrimp Trawl Fishery in Madagascar and Strategies for Bycatch Management*; TCP/MAG/3201–REBYC2 Report; United Nations Food and Agriculture Organization: Rome, Italy, 2010; p. 111.
24. Nzioka, R.M. Observations on the spawning seasons of East African reef fishes. *J. Fish Biol.* **1979**, *14*, 329–342. [CrossRef]
25. Mutagyera, W.B. Distribution of some deepwater prawns and lobster species in Kenya waters. In Proceedings of the NORAD-Kenya Seminar to Review the Marine Fish Stocks in Kenya, Mombasa, Kenya, 13–15 March 1984.
26. Ruwa, R.K.; Muthiga, N.; Esposito, M.; Zanre, R.; Mueni, E.; Muchiri, M. Report of the scientific information and conservation sub-committee-prawn fishery in Kenya. In *Marine Waters. Prawn Trawling Task Force*; Kenya Marine and Fisheries Research Institute: Mombasa, Kenya, 2001; p. 8.
27. Government of Kenya. *The Prawn Fishery Management Plan 2010*; Kenya Gazette; Government of Kenya: Nairobi, Kenya, 2010.
28. Mutagyera, W.B. On *Thenus orientalus* and *Metanephrops andamanicus* (Macrura, Scyllaridae and Nephropidae) off Kenya coast. *East Afr. Agric. For. J.* **1979**, *45*, 142–144. [CrossRef]
29. Ruwa, R.K. *Coastal and Offshore Marine Fisheries of Kenya: Status and Opportunities*; KMFRI Technical Report/2004/FP/1; Kenya Marine and Fisheries Research Institute: Mombasa, Kenya, 2006; p. 44.
30. Abuodha, P. Sediment composition and characteristics in the Malindi-Ungwana bay. In *Current Status of Trawl Fishery of Malindi-Ungwana Bay*; KMFRI Technical Report; Kenya Marine and Fisheries Research Institute: Mombasa, Kenya, 2002; pp. 23–30.
31. Mirera, D.; Kairo, J.; Kimani, E.; Waweru, F. A comparison between fish assemblages in mangrove forests and on intertidal flats at Ungwana Bay, Kenya. *Afr. J. Aquat. Sci.* **2010**, *35*, 165–171. [CrossRef]
32. Kitheka, J.U.; Obiero, M.; Nthenge, P. River discharge, sediment transport and exchange in the Tana estuary, Kenya. *Est. Coast. Shelf. Sci.* **2005**, *63*, 455–468. [CrossRef]
33. Schott, F.A.; McCreary, J.P., Jr. The monsoon circulation of the Indian Ocean. *Prog. Oceangr.* **2001**, *51*, 1–123. [CrossRef]
34. Athayde, T. *SWIOFP Observer Program Data Collection Guide*; South West Indian Ocean Fisheries Project; Regional Management Unit, Kenya Marine and Fisheries Research Institute: Mombasa, Kenya, 2012; p. 79.
35. Smith, M.M.; Heemstra, P.C. (Eds.) *Smith's Sea Fishes*, 6th ed.; Springer: Berlin, Germany, 1986; Volume 1038, p. 144.
36. FAO FishFinder. *Commercially Important Coastal Fishes of Kenya. A Pocket Guide*; FAO: Rome, Italy, 2010; p. 39.
37. Richmond, M.D.E. *A Field Guide to the Seashores of Eastern Africa and the Western Indian Ocean Islands*, 3rd ed.; Sida/WIOMSA; Samaki Consultants Ltd.: Dar es Salaam, Tanzania, 2011; p. 464.
38. Anam, R.; Mostarda, E. Field identification guide to the living marine resources of Kenya. In *FAO Species Identification Guide for Fishery Purposes*; FAO: Rome, Italy, 2012; p. 25.
39. Froese, R.; Pauly, D. (Eds.) FishBase. World Wide Web Electronic Publication. Available online: www.fishbase.org (accessed on 12 December 2019).
40. Clarke, K.R.; Warwick, R.M. *Change in Marine Communities: An Approach to statistical Analysis and Interpretation*, 2nd ed.; PRIMER-E, Ltd.: Plymouth, UK, 2001.
41. Clarke, K.R.; Gorley, R.N. *PRIMER v6: User Manual/Tutorial (Plymouth Routines in Multivariate Ecological Research)*; PRIMER-E, Ltd.: Plymouth, UK, 2006.
42. Munga, C.; Ndegwa, S.; Fulanda, B.; Manyala, J.; Kimani, E.; Ohtomi, J.; Vanreusel, A. Bottom shrimp trawling impacts on species distribution and fishery dynamics; Ungwana bay fishery Kenya before and after the 2006 trawl ban. *Fish. Sci.* **2012**, *78*, 209–219. [CrossRef]

43. Mwatha, G.K. *Stock Assessment and Population Dynamics of Penaeid Prawns in the Prawn Trawling Grounds of Malindi-Ungwana Bay: The Challenges of Managing the Prawn Fishery in Kenya*; WIOMSA/MARG-I/2005-06; Western Indian Ocean Marine Science Association: Zanzibar, Tanzania, 2005; p. 21.

44. Kimani, E.N.; Fulanda, B.; Wambiji, N.; Munga, C.; Okemwa, G.; Thoya, P.; Omukoto, J. *Spatial Mapping of Trawl Fishery Resources of Malindi-Ungwana Bay, Kenya*; SWIOFP Components 2 (Crustaceans); Special Technical Report; SWIOFP Survey Report for SWIOFP 2011C201a; Kenya Marine and Fisheries Research Institute: Mombasa, Kenya, 2012; p. 28.

45. Munga, C.N.; Mwangi, S.; Ong'anda, H.; Ruwa, R.; Manyala, J.; Groeneveld, J.C.; Kimani, E.; Vanreusel, A. Species composition, distribution patterns and population structure of penaeid shrimps in Malindi-Ungwana bay, Kenya, based on experimental bottom trawl surveys. *Fish. Res.* **2013**, *147*, 93–102. [CrossRef]

46. Fennessy, S.T. Incidental capture of elasmobranchs by commercial prawn trawlers on the Tugela Bank, Natal, South Africa. *S. Afr. J. Mar. Sci.* **1994**, *14*, 287–296. [CrossRef]

47. Tonks, M.L.; Griffiths, S.P.; Heales, D.S.; Brewer, D.T.; Dell, Q. Species composition and temporal variation of prawn trawl bycatch in the Joseph Bonaparte Gulf, northwestern Australia. *Fish. Res.* **2008**, *89*, 276–293. [CrossRef]

48. Munga, C.N.; Mwangi, S.; Ong'anda, H.; Ruwa, R.; Manyala, J.; Groeneveld, J.; Kimani, E. Fish catch composition of artisanal and bottom trawl fisheries in Malindi-Ungwana bay, Kenya: A cause for conflict? *WIO J. Mar. Sci.* **2014**, *13*, 31–46.

49. KMFRI. *Current Status of the Shallow Water Industrial Prawn Trawling Fishery in Kenya*; Kenya Marine and Fisheries Research Institute Research Report; OCS/FIS/2018–2019/ C1.1.i&C1.4.i; KMFRI: Mombasa, Kenya, 2019; p. 59.

50. Munga, C.N. Ecological and Socio-Economic Assessment of Kenyan Coastal Fisheries: The Case of Malindi-Ungwana Bay Artisanal Fisheries versus Semi-Industrial Bottom Trawling. Ph.D. Thesis, Ghent University, Ghent, Belgium, 2013.

51. Pauly, D.; Palomares, M.L.; Froese, R.; Saa, P.; Vakily, M.; Preikshot, D.; Wallace, S. Fishing down Canadian aquatic food webs. *Can. J. Fish. Aquat. Sci.* **2001**, *58*, 51–62. [CrossRef]

52. Rochet, M.-J.; Trenkel, V.M. Which community indicators can measure the impact of fishing? A review and proposals. *Can. J. Fish. Aquat. Sci.* **2003**, *60*, 86–99. [CrossRef]

53. Yang, J. A tentative analysis of the trophic levels of North Sea fish. *Mar. Ecol. Prog. Ser.* **1982**, *7*, 47–52. [CrossRef]

54. Hall, M.A.; Alverson, D.L.; Metuzals, K.I. By-Catch: Problems and Solutions. *Mar. Pollut. Bull.* **2000**, *41*, 204–219. [CrossRef]

55. Sánchez, P.; Demestre, M.; Martin, P. Characterisation of the discards generated by bottom trawling in the northwestern Mediterranean. *Fish. Res.* **2004**, *67*, 71–80. [CrossRef]

56. Ramkumar, S.; Ranjith, L.; Jaiswar, A.K.; Vinod, K.; Deshmukhm, V. Does the mechanised trawl target the non-targets from the commercial fishing grounds of northern Maharashtra, eastern Arabian Sea India. *J. Entom. Zool. Stud.* **2019**, *7*, 1133–1140.

57. Ochiewo, J. *Harvesting and Sustainability of Marine Fisheries in Malindi-Ungwana Bay, Northern Kenya Coast*; Report NO: WIOMSA/MARG-I/2006–03; Western Indian Ocean Marine Science Association: Zanzibar, Tanzania, 2006; p. 44.

58. Kaunda-Arara, B.; Munga, C.; Manyala, J.; Kuguru, B.; Igulu, M.; Chande, M.; Kangwe, S.; Mwakiti, S.; Thoya, P.; Mbaru, E.; et al. Spatial variation in benthopelagic fish assemblage structure along coastal East Africa from recent bottom trawl surveys. *Reg. Stud. Mar. Sci.* **2016**, *8*, 201–209. [CrossRef]

59. Kennelly, S.J.; Liggins, G.W.; Broadhurst, M.K. Retained and discarded by-catch from oceanic prawn trawling in New South Wales, Australia. *Fish. Res.* **1998**, *36*, 217–236. [CrossRef]

60. Komoroske, L.; Lewison, R. Addressing fisheries bycatch in a changing world. *Front. Mar. Sci.* **2015**, *2*, 83. [CrossRef]

61. Fulanda, B.; Ohtomi, J.; Mueni, E.; Kimani, E. Fishery trends, resource-use and management system in the Ungwana bay fishery Kenya. *Ocean Coast. Manag.* **2011**, *54*, 401–414. [CrossRef]

62. Mwatha, G.K. Assessment of the prawn fishery, by-catch, resource use conflicts and performance of the turtle excluder device. In *Current Status of Trawl Fishery of Malindi-Ungwana Bay*; KMFR Technical Report; Kenya Marine and Fisheries Research Institute: Mombasa, Kenya, 2002; p. 97.

63. Olbers, J.M.; Fennessy, S.T. A retrospective assessment of the stock status of Otolithes ruber (Pisces: Sciaenidae) as bycatch on prawn trawlers from KwaZulu-Natal, South Africa. *Afr. J. Mar. Sci.* **2007**, *29*, 247–252. [CrossRef]

64. de Freitas, A.J. The Penaeoidea of South Africa IV- The Family Penaeidae: Genus Penaeus. *Oceanogr. Res. Inst. Investig. Res. Rep.* **2011**, *59*, 125.

65. Macia, A. Juvenile penaeid shrimp density, spatial distribution and size composition in four adjacent habitats within a mangrove-fringed bay on Inhaca Island, Mozambique. *WIO J. Mar. Sci.* **2004**, *3*, 163–178. [CrossRef]

66. Rogers, D.R.; de Silva, J.A.; Wright, V.L.; Watson, J.W. Evaluation of shrimp trawls equipped with bycatch reduction devices in inshore waters of Louisiana. *Fish. Res.* **1997**, *33*, 55–72. [CrossRef]

67. Brewer, D.; Rawlinson, N.; Eayrs, S.; Burridge, C. An assessment of Bycatch Reduction Devices in a tropical Australian prawn trawl fishery. *Fish. Res.* **1998**, *36*, 195–215. [CrossRef]

68. Robins, J.B.; Campbell, M.J.; McGilvray, J.G. Reducing Prawn-trawl Bycatch in Australia: An Overview and an Example from Queensland. *Mar. Fish. Rev.* **1999**, *61*, 46–55.

69. Bage, H. *Desk Review of Bycatch in Shrimp Fisheries*; Report; Food and Agriculture Organization of the United Nations: Ebene, Mauritius, 2013; p. 38.

70. Zhang, W.; Liu, M.; de Mitcheson, Y.S.; Cao, L.; Leadbitter, D.; Newton, R.; Little, D.C.; Li, S.; Yang, Y.; Chen, X.; et al. Fishing for feed in China: Facts, impacts and implications. *Fish Fish.* **2020**, *21*, 47–62. [CrossRef]

71. Matsue, N.; Daw, T.; Garrett, L. Women Fish Traders on the Kenyan Coast: Livelihoods, Bargaining Power, and Participation in Management. *Coast. Manag.* **2014**, *42*, 531–554. [CrossRef]
72. Swaleh, K.; Kaunda-Arara, B.; Ruwa, R.; Raburu, P. Ecosystem-based assessment of a prawn fishery in coastal Kenya using ecological indicators. *Ecol. Indic.* **2015**, *50*, 233–241. [CrossRef]

 *diversity*

Article

Over, Under, Sideways and Down: Patterns of Marine Species Richness in Nearshore Habitats off Santa Catalina Island, California

David W. Ginsburg * and Andrew H. Huang

Environmental Studies Program, University of Southern California, 3454 Trousdale Parkway, Los Angeles, CA 90089-0156, USA; huangah@usc.edu
* Correspondence: dginsbur@usc.edu

Abstract: Santa Catalina Island, located off the southern California coast, is home to the Blue Cavern Onshore State Marine Conservation Area (SMCA), which is recognized as a marine protected area. Here, we provide an updated species inventory of nearshore macroalgae, seagrasses, bony and cartilaginous fishes and invertebrates documented inside the Blue Cavern Onshore SMCA. Species richness data were compiled using scuba-based visual surveys conducted in the field, references from the primary and gray literature, museum records, unpublished species lists and online resources. The current checklist consists of 1091 marine species from 18 different taxonomic groups, which represents an ~43% increase in species diversity compared to the value reported previously. These data are indicative of the high biodiversity known from the Southern California Bight (SCB) region. The total number of intertidal and subtidal taxa reported represent approximately 85% and 45% of the documented macroalgae and plants, 41% and 24% invertebrates, and 62% and 20% of fishes from Catalina Island and the SCB, respectively. Among the marine taxa documented, 39 species either have undergone a geographic range shift or were introduced as the result of human activities, while another 4 species are listed as threatened, endangered or critically endangered. Research findings presented here offer an important baseline of species richness in the California Channel Islands and will help improve the efforts by resource managers and policy makers to conserve and manage similar habitats in the coastal waters off southern California.

Keywords: species richness; coastal biodiversity; California Channel Islands; marine protected area

Citation: Ginsburg, D.W.; Huang, A.H. Over, Under, Sideways and Down: Patterns of Marine Species Richness in Nearshore Habitats off Santa Catalina Island, California. *Diversity* 2022, 14, 366. https://doi.org/10.3390/d14050366

Academic Editors: Thomas J. Trott and Michael Wink

Received: 6 April 2022
Accepted: 3 May 2022
Published: 5 May 2022

Publisher's Note: MDPI stays neutral with regard to jurisdictional claims in published maps and institutional affiliations.

1. Introduction

Ecosystem function is integral for effective management in marine systems [1], thus local and global species losses could threaten the stability of the ecosystem services on which humans depend [2]. An ongoing challenge to the documentation and management of biodiversity is that the number of taxa (species richness) in a given area is not typically known. Likewise, the patterns of species richness play an important role in studies of biogeography and conservation biology [3]. For example, while ~300,000 taxa have been described from the global ocean, the total number of marine organisms is estimated to include as many as 10 million species [4]. Not surprisingly, the taxonomic status of many of these organisms has yet to be evaluated [5], even among well-studied species, such as coastal fishes and invertebrates [6].

The California Channel Islands are located in the Southern California Bight (SCB), which stretches along ~700 km of coastline from Point Conception off California to Ensenada, just south of the US–Mexico border, and is known as a marine biodiversity hotspot [7]. The SCB is a dynamic region in which the subtropical, Southern California countercurrent flows nearshore along a northward trajectory, and the subarctic, California current moves offshore in a southerly direction. As these different water masses converge, this unique oceanographic circulation pattern acts as a biological transition zone, making the

SCB one of the most productive and economically valuable coastal regions in the United States [8]. This area hosts nearly 500 species of macroalgae [9] and fishes [10], and more than 5000 species of invertebrates [11]. Concomitantly, the Bight coastal zone is also home to more than 22 million people, the busiest and largest container ports (Los Angeles and Long Beach) in the Western Hemisphere, as well as the second largest naval facility (San Diego) in the US [12].

Santa Catalina, the largest of the Southern Channel Islands (area = 194 km^2), is the only island in the archipelago with a permanent civilian population. Surrounded by nearly ~87 km of rocky cliffs and sheltered bays, Catalina's coastal zone is dominated by rocky reef and kelp forest habitat [13]. Located ~35 km south-southwest of Los Angeles, the island is easily accessible from multiple ports and marinas on the southern California mainland and is a popular tourist destination. Moreover, the diversity of its nearshore habitats and relatively inaccessible coastline makes the island an important resource to a range of stakeholders, including fishers, recreational groups, local residents and scientists. However, with more than one million visitors annually, Catalina's marine biodiversity and ecosystems are under increasing pressure from anthropogenic stressors, such as nutrient pollution, habitat modification and climate change [14].

Recognized as an ecosystem unique for its species diversity, the State of California designated the Blue Cavern Onshore State Marine Conservation Area (SMCA), located on the leeward side of the island, as a marine protected area (MPA) in 2012. This site encompasses ~6.8 km^2 of ocean habitat and is an expansion of the Catalina Marine Life Refuge, which was established by the state as a protected area in 1974. Situated adjacent to the University of Southern California (USC) Philip K. Wrigley Marine Science Center (WMSC), Blue Cavern Onshore SMCA is part of a network of nine MPAs established around the island in which the removal of living resources is either limited or prohibited altogether as outlined by California's Marine Life Protection Act.

Here, we provide an updated species inventory of macroalgae, seagrasses, bony and cartilaginous fishes and invertebrates documented inside Blue Cavern Onshore SMCA. These findings are based on a recently published checklist of marine taxa from this area [15] and include updated records from two additional reef sites along with new data from previous data reports and articles published since 2021. It is worth noting that while the habitats and topographic complexities in and around Catalina are well known, species richness data on the intertidal and subtidal biota are far less complete [16]. Taxonomic data on species from this location are documented in an assortment of scholarly articles, technical reports, unpublished data and marine species databases, and likely represent a fraction of the biodiversity present, much of which remains unknown.

We anticipate that these new findings will provide an important baseline of species richness from Catalina Island, as well as prove useful to resource managers and policy makers for determining the mitigation costs associated with a loss in natural services. In particular, targeted conservation efforts that represent biodiversity in different regions and taxonomic groups require comprehensive inventories of the number of species present in a given area [17]. Data presented here will further improve coastal zone management by characterizing the marine biodiversity in Blue Cavern Onshore SMCA relative to other nearshore habitats in the California Channel Islands, as well as within the larger context of the SCB.

2. Materials and Methods

2.1. Study Sites

Species inventory data were recorded at 7 reef sites on the leeside of Santa Catalina Island in Blue Cavern Onshore SMCA: Big Fisherman's Cove (33°26′43.34″ N, 118°29′11.40″ W), Bird Rock (33°27′01.7″ N, 118°29′11.7″ W), Blue Caverns (33°26′47.70″ N, 118°29′35.62″ W), Habitat Reef (33°26′41.50″ N, 118°29′17.00″ W), Intake Pipes (33°26′48.95″ N, 118°29′05.75″ W), Isthmus Reef (33°26′56.2″ N, 118°29′19.4″ W) and Pumpernickel Cove (33°26′54.08″ N, 118°29′47.90″ W) (Figure 1). These sites were selected because they support a variety of intertidal and subtidal species [13,15,18] and are the most convenient to access and use for research studies based at WMSC. The subtidal habitat structure of Blue Cavern Onshore SMCA is characterized as a major reef complex [19], which includes sandy areas with rock cobble and bedrock escarpments covered with fleshy macroalgae and kelp that provide a forest habitat for invertebrates and fishes [20,21]. Details on the nearshore reef habitat in and around Blue Cavern Onshore SMCA is reported elsewhere [15].

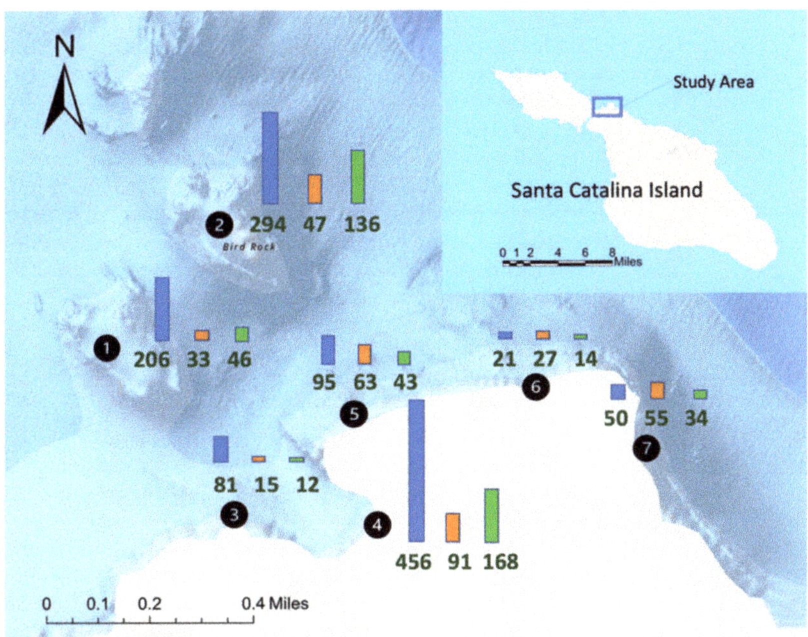

Figure 1. Map indicating location of Blue Cavern Onshore State Marine Conservation Area reef sites off Santa Catalina Island, California. Numbers circled in black correspond to each study site: 1 = Isthmus Reef; 2 = Bird Rock; 3 = Habitat Reef; 4 = Big Fisherman's Cove; 5 = Intake Pipes; 6 = Pumpernickel Cove; 7 = Blue Caverns. Values below each bar represent species count of marine invertebrates (blue bar), macroalgae and seagrasses (orange bar), and bony and cartilaginous fishes (green bar) recorded at each study location (n = 1091 species total).

2.2. Field Surveys

Timed, roving visual surveys were conducted by scuba divers between May 2015 and September 2016 and from May to September 2021 to identify individual species of subtidal marine macroalgae (excluding most crustose coralline algae), seagrasses, bony and cartilaginous fishes and invertebrates. This survey method is applicable for assessing the species richness of a variety of temperate [22] and tropical [23] marine taxa. Individual species were observed and recorded at each reef site based on methods reported previously [15,24]. Visual surveys were conducted while swimming ~1 m above the bottom contour (2–30 m depth). Total bottom time (35–60 min dive^{-1}) varied among reef sites and was dependent on the amount of breathing gas available to a diver at depth, ambient water temperatures, survey area size and visibility beneath the surface. Priority was given to conspicuous (>1 cm) subtidal macrofauna and flora that could be identified to at least the genus level of classification. Survey data on deep-water (>30 m) species, marine parasites and most planktonic organisms (<1 cm) were not recorded in this study.

2.3. Species Records

An updated species inventory from Blue Cavern Onshore SMCA was compiled from the primary and gray literature, museum records and unpublished species lists described elsewhere [15]. The earliest collections of marine taxa from Catalina date back more than 100 years. Unfortunately, these published reports are often not readily available, and in some cases, the historic names used to describe the geographic locations where samples were collected have changed. We chose to focus the current checklist on research studies and collections completed over the last 57 years, beginning with the construction of WMSC at Big Fisherman's Cove in 1965 up to the present year, 2022.

All marine taxa data and documents were either obtained via digital bibliographic resources or as a hard copy from the USC Libraries. Many intertidal and subtidal specimen records were sourced from electronic databases [25–34]. Only species that were explicitly reported as either observed or collected in water <30 m in depth at one of the study sites designated inside Blue Cavern Onshore SMCA were considered in this study. To eliminate synonyms and create a comprehensive list of valid species names for as many taxa reviewed as possible, the scientific nomenclature was confirmed for invertebrates and fishes [35], and for macroalgae and plants [29]. The open nomenclature abbreviations sp. and spp. were used to indicate that an individual or group of species within a genus were either unidentified or have yet to be described.

2.4. Data Analysis

Species richness of marine taxa from Blue Cavern Onshore SMCA was used to create an incidence matrix relating the presence of each species to a specific source citation. Individual references and field survey data used to construct the current checklist were recorded as discrete sampling units following methods established previously [15]. The numbers of bony and cartilaginous fishes, macroalgae, seagrasses and invertebrates were then used to estimate the expected richness using the mean value of 4 non-parametric incidence-based estimators (Chao2, Jack1, Jack2, Bootstrap) of species biodiversity [36–38]. These non-parametric statistical procedures are tractable for the analysis of binary species data [39] and have been demonstrated to be accurate estimators of species richness in other marine biodiversity studies [40–42]. Species richness was also evaluated by measuring the cumulative number of marine taxa documented in this study as a function of the first year in which they were reported by a specific reference. All data were analyzed using R Statistical Software [43] with the vegan [44] package.

3. Results

3.1. Species Biodiversity

A total of 1091 species of valid and unidentified marine taxa from 18 major phylogenetic groups were documented from 7 different reef sites in Blue Cavern Onshore SMCA (see Species checklist, Supplementary Table S1). A comprehensive inventory of 763 species of invertebrates, 225 species of macroalgae, 2 species of seagrass and 101 species of bony and cartilaginous fishes was compiled from a list of 158 discrete citations and observations. These findings include species reports from two phyla (Nemertea and Platyhelminthes) not reported in the previous inventory [15]. Despite a nearly 10.5-fold increase in the inventory of taxa reported in Blue Cavern Onshore SMCA since the construction of WMSC (from 104 in 1965 to a cumulative total of 1091 in 2022), none of the species-accumulation curves produced in this study approached an asymptote (Figure 2). Overall, a total of 33.8 h was spent underwater conducting roving visual surveys (n = 68 total dives) from which 105 species were documented. The majority of scuba-based surveys (47.1%) were performed in Big Fisherman's Cove, with the remainder spread out among the other 6 reef sites (Bird Rock 6.9%, Blue Caverns 4.4%, Habitat Reef 10.3%, Intake Pipes 22.2%, Isthmus Reef 5.9%, Pumpernickel Cove 4.4%).

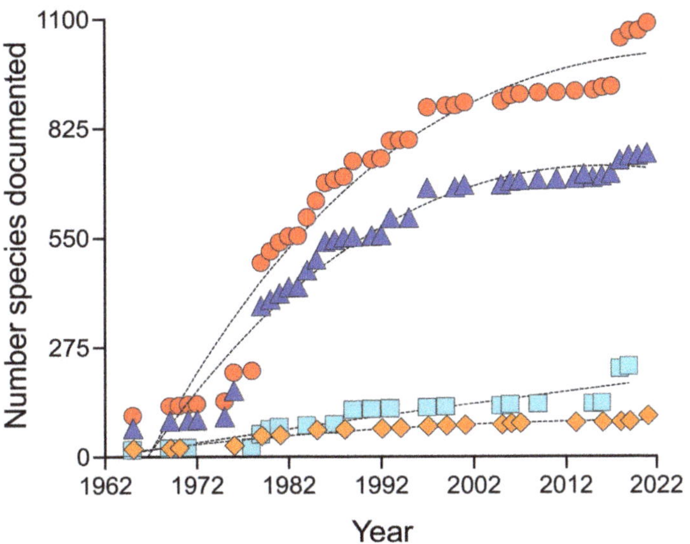

Figure 2. Species-accumulation curve for bony and cartilaginous fishes (diamonds), macroalgae and seagrasses (squares), invertebrates (triangles), as well as all marine taxa (circles) documented from seven different reef sites in Blue Cavern Onshore SMCA from 1965 to 2022.

3.2. Species Richness

The expected species richness of bony and cartilaginous fishes (123.2 ± 0.93), macroalgae and seagrasses (369.4 ± 5.2), invertebrates (1345.7 ± 11.4), as well as all taxa combined (1828.7 ± 11.7) recorded in Blue Cavern State Onshore SMCA were 1.2- to 1.8-times greater than the total number of organisms cataloged from all external sources and scuba-based surveys (Table 1). A heat map analysis was also performed on the number of species recorded within a given taxonomic group at each reef site assessed (Figure 3). Squares highlighted in red indicate taxa with high species richness, while squares colored in yellow represent taxonomic groups with the least number of species present. Marine taxa documented with the highest biodiversity (>40 species per reef site) include the Rhodophyta macroalgae, Vertebrata bony and cartilaginous fishes and the Mollusca and Arthropoda invertebrates, whereas those with the lowest richness (<2 species per site) include the Tracheophyta seagrasses, as well as the Phoronida, Sipuncula, Chaetognatha, Nemertea and Platyhelminthes invertebrates. Finally, the current species inventory for Blue Cavern State Onshore SMCA contains a total of 21 nonindigenous and invasive species, 18 species that have undergone a geographic range shift and 4 taxa listed as either a species of concern, endangered or critically endangered (Supplementary Table S1).

Table 1. Species richness of marine biota documented from Blue Cavern Onshore SMCA. Expected species richness was estimated using the mean value (±1 SE) of four non-parametric incidence-based estimators (Chao and Chiu 2016) of the number of bony and cartilaginous fishes, macroalgae and plants, invertebrates, as well as all marine taxa recorded in this study.

Type	Number Species Recorded	Non-Parametric Incidence-Based Estimators				Expected Species Richness
		Chao2	Jack1	Jack2	Bootstrap	
Bony and Cartilaginous Fishes	101	118.8	126.4	134.5	113.1	123.2 ± 0.93
Macroalgae and Seagrasses	227	441.7	339.8	423.2	272.9	369.4 ± 5.2
Invertebrates	763	1631.3	1228.6	1567.3	955.7	1345.7 ± 11.4
All taxa	1091	2148.9	1696.2	2127.6	1342.3	1828.7 ± 11.7

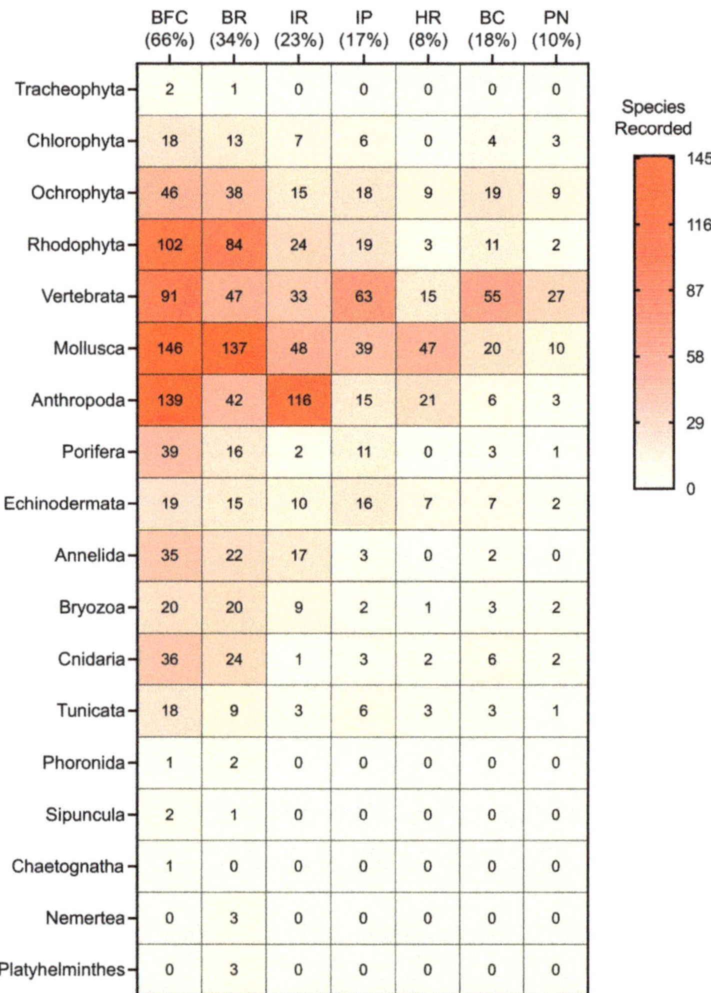

Figure 3. Heat map showing species richness of seagrasses, macroalgae, bony and cartilaginous fishes and invertebrates documented at seven reef sites in Blue Cavern Onshore SMCA. Each square indicates number of species recorded within a taxonomic group. Values in parentheses indicate percentage of studies (n = 158 total; does not include scuba-based surveys from this study) conducted at each reef site (BFC = Big Fisherman's Cove, BR = Bird Rock, IR = Isthmus Reef, IP = Intake Pipes, HR = Habitat Reef, BC = Blue Caverns, PN = Pumpernickel Cove). Color gradient (ranging from yellow to red, as shown in key) signifies changes in the number of species present (rows) among marine taxa recorded at different study sites (columns).

4. Discussion

4.1. Nearshore Biodiversity

The species inventory of macroalgae, seagrasses, bony and cartilaginous fishes and invertebrates presented here provides an important baseline of the nearshore biodiversity in the California Channel Islands and will help improve the efforts to conserve and manage similar habitats in the region. The current checklist consists of 1091 individual marine species from 18 different taxonomic groups, which represents an ~43% increase in species diversity compared to the value reported previously [15]. This increase is due in part to advances in environmental DNA (eDNA) techniques (i.e., genetic material obtained from seawater or soil samples rather than directly from an individual organism), which has increasingly been used to measure the biodiversity of marine ecosystems [45,46]. Specifically, eDNA methods were recently used [47] to survey more than 50 different species of fishes in Blue Cavern SMCA. Nonetheless, the total number of marine taxa documented here continues to represent only 60% of the estimated species richness from this area (Table 1). As previously suggested [15], this value might indicate that as much as 40% of the macrofauna and flora from this area have yet to be reported.

These findings corroborate our species-accumulation curves (Figure 2), which are regarded as well-established predictive tools for estimating species richness and sampling effort [38]. Species-accumulation data infer that the current species inventory represents only a fraction of a more highly diverse ecosystem. This underrepresentation of true species richness is likely due in part to the under-sampling of some habitats compared to others [6]. Among the many reef sites within Blue Cavern Onshore SMCA, those closest and with the most convenient access for conducting research studies at WMSC have received the most attention. Based on 158 discrete references and collections used to evaluate biodiversity data in this study, Big Fisherman's Cove, Bird Rock and Isthmus Reef were selected as survey sites 2- to 8-times more often than other locations and were host to the greatest numbers of marine taxa documented (Figure 3). Future efforts to investigate patterns of species richness in this area might close the apparent gap in the estimated number of taxa (compared to the number observed) by focusing on reef sites that are less frequented, such as Intake Pipes, Habitat Reef, Blue Caverns, Pumpernickel Cove, as well as other areas in Blue Cavern Onshore SMCA, which covers nearly 7 km^2 of nearshore habitat.

Still, the current inventory of marine taxa is reflective of the high biodiversity recognized in the coastal waters off southern California [11,18]. Overall, these data are beneficial for improving marine biodiversity conservation actions in the California Channel Islands, as well as within the larger context of the SCB [48]. The total number of species reported from Blue Cavern Onshore SMCA represents approximately 85% and 45% of the documented macroalgae and plants, 41% and 24% invertebrates and 62% and 20% of fishes known from Catalina Island [49] and the SCB [9,10], respectively. Although more than 5000 species of marine invertebrates are known from the SCB [11], the taxonomic status of many of these organisms is not yet known. Thus, a more conservative assessment of 3250 valid benthic taxa [32] was used here to estimate the number of invertebrate species documented off Catalina relative to the SCB.

4.2. Colonizers and Nonindigenous Species

Species checklists provide a means to track and monitor marine communities over time and are beneficial for detecting changes in the presence and condition of select organisms [50]. Findings from this study highlight the temporal occurrence and spatial distribution of a variety of marine taxa recorded inside Blue Cavern Onshore SMCA. In particular, the eelgrass *Zostera marina* (one of two species of seagrasses known from this area) was not documented in Big Fisherman's Cove until 1996 and is likely the result of natural colonization from populations in the Northern Channel Islands [51]. Eelgrass beds are an important refuge for marine invertebrates and fishes and provide ecosystem services whose economic value, in terms of their overall abundance and density of eelgrass habitats as a whole, outweighs their ecological function [52]. Along the Pacific coast of the United States, state and federal resource agencies recognize eelgrass as habitat areas of particular concern that provide ecologically important habitat for species to survive and reproduce and are high priorities for conservation. Once established, the presence of *Z. marina* likely attracted a variety of conspicuous species associated with eelgrass habitats to the Big Fisherman's Cove reef site, such as the California sea cucumber *Apostichopus californicus*, orangethroat pikeblenny *Chaenopsis alepidota*, Pacific angelshark *Squatina californica*, as well as the rays *Myliobatis californica* and *Urolophus halleri*.

Additionally, a total of 39 species from Blue Cavern Onshore SMCA were either introduced or have undergone a geographic range shift, while another 4 species are listed as threatened, endangered or critically endangered (Supplementary Table S1). For example, the invasive brown seaweed *Sargassum horneri* has raised concerns about its impact on native ecosystems in southern California, particularly off the Channel Islands [29,53]. Populations of *S. horneri* can cause a decrease in the abundance of fleshy macroalgae and kelp that provide a refuge for invertebrates and fishes [54], as well as significant economic losses to a variety of commercial industries ranging from fisheries and boating to tourism [55].

4.3. Survivors, Visitors and Missing in Action

Among the 864 species of invertebrates and fishes documented in Blue Cavern Onshore SMCA, 18 taxa have experienced either a geographic range expansion or contraction. For example, the range of the California dorid nudibranch *Felimare californiensis* was once widespread throughout the SCB; however, by the mid-1980s, this species was extirpated from the region [56]. After disappearing for nearly 20 years, the first sightings of *F. californiensis* were reported in 2003. Currently, only a handful of populations are known to exist off the southern California mainland and Channel Islands, which include the Big Fisherman's Cove reef site on Catalina Island. The marked decline of *F. californiensis* populations is likely due to a variety of factors, including significant increases in coastal eutrophication, loss of essential habitat and historical overharvesting by the aquarium trade [56].

Over the past 200 years, 133 local- to global-scale marine extinctions are known to have taken place [57]. Therefore, our findings are in agreement with previous studies, which suggest that that species loss and ecosystem change have become more widespread over shorter ecological timescales [58–60]. For instance, a northward shift in the species ranges of several subtropical fishes from the Pacific coast of Mexico has resulted in the frequent occurrence of both the finescale triggerfish *Balistes polylepis* and largemouth blenny *Labrisomus xanti* in Big Fisherman's Cove [15,61]. Other species of fish whose northern ranges have expanded into Blue Cavern Onshore SMCA include the Rainbow scorpionfish *Scorpaenodes xyris* [62], as well as the cardinalfishes *Apogon guadalupensis* and *A. pacificus* [63]. While the presence of these species in the SCB was once considered a relatively rare occurrence, such sightings have become more common and are likely attributed to the increasing frequency and spatial extent of marine heatwaves [64,65] and El Niño events [66,67].

Furthermore, several widespread, cryptogenic invertebrate species reported from Blue Cavern Onshore SMCA, such as the bryozoans *Bugula neretina*, *Watersipora subatra*, *W. subtorquata*, the colonial ascidian *Diplosoma listerianum* and the sea anemone *Bunodeopsis* sp., are cause for concern, given their ability to quickly settle and encrust hard substrates [26,68,69]. These species can alter the diversity of benthic ecosystems by competing with native biota for space, facilitate the spread of other nonindigenous taxa and cause significant damage to marine ecosystem services [70,71]. Other marine invertebrates, however, once commonly found in Blue Cavern Onshore SMCA, are now extremely rare. For example, two different species of echinoderms, the sea stars *Patiria miniata* and *Pisaster giganteus*, were documented in at least 15 different studies performed between 1965 and 1988 at the reef sites evaluated in this study. The sudden disappearance of these species from subtidal habitats (stretching from Alaska to Mexico) is linked to an infectious pathogen known as sea star-associated densovirus (SSaDV) [72], which caused a mass die-off of both species from Catalina and other locations in the Channel Islands [73]. Although SSaDV is not fully understood, mortality events appear to be most prevalent when sea surface temperatures are anomalously warmer than usual in the East Pacific Ocean [74]. Since 1988, there have been no reports of *P. miniata* in Blue Cavern Onshore SMCA, while *P. giganteus* has been documented twice (in 1997 and 2004).

4.4. Vulnerable and Endangered Species

Currently, 13 species of marine invertebrates and fishes living in California's nearshore waters are in danger of extinction, as outlined by the Endangered Species Act (ESA). Among the animals listed, three species of gastropod mollusks are known from Blue Cavern Onshore

SMCA (Supplementary Table S1). One of these, the endangered black abalone *Haliotis cracherodii* is a large herbivorous sea snail that inhabits both intertidal and subtidal habitat. It is worth noting that *H. cracherodii* was last documented in Big Fisherman's Cove more than 40 years ago [75]. Two additional species, the pink (*H. corrugata*) and green abalones (*H. fulgens*), are recognized as species of concern. Interestingly, *H. corrugata* and *H. fulgens* are frequently observed in Blue Cavern Onshore SMCA, serving as a reminder that MPAs are important refuges for populations facing multiple threats, such as overfishing, habitat degradation and climate change [76]. Nevertheless, despite the closure of the regional fishery in 1997, as well as the implementation of numerous restoration programs over the past three decades, abalone populations throughout the SCB are still recovering from a combination of natural and human-induced stock collapses [77] and disease events [78].

Other vulnerable species recorded in Blue Cavern Onshore SMCA include the giant sea bass *Stereolepis gigas*, the largest bony fish known from California's kelp forest habitat and classified as critically endangered by the IUCN Red List. Truly a behemoth fish, giant sea bass are members of the wreckfish family and can grow to more than 2 m in total length and 255 kg in weight [79]. One of the largest individuals documented was observed off Catalina Island at Goat Harbor (~5 km southeast of Blue Cavern Onshore SMCA) and measured 2.75 m in total length with an estimated weight of 380 kg [80]. Although never listed as endangered under the ESA [81], *S. gigas* stocks along the California coast were so severely impacted by overfishing that a moratorium was declared in 1982 [82]. Remarkably though, the number of giant sea bass documented in Blue Cavern Onshore SMCA has become more common in recent years, in which encounters with these fish are known from four of the reef sites (Blue Caverns, Big Fisherman's Cove, Intake Pipes and Isthmus Reef) assessed in this study. Finally, East Pacific green turtles (*Chelonia mydas*), listed as threatened under the ESA, regularly visit the highly productive coastal waters off southern California from their nesting beaches in Mexico. Although not one of the species included in this study, one of the authors (Ginsburg) recently observed an adult green turtle in Big Fisherman's Cove, which is the first report to our knowledge to document their presence at this location.

5. Conclusions

Further investigations of species richness from Blue Cavern Onshore SMCA and other nearby protected areas on Catalina Island (and the California Channel Islands archipelago altogether) are required to provide new insights into the mechanisms that contribute to both the spatial and temporal connectivity among populations in the region. Such studies will help communicate the significance of conserving marine biodiversity for future generations with both the general public and stakeholder groups.

Supplementary Materials: The following supporting information can be downloaded at: https://www.mdpi.com/article/10.3390/d14050366/s1, Supplementary Table S1. Species checklist of (I) Macroalgae (excluding most crustose coralline algae), (II) Seagrasses, (III) Invertebrates and (IV) Bony and Cartilaginous Fishes documented from Blue Cavern Onshore SMCA (n = 1091 species total). The abbreviations sp. and spp. indicate that an individual or group of species within a genus were either unidentified or have yet to be described. Taxa organized alphabetically by phylum and class (columns organized top to bottom, left to right). Superscripted letters indicate the following: [a] Nonindigenous species; [b] Species range shift; [c] Species of concern; d Endangered species; [e] Critically endangered species; [f] Identification based on morphology; [g] New taxon added to checklist.

Author Contributions: Conceptualization, D.W.G. and A.H.H.; Data curation, D.W.G. and A.H.H.; Formal analysis, D.W.G. and A.H.H.; Methodology, D.W.G.; Validation, D.W.G. and A.H.H.; Writing—original draft, D.W.G.; Writing—review and editing, D.W.G. and A.H.H. All authors have read and agreed to the published version of the manuscript.

Funding: Funding and resources for this study provided by NSF-OCE (#1559941), USC Sea Grant, USC Dornsife College, the Zinsmeyer Family Endowed Undergraduate Research Fund and the Wrigley Institute for Environmental Studies.

Institutional Review Board Statement: Not applicable.

Informed Consent Statement: Not applicable.

Data Availability Statement: Data available from authors upon reasonable request.

Acknowledgments: We would like to thank the USC Wrigley Institute for Environmental Studies and Wrigley Marine Science Center staff for their assistance with boat and dive operations for this project. Special thanks to A. Looby for getting this project started and to J. Engle, M. Fourriére, G. Hendler and K.A. Miller for their suggestions regarding the analysis of species inventory data. The authors would also like to acknowledge C. Dreja, J. McCarty, J. Beck, K. Relf, and P. Samwell-Smith for inspiring the title of this article. This study utilized data collected by the Multi-Agency Rocky Intertidal Network (MARINe): a long-term ecological consortium primarily supported by BOEM, NPS, The David & Lucile Packard Foundation and USN, D. Pondella and J. Williams from the Vantuna Research Group at Occidental College and J. Freiwald from Reef Check California. This is contribution no. 258 from the USC Wrigley Marine Science Center on Santa Catalina Island.

Conflicts of Interest: The authors declare no conflict of interest.

References

1. Caldow, C.; Monaco, M.E.; Pittman, S.J.; Kendall, M.S.; Goedeke, T.L.; Menza, C.; Kinlan, B.P.; Costa, B.M. Biogeographic assessments: A framework for information synthesis in marine spatial planning. *Mar. Policy* **2015**, *51*, 423–432. [CrossRef]
2. McCann, K.S. The diversity-stability debate. *Nature* **2000**, *405*, 228–233. [CrossRef]
3. Gotelli, N.J.; Colwell, R.K. Quantifying biodiversity: Procedures and pitfalls in the measurement and comparison of species richness. *Ecol. Lett.* **2001**, *4*, 379–391. [CrossRef]
4. Briggs, J.C. Species diversity: Land and sea compared. *Syst. Biol.* **1994**, *43*, 130–135. [CrossRef]
5. Sala, E.; Knowlton, N. Global marine biodiversity trends. *Annu. Rev. Environ. Resour.* **2006**, *31*, 93–122. [CrossRef]
6. Bearham, D.; Strzelecki, J.; Hara, A.; Hosie, A.; Kirkendale, L.; Richards, Z.; Huisman, J.M.; Liu, D.; McLaughlin, J.; Naughton, K.M.; et al. Habitats and benthic biodiversity across a tropical estuarine–marine gradient in the eastern Kimberley region of Australia. *Reg. Stud. Mar.* **2022**, *49*, 102039. [CrossRef]
7. Spalding, M.D.; Fox, H.E.; Allen, G.R.; Davidson, N.; Ferdaña, Z.A.; Finlayson, M.A.X.; Halpern, B.S.; Jorge, M.A.; Lombana, A.L.; Lourie, S.A.; et al. Marine ecoregions of the world, a bioregionalization of coastal and shelf areas. *BioScience* **2007**, *57*, 573–583. [CrossRef]
8. Fautin, D.; Dalton, P.; Incze, L.S.; Leong, J.A.C.; Pautzke, C.; Rosenberg, A.; Sandifer, P.; Sedberry, G.; Tunnell, J.W., Jr.; Abbott, I.; et al. An overview of marine biodiversity in United States waters. *PLoS ONE* **2010**, *5*, e11914. [CrossRef]
9. Murray, S.N.; Bray, R.N. Benthic macrophytes. In *Ecology of the Southern California Bight, a Synthesis and Interpretation*; Dailey, M.D., Reish, D.J., Anderson, J.W., Eds.; University of California Press: Berkeley, CA, USA, 1993; pp. 304–368. [CrossRef]
10. Cross, J.N.; Allen, L.G. Fishes. In *Ecology of the Southern California Bight, a Synthesis and Interpretation*; Dailey, M.D., Reish, D.J., Anderson, J.W., Eds.; University of California Press: Berkeley, CA, USA, 1993; pp. 459–540. [CrossRef]
11. Dailey, M.D.; Anderson, J.W.; Reish, D.J.; Gorsline, D.S. The Southern California Bight: Background and Setting. In *Ecology of the Southern California Bight, a Synthesis and Interpretation*; Dailey, M.D., Reish, D.J., Anderson, J.W., Eds.; University of California Press: Berkeley, CA, USA, 1993; pp. 1–18. [CrossRef]
12. Schiff, K.; Greenstein, D.; Dodder, N.; Gillett, D.J. Southern California bight regional monitoring. *Reg. Stud. Mar. Sci.* **2016**, *4*, 34–46. [CrossRef]
13. Zahn, L.A.; Claisse, J.T.; Williams, J.P.; Williams, C.M.; Pondella, D.J. The biogeography and community structure of kelp forest macroinvertebrates. *Mar. Ecol.* **2016**, *37*, 770–785. [CrossRef]
14. Harley, C.D.; Rogers-Bennett, L. The potential synergistic effects of climate change and fishing pressure on exploited invertebrates on rocky intertidal shores. *Calif. Coop. Ocean. Fish. Investig. Rep.* **2004**, *45*, 98.
15. Looby, A.; Ginsburg, D.W. Nearshore species biodiversity of a marine protected area off Santa Catalina Island, California. *West. N. Am. Nat.* **2021**, *81*, 113–130. [CrossRef]
16. Engle, J.M. Distribution patterns of rocky subtidal fishes around the California Islands. In *Proceedings of the Third California Islands Symposium*; Hochberg, F.G., Ed.; Santa Barbara Museum of Natural History: Santa Barbara, CA, USA, 1993; pp. 475–484.
17. Luypaert, T.; Hagan, J.G.; McCarthy, M.L.; Poti, M. Status of Marine Biodiversity in the Anthropocene. In *YOUMARES 9—The Oceans: Our Research, Our Future*; Jungblut, S., Liebich, V., Bode-Dalby, M., Eds.; Springer: Cham, Switzerland, 2020; pp. 57–82. [CrossRef]
18. Claisse, J.T.; Blanchette, C.A.; Dugan, J.E.; Williams, J.P.; Freiwald, J.; Pondella, D.J.; Schooler, N.K.; Hubbard, D.M.; Davis, K.; Zahn, L.A. Biogeographic patterns of communities across diverse marine ecosystems in southern California. *Mar. Ecol.* **2018**, *39*, e12453. [CrossRef]
19. Pondella, D.J.; Williams, J.; Claisse, J.; Schaffner, B.; Ritter, K.; Schiff, K. The physical characteristics of nearshore rocky reefs in the Southern California Bight. *Bull. South. Calif. Acad. Sci.* **2015**, *114*, 105–122. [CrossRef]
20. Abbott, I.A.; Hollenberg, G.J. *Marine Algae of California*; Stanford University Press: Stanford, CA, USA, 1992; p. 844. [CrossRef]

21. Parnell, P.E.; Miller, E.F.; Lennert-Cody, C.E.; Dayton, P.K.; Carter, M.L.; Stebbins, T.D. The response of giant kelp (*Macrocystis pyrifera*) in southern California to low-frequency climate forcing. *Limnol. Oceanogr.* **2010**, *55*, 2686–2702. [CrossRef]
22. Davis, G.E.; Kushner, D.J.; Mondragon, J.M.; Mondragon, J.E.; Lerma, D.; Richards, D.V. Sampling protocol. In *Kelp Forest Monitoring Handbook*; Channel Islands National Park: Ventura, CA, USA, 1997; Volume 1, pp. 8–33.
23. Holt, B.G.; Rioja-Nieto, R.; MacNeil, M.A.; Lupton, J.; Rahbek, C. Comparing diversity data collected using a protocol designed for volunteers with results from a professional alternative. *Methods Ecol. Evol.* **2013**, *4*, 383–392. [CrossRef]
24. Schmitt, E.F.; Sluka, R.D.; Sullivan-Sealey, K.M. Evaluating the use of roving diver and transect surveys to assess the coral reef fish assemblage off southeastern Hispaniola. *Coral Reefs.* **2002**, *21*, 216–223. [CrossRef]
25. Catania, D. CAS Ichthyology (ICH). v150.315. California Academy of Sciences. Available online: http://ipt.calacademy.org:8080/resource?r=ich&v=150.315 (accessed on 30 April 2022).
26. Fofonoff, P.W.; Ruiz, G.M.; Steves, B.; Simkanin, C.; Carlton, J.T. National Exotic Marine and Estuarine Species Information System. Available online: http://invasions.si.edu/nemesis (accessed on 4 April 2022).
27. marine.ucsc.edu. Long-Term Monitoring Program. Available online: https://marine.ucsc.edu/sitepages/bigfisherman-bio.html (accessed on 30 April 2022).
28. National Museum of Natural History, Smithsonian Institution Invertebrate Zoology Collection Database. Available online: https://collections.nmnh.si.edu/search/iz/ (accessed on 30 April 2022).
29. Miller, K.A. California Seaweeds eFlora. Available online: http://ucjeps.berkeley.edu/seaweedflora/ (accessed on 30 April 2022).
30. Natural History Museum of Los Angeles County Vertebrate Collection. Available online: http://ipt.vertnet.org:8080/ipt/resource.do?r=lacm_verts (accessed on 30 April 2022).
31. Reef Check California, Global Reef Tracker. Available online: http://data.reefcheck.org/ (accessed on 30 April 2022).
32. Southern California Association of Marine Invertebrate Taxonomists: Taxonomic Listing of Macro- and Megainvertebrates from Infaunal and Epibenthic Programs in the Southern California Bight, 12th Edition. Available online: https://www.scamit.org/publications/SCAMIT%20Ed%2012-2018.pdf (accessed on 30 April 2022).
33. Scripps Institution of Oceanography Marine Vertebrate Collection. Available online: http://ipt.vertnet.org:8080/ipt/resource.do?r=sio_marine_vertebrates (accessed on 30 April 2022).
34. Pondella, D.J.; Caselle, J.E.; Claisse, J.T.; Williams, J.P.; Davis, K.; Williams, C.M.; Zahn, L.A. South Coast Baseline Program Final Report: Kelp and Shallow Rock Ecosystems. Available online: https://caseagrant.ucsd.edu/sites/default/files/SCMPA-27-Final-Report_0.pdf (accessed on 30 April 2022).
35. World Register of Marine Species. Available online: https://www.marinespecies.org (accessed on 30 April 2022).
36. Burnham, K.P.; Overton, W.S. Estimation of the size of a closed population when capture probabilities vary among animals. *Biometrika* **1978**, *65*, 625–633. [CrossRef]
37. Chao, A. Estimating the population size for capture-recapture data with unequal catchability. *Biometrics* **1987**, *43*, 783–791. [CrossRef]
38. Colwell, R.K.; Coddington, J.A. Estimating terrestrial biodiversity through extrapolation. *Philos. Trans. R. Soc. Lond. B Biol. Sci.* **1994**, *345*, 101–118. [CrossRef]
39. Chao, A.; Chiu, C.-H. *Nonparametric Estimation and Comparison of Species Richness*; John Wiley and Sons, Ltd.: Hoboken, NJ, USA, 2016; pp. 1–11. [CrossRef]
40. Foggo, A.; Attrill, M.J.; Frost, M.T.; Rowden, A.A. Estimating marine species richness: An evaluation of six extrapolative techniques. *Mar. Ecol. Prog. Ser.* **2003**, *248*, 15–26. [CrossRef]
41. Drew, J.A.; Buxman, C.L.; Holmes, D.D.; Mandecki, J.L.; Mungkaje, A.J.; Richardson, A.C.; Westneat, M.W. Biodiversity inventories and conservation of the marine fishes of Bootless Bay, Papua New Guinea. *BMC Ecol.* **2012**, *12*, 15. [CrossRef] [PubMed]
42. Fourriére, M.; Reyes-Bonilla, H.; Rodríguez-Zaragoza, F.A.; Nicole, N. Fishes of Clipperton Atoll, Eastern Pacific: Checklist, endemism, analysis of completeness of the inventory. *Pac. Sci.* **2014**, *68*, 375–395. [CrossRef]
43. R Core Team. R: A Language and Environment for Statistical Computing, Vienna, Austria. Available online: https://www.R-project.org/ (accessed on 10 February 2022).
44. Oksanen, J.; Blanchet, F.G.; Friendly, M.; Kindt, R.; Legendre, P.; McGlinn, D.; Minchin, P.R.; O'Hara, R.B.; Simpson, G.L.; Solymos, P.; et al. Vegan: Community Ecology Package. R Package Version 2.5-6. Available online: https://CRAN.R-project.org/package=vegan (accessed on 10 February 2022).
45. Thomsen, P.F.; Kielgast, J.; Iversen, L.L.; Møller, P.R.; Rasmussen, M.; Willerslev, E. Detection of a Diverse Marine Fish Fauna Using Environmental DNA from Seawater Samples. *PLoS ONE* **2012**, *7*, e41732. [CrossRef] [PubMed]
46. Bohmann, K.; Evans, A.; Gilbert, M.T.P.; Carvalho, G.R.; Creer, S.; Knapp, M.; Yu, D.W.; Bruyn, M. Environmental DNA for wildlife biology and biodiversity monitoring. *Trends Ecol. Evol.* **2014**, *29*, 358–367. [CrossRef] [PubMed]
47. Ely, T.; Barber, P.H.; Man, L.; Gold, Z. Short-lived detection of an introduced vertebrate eDNA signal in a nearshore rocky reef environment. *PLoS ONE* **2021**, *16*, e0245314. [CrossRef]
48. Costello, M.J.; Ballantine, B. Biodiversity conservation should focus on no-take marine reserves: 94% of marine protected areas allow fishing. *Trends Ecol. Evol.* **2015**, *30*, 507–509. [CrossRef]
49. Engle, J.M. *Provisional Checklist of the Marine Species of Santa Catalina Island*; Unpublished Report; University of Southern California, Wrigley Marine Center: Santa Catalina Island, CA, USA, 1978; p. 259.

50. Hammond, P. Species inventory. In *Global Biodiversity: Status of the Earth's Living Resources*; Groombridge, B., Ed.; Chapman and Hall: London, UK, 1992; pp. 17–39. [CrossRef]

51. Engle, J.M.; Miller, K.A. *Distribution and Morphology of Eelgrass* (Zostera marina *L.) at the California Channel Islands*; Garcelon, D.K., Schween, C.A., Eds.; Institute for Wildlife Studies: Arcata, CA, USA, 2005; pp. 405–414.

52. Barbier, E.B.; Hacker, S.D.; Kennedy, C.; Koch, E.W.; Stier, A.C.; Silliman, B.R. The value of estuarine and coastal ecosystem services. *Ecol. Monogr.* **2011**, *81*, 169–193. [CrossRef]

53. Marks, L.; Reed, D.; Obaza, A. Assessment of control methods for the invasive seaweed *Sargassum horneri* in California, USA. *Manag. Biol. Invasions* **2017**, *8*, 205–213. [CrossRef]

54. Ginther, S.C.; Steele, M.A. Limited recruitment of an ecologically and economically important fish *Paralabrax clathratus* to an invasive alga. *Mar. Ecol. Prog. Ser.* **2018**, *602*, 213–224. [CrossRef]

55. Williams, S.L.; Smith, J.E. A global review of the distribution, taxonomy, and impacts of introduced seaweeds. *Annu. Rev. Ecol. Evol. Syst.* **2007**, *38*, 327–359. [CrossRef]

56. Goddard, J.H.R.; Schaefer, M.C.; Hoover, C.; Valdés, Á. Regional extinction of a conspicuous dorid nudibranch (Mollusca: Gastropoda) in California. *Mar. Biol.* **2013**, *160*, 1497–1510. [CrossRef]

57. Dulvy, N.K.; Sadovy, Y.; Reynolds, J.D. Extinction vulnerability in marine populations. *Fish Fish.* **2003**, *4*, 25–64. [CrossRef]

58. Harnik, P.G.; Lotze, H.K.; Anderson, S.C.; Finkel, Z.V.; Finnegan, S.; Lindberg, D.R.; Liow, L.H.; Lockwood, R.; McClain, C.R.; McGuire, J.L.; et al. Extinctions in ancient and modern seas. *Trends Ecol. Evol.* **2012**, *27*, 608–617. [CrossRef] [PubMed]

59. McCauley, D.J.; Pinsky, M.L.; Palumbi, S.R.; Estes, J.A.; Joyce, F.H.; Warner, R.R. Marine defaunation: Animal loss in the global ocean. *Science* **2015**, *347*, 1255641. [CrossRef]

60. Johnson, C.N.; Balmford, A.; Brook, B.W.; Buettel, J.C.; Galetti, M.; Guangchun, L.; Wilmshurst, J.M. Biodiversity losses and conservation responses in the Anthropocene. *Science* **2017**, *356*, 270–275. [CrossRef]

61. Love, M.S.; Passarelli, J.K.; Cantrell, B.; Hastings, P.A. The largemouth blenny, *Labrisomus xanti*, new to the California marine fauna with a list of and key to the species of Labrisomidae, Clinidae, Chaenopsidae found in California waters. *Bull. South. Calif. Acad. Sci.* **2016**, *115*, 191–197. [CrossRef]

62. Lea, R.N.; Rosenblatt, R.H. Observations on fishes associated with the 1997-98 El Niño off California. *Calif. Coop. Ocean. Fish. Investig. Rep.* **2000**, *41*, 117–129.

63. Lea, R.N.; Fraser, T.H.; Baldwin, C.C.; Craig, M.T. Five Valid Species of Cardinalfishes of the Genus *Apogon* (Apogonidae) in the Eastern Pacific Ocean, with a Redescription of *A. atricaudus* and Notes on the Distribution of *A. atricaudus* and *A. atradorsatus*. *Ichthyol. Herpetol.* **2022**, *110*, 106–114. [CrossRef]

64. Hobday, A.J.; Alexander, L.V.; Perkins, S.E.; Smale, D.A.; Straub, S.C.; Oliver, E.C.; Benthuysen, J.A.; Burrows, M.T.; Donat, M.G.; Feng, M.; et al. A hierarchical approach to defining marine heatwaves. *Prog. Oceanogr.* **2016**, *141*, 227–238. [CrossRef]

65. Sanford, E.; Sones, J.L.; García-Reyes, M.; Goddard, J.H.; Largier, J.L. Widespread shifts in the coastal biota of northern California during the 2014–2016 marine heatwaves. *Sci. Rep.* **2019**, *9*, 4216. [CrossRef] [PubMed]

66. Rosales-Casián, J.A. Finescale triggerfish (*Balistes polylepis*) and roosterfish (*Nematistius pectoralis*) presence in temperate waters off Baja California, México, evidence of El Niño conditions. *Calif. Cooperative Ocean. Fish. Investig. Rep.* **2013**, *54*, 81–84.

67. Walker, H.J.; Hastings, P.A.; Hyde, J.R.; Lea, R.N.; Snodgrass, O.E.; Bellquist, L.F. Unusual occurrences of fishes in the southern California current system during the warm water period of 2014–2018. *Estuar. Coast. Shelf Sci.* **2020**, *236*, 106634. [CrossRef]

68. Obaza, A.K.; Williams, J.P. Spatial and temporal dynamics of the overwater structure fouling community in southern California. *Mar. Freshw. Res.* **2018**, *69*, 1771–1783. [CrossRef]

69. Susick, K.; Scianni, C.; Mackie, J.A. Artificial structure density predicts fouling community diversity on settlement panels. *Biol. Invasions.* **2019**, *22*, 271–292. [CrossRef]

70. Molnar, J.L.; Gamboa, R.L.; Revenga, C.; Spalding, M.D. Assessing the global threat of invasive species to marine biodiversity. *Front. Ecol. Environ.* **2008**, *6*, 485–492. [CrossRef]

71. Mackie, J.A.; Darling, J.A.; Geller, J.B. Ecology of cryptic invasions: Latitudinal segregation among *Watersipora* (Bryozoa) species. *Sci. Rep.* **2012**, *2*, 871. [CrossRef] [PubMed]

72. Hewson, I.; Button, J.B.; Gudenkauf, B.M.; Miner, B.; Newton, A.L.; Gaydos, J.K.; Wynne, J.; Groves, C.L.; Hendler, G.; Murray, S.; et al. Densovirus associated with sea-star wasting disease and mass mortality. *Proc. Natl. Acad. Sci. USA* **2014**, *111*, 17278–17283. [CrossRef]

73. Eckert, G.L.; Engle, J.M.; Kushner, D.J. Sea star disease and population declines at the Channel Islands. In *Proceedings of the Fifth California Islands Symposium*; Browne, D.R., Mitchell, K.L., Chaney, H.W., Eds.; U.S. Department of the Interior, Minerals Management Service: Camarillo, CA, USA, 2000; pp. 390–393.

74. Harvell, C.D.; Montecino-Latorre, D.; Caldwell, J.M.; Burt, J.M.; Bosley, K.; Keller, A.; Heron, S.F.; Salomon, A.K.; Lee, L.; Pontier, O.; et al. Disease epidemic and a marine heat wave are associated with the continental-scale collapse of a pivotal predator (*Pycnopodia helianthoides*). *Sci. Adv.* **2019**, *5*, eaau7042. [CrossRef]

75. Given, R.R.; Robertson, D. *Biological Studies on the Rock Mole-Pier Complex*; Unpublished Report; University of Southern California, Wrigley Marine Center: Santa Catalina Island, CA, USA, 1981; p. 134.

76. Nickols, K.J.; White, J.W.; Malone, D.; Carr, M.H.; Starr, R.M.; Baskett, M.L.; Hastings, A.; Botsford, L.W. Setting ecological expectations for adaptive management of marine protected areas. *J. Appl. Ecol.* **2019**, *56*, 2376–2385. [CrossRef]

77. Karpov, K.A.; Haaker, P.L.; Taniguchi, I.K.; Rogers-Bennett, L. Serial depletion and the collapse of the California abalone fishery (*Haliotis* spp.) fishery. *Can. Spec. Publ. Fish. Aquat. Sci.* **2000**, *130*, 11–24.

78. Harvell, C.D.; Lamb, J.B. Disease outbreaks can threaten marine biodiversity. In *Marine Disease Ecology*; Oxford University Press: Oxford, UK, 2020; pp. 141–158. [CrossRef]

79. Allen, L.G. GIANTS! Or . . . The Return of the Kelp Forest King. *Copeia* **2017**, *105*, 10–13. [CrossRef]

80. House, P.H.; Clark, B.L.; Allen, L.G. The return of the king of the kelp forest: Distribution, abundance, biomass of giant sea bass (*Stereolepis gigas*) off Santa Catalina Island, California, 2014–2015. *Bull. South. Calif. Acad. Sci.* **2016**, *115*, 1–14. [CrossRef]

81. Ramírez-Valdez, A.; Rowell, T.J.; Dale, K.E.; Craig, M.T.; Allen, L.G.; Villaseñor-Derbez, J.C.; Cisneros-Montemayor, A.M.; Hernández-Velasco, A.; Torre, J.; Hofmeister, J.; et al. Asymmetry across international borders: Research, fishery and management trends and economic value of the giant sea bass (*Stereolepis gigas*). *Fish Fish.* **2021**, *22*, 1392–1411. [CrossRef]

82. Hawk, H.A.; Allen, L.G. Age and growth of the giant sea bass, *Stereolepis gigas*. *Coop. Ocean. Fish. Investig. Rep.* **2014**, *55*, 128–134.

Article

Gene Transfer Agent *g5* Gene Reveals Bipolar and Endemic Distribution of *Roseobacter* Clade Members in Polar Coastal Seawater

Yin-Xin Zeng [1,2,3,*], **Hui-Rong Li** [1] and **Wei Luo** [1]

[1] Key Laboratory for Polar Science, Polar Research Institute of China, Ministry of Natural Resources, Shanghai 200136, China; lihuirong@pric.org.cn (H.-R.L.); luowei@pric.org.cn (W.L.)

[2] School of Oceanography, Shanghai Jiao Tong University, Shanghai 200030, China

[3] Antarctic Great Wall Ecology National Observation and Research Station, Polar Research Institute of China, Shanghai 201209, China

[*] Correspondence: zengyinxin@pric.org.cn; Tel.: +86-21-5871-7207

Abstract: The *Roseobacter* clade represents one of the most abundant groups of marine bacteria and plays important biogeochemical roles in marine environments. *Roseobacter* genomes commonly contain a conserved gene transfer agent (GTA) gene cluster. A major capsid protein-encoding GTA (*g5*) has been used as a genetic marker to estimate the diversity of marine roseobacters. Here, the diversity of roseobacters in the coastal seawater of Arctic Kongsfjorden and Antarctic Maxwell Bay was investigated based on *g5* gene clone library analysis. Four *g5* gene clone libraries were constructed from microbial assemblages representing Arctic and Antarctic regions. The genus *Phaeobacter* was exclusively detected in Arctic seawater, whereas the genera *Jannaschia*, *Litoreibacter* and *Pacificibacter* were only observed in Antarctic seawater. More diverse genera within the *Roseobacter* clade were observed in Antarctic clones than in Arctic clones. The genera *Sulfitobacter*, *Loktanella* and *Yoonia* were dominant (higher than 10% of total clones) in both Arctic and Antarctic samples, implying their roles in polar marine environments. The results not only indicated a bipolar or even global distribution of roseobacters in marine environments but also showed their endemic distribution either in the Arctic or Antarctic. Endemic phylotypes were more frequently observed in polar regions than cosmopolitan phylotypes. In addition, endemic phylotypes were more abundant in Arctic samples (84.8% of Arctic sequences) than in Antarctic samples (54.3% of Antarctic sequences).

Keywords: gene transfer agent; *Roseobacter*; diversity; distribution; Kongsfjorden; Maxwell Bay

Citation: Zeng, Y.-X.; Li, H.-R.; Luo, W. Gene Transfer Agent *g5* Gene Reveals Bipolar and Endemic Distribution of *Roseobacter* Clade Members in Polar Coastal Seawater. *Diversity* **2022**, *14*, 392. https://doi.org/10.3390/d14050392

Academic Editors: Michael Wink and Thomas J. Trott

Received: 12 March 2022
Accepted: 12 May 2022
Published: 14 May 2022

Publisher's Note: MDPI stays neutral with regard to jurisdictional claims in published maps and institutional affiliations.

1. Introduction

The marine *Roseobacter* clade is the largest lineage within the family Rhodobacteraceae and is metabolically, phenotypically, and ecologically diverse [1–3]. Members of the roseobacters can comprise up to 25% of total marine bacterioplankton [1], making it one of the most abundant groups of marine bacteria [4]. In addition, the *Roseobacter* clade contains isolates capable of dimethylsulfoniopropionate (DMSP) degradation [5], carbon monoxide oxidation [6], anoxygenic photosynthesis [7], and quorum sensing [8], which play crucial ecological and environmental roles [9,10]. Moreover, some roseobacters have been found to be dominant in polar marine environments [11–13].

Gene transfer agents (GTAs) are DNA-containing phage-like particles encoded and produced by certain bacteria and archaea [14]. They can package fragments of the host genome and transfer the encapsidated genetic material to the recipient [15]. GTA-related gene transfer is regarded as a potential adaptive mechanism for microbes to maintain metabolic flexibility in changing environments. GTAs are common in genomes of the *Roseobacter* clade [5,16]. Among the GTA genes, a major capsid gene (*g5*) is highly conserved among those bacteria and has been widely used as a gene marker to estimate the diversity of marine roseobacters [17,18]. However, information on the diversity of GTA *g5* genes in marine roseobacters at high latitudes is still limited.

Previous studies have revealed both bipolar and endemic distributions of marine *Roseobacter* clade members and their functional genes (e.g., *pufM*, *dmdA* and *dddP*) in polar regions [19,20]. It is hypothesized that a similar phenomenon can be found in the distribution of the *g5* gene in roseobacters in polar marine environments. In this study, the GTA *g5* gene was directly obtained from the marine environment to explore the diversity of roseobacters and the difference in the phylogenetic diversity of *Roseobacter* clade bacteria between Arctic and Antarctic coastal seawaters. The results will help improve our understanding of the distribution of *Roseobacter* clade bacteria and their adaptation to unique polar environments.

2. Materials and Methods

2.1. Sample Collection and DNA Extraction

Two Arctic surface seawater samples were collected from an outer station, St1, and inner station, St5, in Kongsfjorden in July 2011, whereas two Antarctic surface seawater samples were collected from Station A5 in Ardley Cove and Station G5 in Great Wall Cove of Maxwell Bay in December 2011 (Figure 1). The locations and sampling dates are summarized in Table 1 [12,21]. Microorganisms were collected by filtering water through a 0.2 μm-pore-size filter. DNA extraction was performed as previously described [22,23]. The purity and concentration of the extracted DNA were estimated using a Nanodrop 2000 spectrophotometer (Thermo Fischer Scientific, Hvidovre, Denmark).

Arctic Ny-Ålesund

Figure 1. *Cont.*

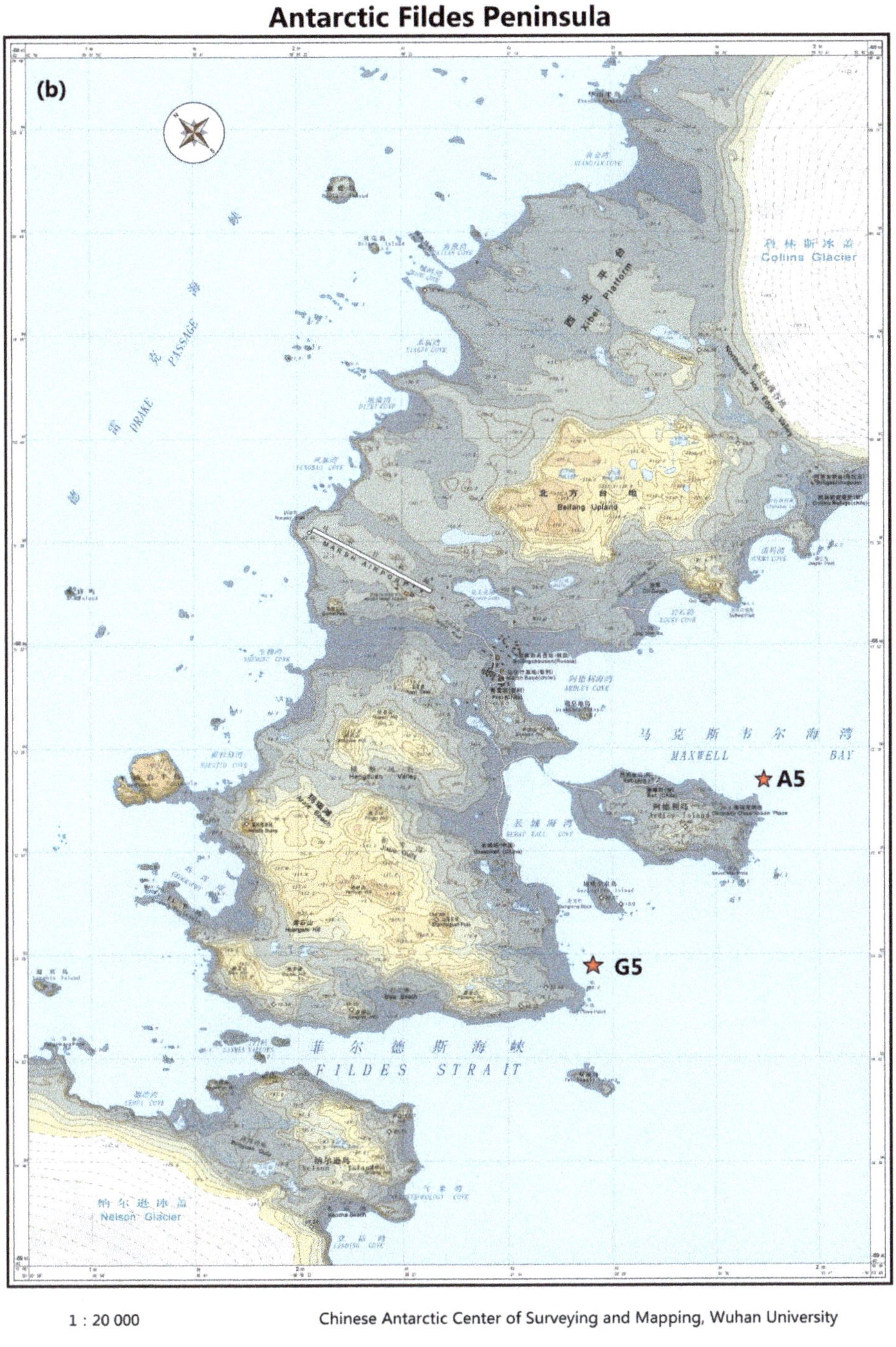

1 : 20 000   Chinese Antarctic Center of Surveying and Mapping, Wuhan University

Figure 1. Maps of sampling sites in Arctic Kongsfjorden (**a**) and Antarctic Maxwell Bay (**b**) in 2011.

Table 1. Estimation of sequence diversity and genotype coverage of gene transfer agent *g5* clone libraries.

Parameter	Value of Arctic Kongsfjorden		Value of Antarctic Maxwell Bay	
Sampling station	St1	St5	A5	G5
Sampling site	78°59′17″ N, 11°39′34″ E	78°54′20″ N, 12°17′34″ E	62°12′42″ S, 58°54′41″ W	62°13′37″ S, 58°56′37″ W
Number of clones sequenced	297	107	157	154
Number of OTUs	112	28	89	79
Coverage (%)	75.3	90.7	68.8	68.8
Shannon–wiener index	3.984	2.752	4.329	4.107
Simpson index	0.042	0.109	0.016	0.022
Evenness	0.844	0.826	0.964	0.940
Species richness (Chao1)	289	35	141	161

2.2. Clone Library Construction and Sequencing of the GTA g5 Gene

Four *g5* gene clone libraries were constructed using nucleic acids obtained from surface water samples. Degenerate PCR primers MCP-109F (5′-GGC TAY CTG GTS GAT CCS CAR AC-3′) and MCP-368R (5′-TAG AAC AGS ACR TGS GGY TTK GC-3′) were used to amplify GTA *g5* genes in the present study [18]. Amplification was conducted using a total volume of 25 µL containing 200 ng of template DNA, 12.5 µL of DreamTaq Green PCR Master Mix (2×; Thermo Scientific, Waltham, USA), 2 µg of bovine serum albumin (BSA), and 0.4 µmol/L of each primer. The thermocycling conditions were listed as follows: 30 s at 98 °C; 35 cycles at 98 °C for 10 s, 60 °C for 30 s, and 72 °C for 30 s; and a final extension step at 72 °C for 7 min. The success of PCR was determined through electrophoresis of 4 µL of the reaction mixture in 0.8% (*w*/*v*) agarose gels. Genomic DNA from one reference strain *Ruegeria pomeroyi* DSS-3, which possesses a *g5* gene (CP000031), was used as a template for the positive control. Both distilled water and genomic DNA from *Escherichia coli* DH5α were used as a template for negative controls. The *g5* gene PCR products were ligated to the vector pMD18-T (Takara, Dalian, China) and used to transform *E. coli* DH5α competent host cells. All clones were screened for inserts through colony PCR with the M13 primer sequences flanking the pMD18-T cloning site as described by Zeng et al. [24]. After verification through gel electrophoresis, all positive clones were sent to Majorbio Biopharm Technology Co., Ltd. (Shanghai, China) for Sanger sequencing.

2.3. Data Analysis

The obtained *g5* gene sequences were checked for chimeras using the Bellerophon program [25]. A 98% identity in the putative *g5* sequence was employed to group sequences into the same operational taxonomic unit (OTU) or genotype in this study [18]. From each OTU, a single sequence was selected as a representative. Coverage, Shannon's diversity and Simpson's diversity indices of the clone library were estimated using the SpadeR (https://chao.shinyapps.io/SpadeR; accessed on 16 January 2022). The DNA sequences were translated into amino acid sequences and then aligned and compared with reference sequences from bacterial taxa in the GenBank database. Amino acid sequences were grouped at the 90% similarity level [18]. Neighbor-joining phylogenetic trees were constructed using the MEGA 5.1 software. The evolutionary distances were calculated under the Jones–Taylor–Thornton model. Neighbor-joining bootstrap tests of phylogeny were performed using 1000 replicates.

2.4. Nucleotide Sequence Accession Numbers

The *g5* nucleotide sequences of representative from each OTU determined in the present study have been deposited in GenBank under accession numbers KC906230 to KC906242, KC951540 to KC951570, KF018557 to KF018562, KF537694 to KF537765, and KF686584 to KF686735.

3. Results and Discussion

3.1. Statistical Analysis of the g5 Gene Library

A total of 715 clones were sequenced, including 404 and 311 clones from Arctic and Antarctic samples, respectively. The coverage of each clone library ranged from 68% to 90% (Table 1). Arctic samples showed higher coverage than Antarctic samples. However, Antarctic samples exhibited higher *g5* gene diversity than Arctic samples: the cloned sequences fell into 274 OTUs, including 122 and 152 OTUs from Arctic and Antarctic regions, respectively. In addition, both Shannon–Wiener and Simpson indices suggested that the Antarctic samples had more diverse *g5* gene genotypes than the Arctic samples (Table 1). The results implied that more diverse roseobacters existed in Antarctic coastal waters than in Arctic coastal waters, consistent with the findings of previous studies [19,20].

3.2. Diversity and Distribution of Arctic g5 Genes

All *g5* gene sequences were subsequently translated into amino acid sequences. In congruence with the diversity of DNA sequences, Arctic *g5* gene product amino acid sequences fell into eight phylogenetic groups, including the genera *Ascidiaceihabitans*, *Loktanella*, *Octadecabacter*, *Phaeobacter*, *Sulfitobacter*, *Tateyamaria* and *Yoonia* (Figure 2). *Sulfitobacter* was the most abundant genus (49.0%) in the Arctic *g5* clones, followed by *Loktanella* (29.5%) and *Yoonia* (11.1%). Phylogenetic analysis of the bacterial *g5* gene clones (Figure 2) showed that there were seven cloned sequences in total (1.7% of Arctic clones) exhibiting a close relationship (>90% amino acid identity) to both *Loktanella* sp. 1ANDIMAR09 (KQB98268) and *Yoonia rosea* (WP076658629); thus, the seven clones were placed in the *Loktanella*/*Yoonia* phylogenetic group. In fact, *Loktanella* sp. 1ANDIMAR09 exhibited identical *g5* gene product amino acid sequence to *Yoonia rosea* (Figure 2). A similar phenomenon was observed in another 26 clones (6.4%), which showed a close relationship (>83% amino acid identity) to *Ascidiaceihabitans donghaensis* (WP108828037) and *Tateyamaria omphalii* (WP 076630136). Both *Ascidiaceihabitans donghaensis* and *Tateyamaria omphalii* were first reported from marine organisms in eastern Asia [26,27].

Although showing lower coverage for the clone library, the outer station sample St1 exhibited higher *g5* gene diversity than the inner station sample St5 (Table 1). Members within the phylogenetic groups *Ascidiaceihabitans*/*Tateyamaria* and *Phaeobacter* were exclusively detected at outer station St1. In addition, represented by groups from Outer1 to Outer11 (Figure 2), genotypes exclusively found at outer station St1 comprised 28.7% of the total Arctic clones. In contrast, genotypes exclusively detected at inner station St5 accounted for only 0.5% of the total clones. Represented by groups from All1 to All4, genotypes shared by the outer and inner stations comprised most of the Arctic clones. Kongsfjorden is a typical glacier fjord in the European Arctic and is influenced by freshwater inputs mainly from glacier meltwater in summer. Consistent with previous studies [13,28], the results of this study indicate an influence of freshwater input on the diversity and distribution of marine roseobacters in Kongsfjorden during Arctic summers.

Genotype K1g5-64 (KF537731) within the genus *Sulfitobacter* accounted for 32.3% and 51.1% at outer station St1 and inner station St5, respectively. Gene product of this genotype showed 100% amino acid identity to strain *Sulfitobacter* sp. BSw21498 (WP138923583), which was isolated from Kongsfjorden and possessed the dimethylsulfoniopropionate (DMSP) lyase gene *dddL* [29]. Members of the genus *Sulfitobacter* can exhibit algicidal effects against microalgae [30] and degrade algae-derived DMSP [31]. This result is not only consistent with those of previous studies [20,32], indicating a role of *Sulfitobacter* species in sulfur metabolism, but also, implies that GTA-related gene transfer can be helpful for the environmental adaptation and ecological function of *Sulfitobacter* species in the Arctic coastal region.

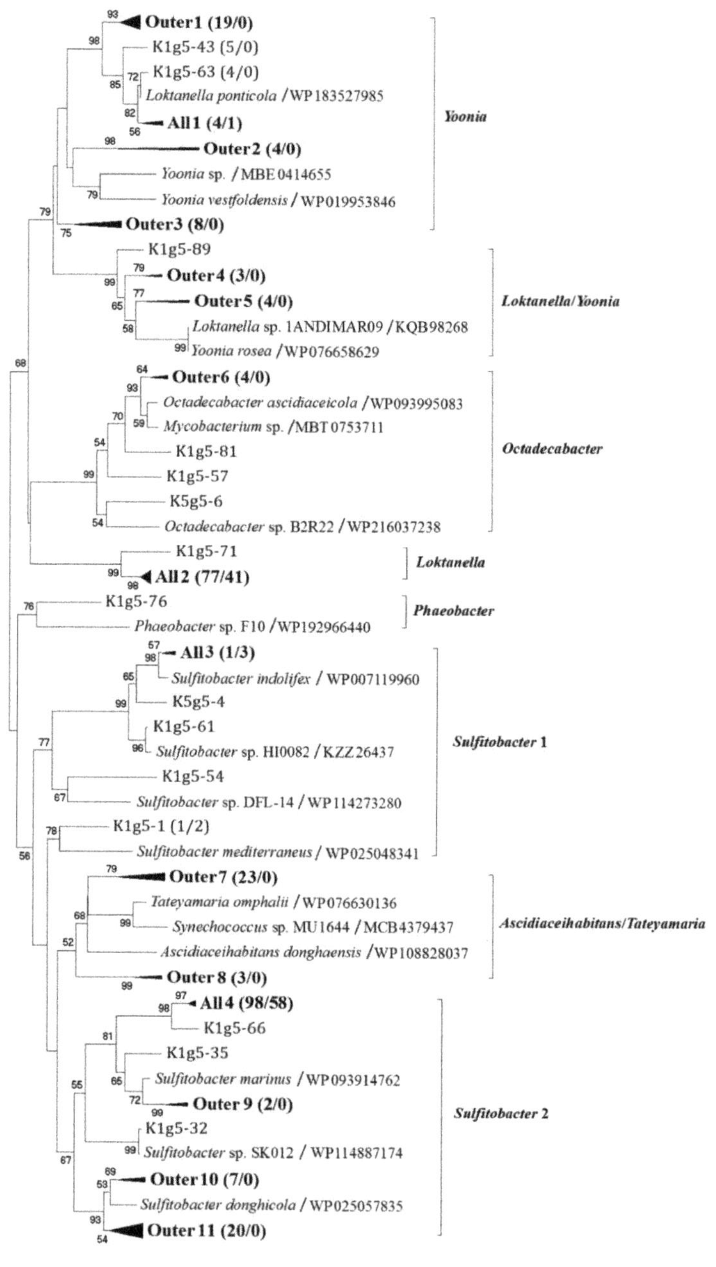

Figure 2. Neighbor-joining phylogenetic tree based on partial *g5* gene product amino acid sequences (ca. 260 aa) showing the phylogenetic diversity of *g5* from Arctic Kongsfjorden bacterial communities. Groups from Outer1 to Outer11 represent closely related clones exclusively found at the outer station St1, whereas groups from All1 to All4 represent closely related clones shared by the outer station St1 and inner station St5. The numbers in parentheses following clone names or group names indicate the number of sequences found at stations St1 and St5, respectively. Bootstrap numbers are shown as percentages based on 1000 replicates, and values of less than 50 were omitted. The scale bar indicates evolutionary distance.

Genotypes K1g5-34 (KF537711) and K1g5-39 (KF537714) within the genus *Loktanella* together comprised 19.2% and 30.6% of the St1 and St5 clones, respectively. They show higher than 98% *g5* gene product amino acid similarity to strain *Loktanella salsilacus* (WP090187762). Members of *Loktanella* can participate in DMSP degradation and aerobic anoxygenic photosynthesis in polar marine environments [19,20]. Except for group All1 (Figure 2) containing genotypes shared by the outer and inner stations, sequences within the *Loktanella/Yoonia* and *Yoonia* phylogenetic groups were exclusively detected at outer station St1, together accounting for 15.8% of the St1 clones. The genus *Yoonia* contains many species (e.g., *Loktanella litorea*, *Loktanella vestfoldensis* and *Roseobacter* sp. CCS2) formerly classified within the genera *Loktanella* and *Roseobacter* [33]. These *Yoonia* species have also been found to possess the aerobic anoxygenic phototrophy gene *pufM* and DMSP degradation genes [19,20,32]. Most species of *Yoonia* require NaCl to grow, indicating that *Yoonia*-related sequences detected in this study originated from marine environments.

3.3. Diversity and Distribution of Antarctic g5 Genes

In congruence with the diversity of DNA sequences, Antarctic *g5* gene product amino acid sequences fell into ten phylogenetic groups, including the genera *Ascidiaceihabitans*, *Jannaschia*, *Litoreibacter*, *Loktanella*, *Octadecabacter*, *Pacificibacter*, *Sulfitobacter*, *Tateyamaria* and *Yoonia* (Figure 3). Among them, *Sulfitobacter* was the most abundant genus (34.0%) in the Antarctic *g5* clones. However, different from Arctic samples, *Ascidiaceihabitans/Tateyamaria*-related sequences were unexpectedly abundant, accounting for 19.2% of the Antarctic clones. The genera *Loktanella* and *Yoonia* were also dominant in Antarctic *g5* sequences, comprising 15.1% and 12.8% of the Antarctic clones, respectively. In addition, sequences within the *Loktanella/Yoonia* phylogenetic group comprised 10.6% of the Antarctic clones.

Similar values of library coverage and diversity indices were observed between Ardley Cove sample A5 and Great Wall Cove sample G5 (Table 1). Represented by groups from Both1 to Both17 (Figure 3) containing closely related clones shared by the two coves, genotypes found in both Antarctic libraries comprised 89.0% of the total Antarctic clones. Mixing exists between the bacterioplankton communities in the two coves due to seawater exchange when high tide occurs, resulting in genotypes shared by the two adjacent coves can comprise most of the Antarctic clones. Represented by group A1, genotypes exclusively found in Ardley Cove accounted for only 5.5% of the total Antarctic clones. The same value (5.5%) was observed for the genotypes (e.g., groups from G1 to G3) exclusively found in Great Wall Cove.

Represented by genotype A-gta103 (KF686586), group Both11 accounted for 18.4% and 15.8% of the A5 and G5 clones, respectively, and showed higher than 88% *g5* gene product amino acid identity to *Tateyamaria omphalii* (WP076630136) and *Ascidiaceihabitans donghaensis* (WP108828037). Members of *Tateyamaria* species show antimicrobial activity and are abundant in healthy marine organisms [34,35]. Moreover, *Ascidiaceihabitans* species have been connected to the health of fish and shrimp in marine environments [36,37]. However, the reason for the relatively high abundance of *Ascidiaceihabitans/Tateyamaria*-related sequences detected in Antarctic seawater remains uncertain.

Represented by genotypes A-gta153 (KF686627), group Both17 comprised 9.5% and 9.7% of the A5 and G5 clones, respectively, showing higher than 82% *g5* gene product amino acid identity to *Sulfitobacter mediterraneus* (WP203198950). Cells of the strain type *S. mediterraneus* can undergo a morphological change during adsorption on polymeric surfaces [38], which may aid *Sulfitobacter* cells in exhibiting algicidal effects. Represented by genotype A-gta131 (KF686603), group Both5 accounted for 12.1% and 16.8% of the A5 and G5 clones, respectively. Sequences of the clones showed higher than 91% *g5* gene product amino acid identity to *Loktanella ponticola* (WP183527985). In addition, represented by genotype A-gta174 (KF686599), group Both6 comprised 6.3% and 14.9% of the A5 and G5 clones, respectively, showing higher than 90% *g5* gene product amino acid similarity to strain *Loktanella* sp. 1ANDIMAR09 (KQB98268) and *Yoonia rosea* (WP076658629). Combined with the Arctic clone libraries described above, the results revealed that *Sulfitobacter*,

Loktanella and *Yoonia* were dominant groups of marine roseobacters in both Arctic and Antarctic coastal regions, playing similar ecological roles in bipolar marine environments.

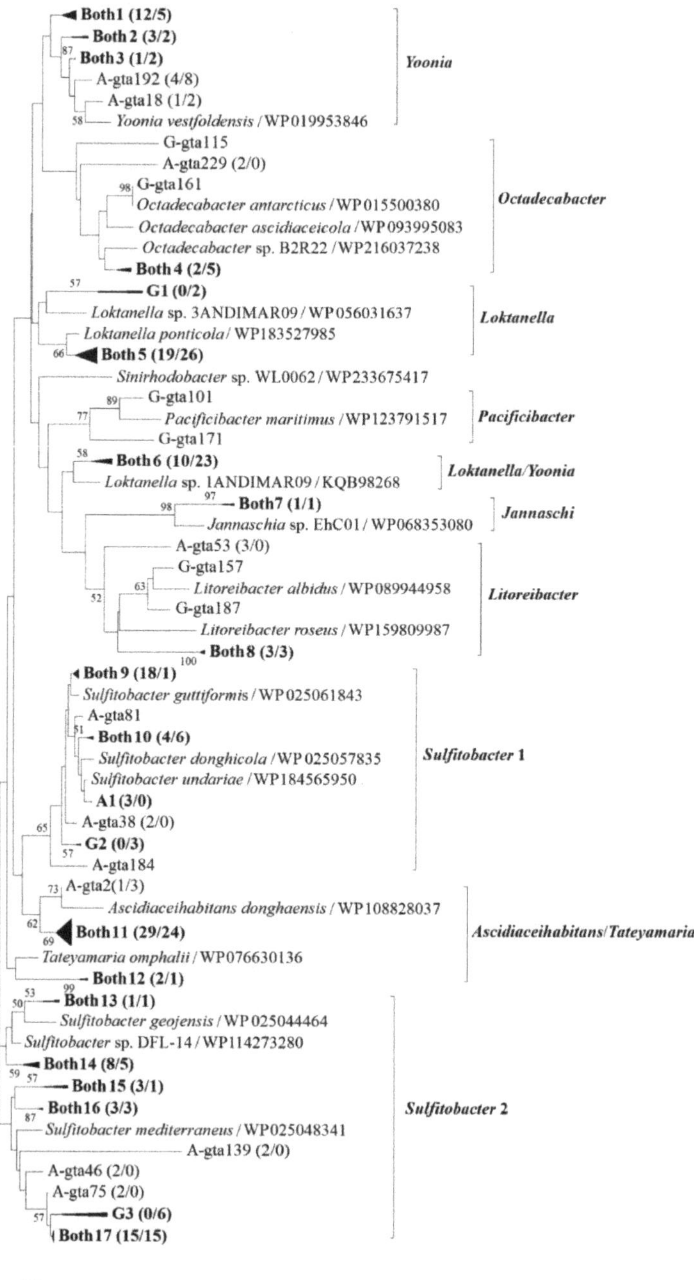

Figure 3. Neighbor-joining phylogenetic tree based on partial *g5* gene product amino acid sequences (ca. 260 aa) showing the phylogenetic diversity of *g5* from Antarctic Maxwell Bay bacterial communities. Group A1 represents closely related clones exclusively found in Ardley Cove, whereas groups

from G1 to G3 represent closely related clones exclusively found in Great Wall Cove. Groups from Both1 to Both17 are closely related clones shared by the two coves. The numbers in parentheses following clone names or group names indicate the number of sequences found at stations A5 (Ardley Cove) and G5 (Great Wall Cove), respectively. Bootstrap numbers are shown as percentages based on 1000 replicates, and values of less than 50 were omitted. The scale bar indicates evolutionary distance.

3.4. Relationships between Arctic and Antarctic g5 Genes

Phylogenetic analysis based on *g5* gene product amino acid sequences was conducted to investigate the similarity or difference between Arctic and Antarctic *g5* genotypes. Compared with *Phaeobacter* (represented by genotype K1g5-76), which was exclusively detected in Arctic seawater, the genera *Jannaschia* (represented by A-gta48), *Litoreibacter* (represented by G-gta157) and *Pacificibacter* (represented by G-gta101) were only observed in Antarctic seawater in this study. In addition, phylogenetic groups Ant1 within the genus *Octadecabacter*, Ant2 within the genus *Yoonia*, Ant5 and Ant6 within the genus *Loktanella*, Ant7 and Ant8 within *Ascidiaceihabitans / Tateyamaria*, and Ant9, Ant10, Ant11, Ant12, Ant13 and Ant14 within the genus *Sulfitobacter* were exclusively detected in Antarctic samples (Figure 4). In contrast, phylogenetic groups from Arc1 to Arc7 were exclusively observed in Arctic seawater. Much more abundant *Ascidiaceihabitans / Tateyamaria*-related sequences were detected in Antarctic seawater (19.2% of Antarctic clones) than in Arctic seawater (6.4% of Arctic clones). Differences in environmental parameters (e.g., water temperature, salinity and nutrients) and bacterial abundance [12,21] can be observed between the two investigated Arctic and Antarctic regions, suggesting niche adaptation of specific roseobacters to unique environments.

Although the genera *Sulfitobacter*, *Loktanella* and *Yoonia* were dominant in both Arctic and Antarctic coastal seawater, an endemic distribution of the genotypes was observed in this study. For example, represented by genotype K1g5-64 (KF537731) showing identical to *Sulfitobacter* sp. BSw21498, the phylogenetic group Arc4 (Figure 4) was dominant (39.0% of Arctic clones) and exclusively detected in Arctic clone libraries. On the other hand, represented by genotype A-gta40 (KF686607) showing higher than 96% sequence similarity to strain *Sulfitobacter guttiformis* (WP025061843), the phylogenetic group Ant13 was abundant (5.4% of Antarctic clones) and exclusively detected in Antarctic seawater. Similar to *Sulfitobacter mediterraneus*, *S. guttiformis* (formerly classified as *Staleya guttiformis*) can also attach to polymeric surfaces and produce extracellular polymeric substances [39]. The strain type *S. guttiformis* was first isolated from an Antarctic hypersaline lake [40]. Within the genus *Yoonia*, genotypes K1g5-45 (KF537719; 0.2% of Arctic clones) and A-gta18 (KF686595; 1.0% of Antarctic clones) were exclusively detected in Arctic and Antarctic seawater, respectively. The two genotypes showed 88.1% *g5* gene product amino acid sequence similarity between each other, both exhibiting higher than 87% similarity to *Yoonia vestfoldensis* (WP019953846). The strain type *Yoonia vestfoldensis* was isolated from an Antarctic salt lake [41]. Endemic distributions of PufM, DmdA, and DddP genotypes have been observed in the Arctic or Antarctic region [19,20,32]. *Sulfitobacter*-related *dddL* gene sequences are detected in Kongsfjorden but absent in Maxwell Bay, whereas *Loktanellar*-related *dddP* gene sequences are abundant in Maxwell Bay but absent in Kongsfjorden [20,32].

Represented by phylogenetic groups from Arc1 to Arc7 (Figure 4), genotypes limited in Arctic samples accounted for 84.8% of total Arctic clones, whereas the proportion of genotypes (represented by groups from Ant1 to Ant14) exclusively found in Antarctic samples comprised 54.3% of Antarctic sequences. The results suggested that the proportion of endemic phylotypes of roseobacters distributed in the Arctic was higher than that in the Antarctic. The endemic phylotypes accounted for a larger share (71.5% on average) of all *g5* clones, suggesting that marine roseobacters in polar regions comprise mainly endemic phylotypes, which are distributed either in the Arctic or Antarctica. Though Louca [42] reports that most prokaryotic species and even closely related strains are globally distributed, geographic endemism at the species or strain level is observed in thermophilic microorganisms. Global-scale microbial distribution patterns are likely the result of recent or current environmental filtering rather than geographic endemism [42]. Polar region is colonized principally by psychrophilic and psychrotrophic microorganisms. Dispersal between poles is problematic for psychrophilic bacteria because of the long distances and the difficulty of transporting across the equator [43]. Whether the high proportion of endemic phylotypes in roseobacters observed in this study is connected to phenotypical characteristics of the bacteria requires further study.

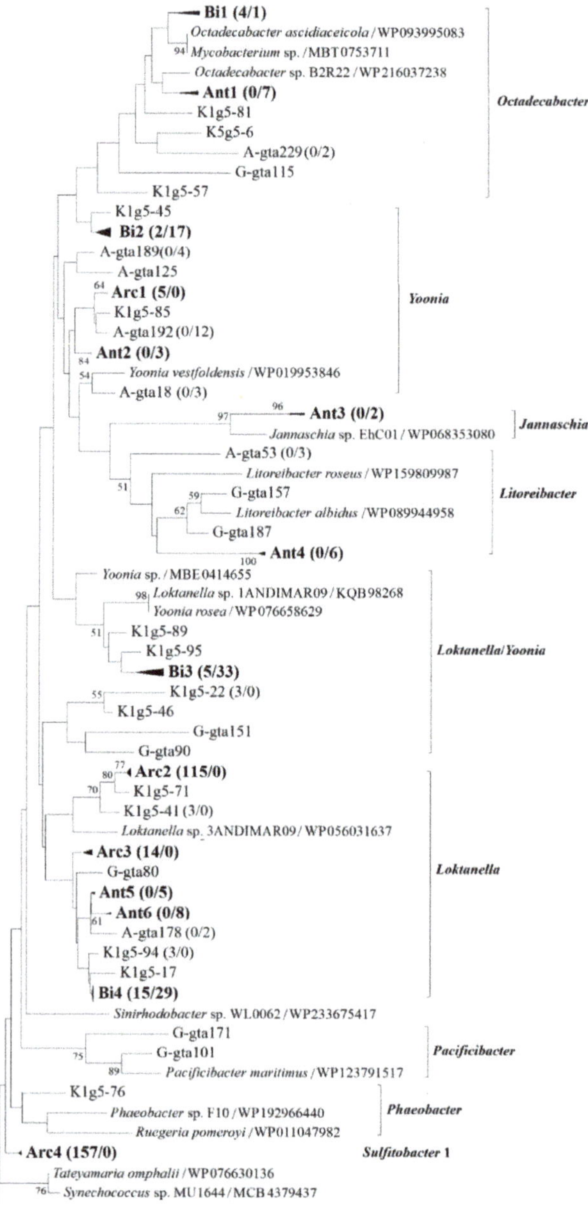

Figure 4. *Cont.*

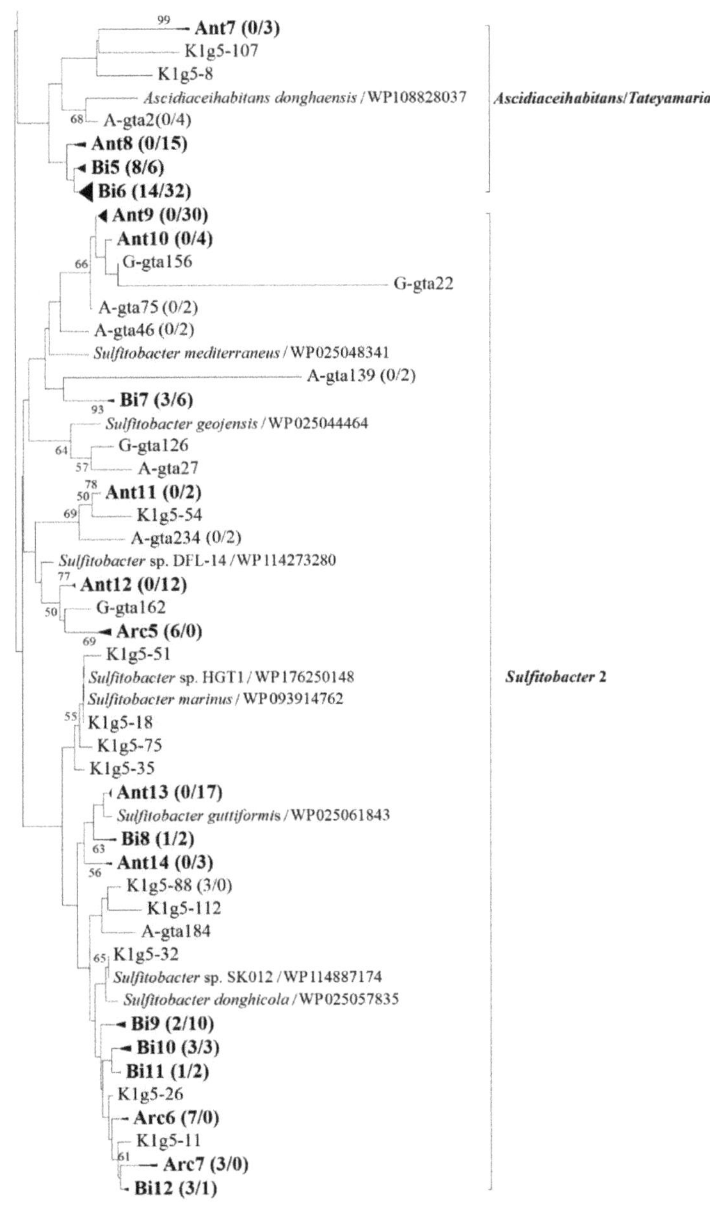

Figure 4. Neighbor-joining phylogenetic tree based on partial *g5* gene product amino acid sequences (ca. 260 aa) showing the phylogenetic diversity of *g5* genes from Arctic and Antarctic bacterial communities. Groups from Arc1 to Arc7 represent closely related clones exclusively found in Arctic region, whereas groups from Ant1 to Ant14 represent closely related clones exclusively found in Antarctic region. Groups from Bi1 to Bi12 are closely related clones detected in bipolar regions. The numbers in parentheses following clone names or group names indicate the number of sequences found in Arctic and Antarctic samples, respectively. Bootstrap numbers are shown as percentages based on 1000 replicates, and values of less than 50 were omitted. The scale bar indicates evolutionary distance.

Represented by genotype K1g5-34 (KF537711) showing identical to *Loktanella salsilacus* (WP 090187762), which was first isolated from an Antarctic salt lake [41], phylogenetic group Arc2 comprised 29.6% of Arctic clones but was not detected in Antarctic samples. In addition, represented by genotypes A-gta166 (KF686615) and A-gta221 (KF686666) showing higher than 93% similarity to *Loktanella ponticola* (WP183527985), groups Ant5 and Ant6 were exclusively observed in Antarctic samples and accounted for 4.1% of the total Antarctic clones. However, the type strain of *Loktanella ponticola* was isolated from seawater in the Korean South Sea [44]. Therefore, *Loktanella*-related genotypes within the phylogenetic groups Arc2, Ant5 and Ant6 can actually be considered to have a bipolar distribution.

The bipolar distribution of roseobacter phylotypes has been reported in Arctic and Antarctic water samples [12,43]. In the present study, phylogenetic groups from Bi1 to Bi12 accounted for 15.2% and 45.7% of the Arctic and Antarctic *g5* gene clones, respectively. Groups from Bi1 to Bi12 (Figure 4) represented closely related clones detected in bipolar regions. Represented by OTU A-gta222 (KF686644), which showed 93.4% similarity to *Loktanella ponticola* (WP183527985), bipolar group Bi4 comprised 3.7% and 9.3% of the Arctic and Antarctic clones, respectively. Moreover, including groups from Bi7 to Bi12, *Sulfitobacter*-related phylotypes accounted for 3.2% and 7.7% of the Arctic and Antarctic clones, respectively. Represented by OTU A-gta103 (KF686586), which showed 89.2% similarity to *Tateyamaria omphalii* (WP076630136), bipolar group Bi6 comprised 3.4% and 10.2% of the Arctic and Antarctic clones, respectively. The results support the bipolar or even global distribution of roseobacters within the *Loktanella*, *Sulfitobacter* and *Ascidiaceihabitans*/*Tateyamaria* groups. Different from previous studies, in this study *Ascidiaceihabitans*/*Tateyamaria* group was abundant in both Arctic and Antarctic coastal seawaters. Neither *Ascidiaceihabitans* nor *Tateyamaria* are reported in previous studies based on analysis of functional genes [19,20,32]. The genera *Sulfitobacter* and *Loktanella* are the dominant members of the Alphaproteobacteria in bacterioplankton community in both Arctic Kongsfjorden [21] and Antarctic Maxwell Bay [12]. *Sulfitobacter*-related sequences are also abundant in Arctic and Antarctic coastal seawaters based on analysis of aerobic anoxygenic phototrophy gene *pufM* [19] and DMSP demethylase gene *dmdA* [20]. In addition, *Loktanella*-related sequences are detected in Kongsfjorden and Maxwell Bay based on *pufM* and *dmdA* genes [19,20]. The bipolar distribution of *g5* genotypes suggests that GTA transduction may contribute to the adaptation of some roseobacters (e.g., *Loktanella* and *Sulfitobacter*) to bipolar marine environments.

The GTA *g5* gene has served as a genetic marker to investigate the diversity of roseobacters in marine environments due to the congruence between *g5* and 16S rRNA gene phylogenies [17,18,45]. However, in the present study, high *g5* gene product amino acid sequence similarities were observed between members of different genera (e.g., *Loktanella*/*Yoonia*, and *Ascidiaceihabitans*/*Tateyamaria*) within the *Roseobacter* clade. It could result in a difficulty in clarifying the taxonomic status of closely related members of the *Roseobacter* clade. Therefore, aside from the *g5* gene focusing on roseobacters harboring GTA, the 16S rRNA gene should be used simultaneously to estimate the diversity of the *Roseobacter* clade in natural environments. It could be helpful for us to obtain more insight into the role of GTA transduction in adaptation and evolution of the *Roseobacter* clade in polar marine environments.

4. Conclusions

Based on analysis of the major capsid protein-encoding GTA *g5* gene, the present study supports previous findings of both bipolar and endemic distributions of roseobacters and their functional genes in polar marine environments [19,20]. Endemic phylotypes were more frequently observed in polar regions than cosmopolitan phylotypes. The genera *Sulfitobacter*, *Loktanella* and *Yoonia* were dominant in the *Roseobacter* clade detected in Arctic and Antarctic coastal seawaters. This may be attributed to GTAs that are maintained in bacterial genomes due to the advantages associated with gene exchange in stressful conditions [14]. The results also indicate that, similar to temperate oceans [46–48], GTAs are common in the *Roseobacter* clade in polar marine environments. However, identical *g5* gene product amino acid sequences were observed between *Loktanella* sp. 1ANDIMAR09 (KQB98268) and *Yoonia rosea* (WP076658629) in this study, suggesting a potential problem for clarifying the taxonomic status of closely related members of the *Roseobacter* clade based on the *g5* gene. Further research on the characterization of bacterial isolates of the *Roseobacter* clade should be performed to improve our understanding of the distribution, environmental adaptation and ecological role of marine roseobacters in polar regions.

Author Contributions: Conceptualization, Y.-X.Z.; methodology, Y.-X.Z. and H.-R.L.; formal analysis, Y.-X.Z.; writing—original draft preparation, Y.-X.Z.; supervision, Y.-X.Z.; project administration, Y.-X.Z.; funding acquisition, Y.-X.Z. and W.L. All authors have read and agreed to the published version of the manuscript.

Funding: This work was supported by the National Key Research and Development Program of China (Grant No. 2018YFC1406903) and the National Natural Science Foundation of China (Grant Nos 91851201 and 41476171).

Institutional Review Board Statement: Not applicable.

Informed Consent Statement: Not applicable.

Data Availability Statement: The *g5* nucleotide sequences determined in the present study have been deposited in GenBank under accession numbers KC906230 to KC906242, KC951540 to KC951570, KF018557 to KF018562, KF537694 to KF537765, and KF686584 to KF686735.

Acknowledgments: We appreciate the assistance of Chinese Antarctic Center of Surveying and Mapping, Wuhan University, for providing maps. We also are grateful to the anonymous reviewers for their helpful comments.

Conflicts of Interest: The authors declare no conflict of interest.

References

1. Buchan, A.; Gonzaìlez, J.M.; Moran, M.A. Overview of the marine roseobacter lineage. *Appl. Environ. Microbiol.* **2005**, *71*, 5665–5677. [CrossRef]
2. Garrity, G.M.; Bell, J.A.; Lilburn, T.; Family, I. Rhodobacteraceae fam. nov. In *Bergey's Manual of Systematic Bacteriology*, 2nd ed.; Volume 2 (The Proteobacteria), Part C (The Alpha-, Beta-, Delta-, and Epsilonproteobacteria); Brenner, D.J., Krieg, N.R., Staley, J.T., Garrity, G.M., Eds.; Springer: New York, NY, USA, 2005; p. 161.
3. Liang, K.Y.H.; Orata, F.D.; Boucher, Y.F.; Case, R.J. Roseobacters in a sea of poly- and paraphyly: Whole genome-based taxonomy of the family *Rhodobacteraceae* and the proposal for the split of the"Roseobacter clade" into a novel family, *Roseobacteraceae* fam. nov. *Front. Microbiol.* **2021**, *12*, 683109. [CrossRef]
4. Moran, M.A.; Belas, R.; Schell, M.A.; Gonzaìlez, J.M.; Sun, F.; Sun, S.; Binder, B.J.; Edmonds, J.; Ye, W.; Orcutt, B.; et al. Ecological genomics of marine roseobacters. *Appl. Environ. Microbiol.* **2007**, *73*, 4559–4569. [CrossRef]
5. Luo, H.; Moran, M.A. Evolutionary ecology of the marine Roseobacter clade. *Microbiol. Mol. Biol. Rev.* **2014**, *78*, 573–587. [CrossRef]
6. Tolli, J.D.; Sievert, S.M.; Taylor, C.D. Unexpected diversity of bacteria capable of carbon monoxide oxidation in a coastal marine environment, and contribution of the Roseobacter-associated clade to total CO oxidation. *Appl. Environ. Microbiol.* **2006**, *72*, 1966–1973. [CrossRef]
7. Imhoff, J.F.; Rahn, T.; Künzel, S.; Neulinger, S.C. Phylogeny of anoxygenic photosynthesis based on sequences of photosynthetic reaction center proteins and a key enzyme in bacteriochlorophyll biosynthesis, the chlorophyllide reductase. *Microorganisms* **2019**, *7*, 576. [CrossRef]
8. Cude, W.N.; Buchan, A. Acyl-homoserine lactone-based quorum sensing in the roseobacter clade: Complex cell-to-cell communication controls multiple physiologies. *Front. Microbiol.* **2013**, *4*, 336. [CrossRef]
9. Moran, M.A.; González, J.M.; Kiene, R.P. Linking a bacterial taxon to sulfur cycling in the sea: Studies of the marine roseobacter group. *Geomicrobiol. J.* **2003**, *20*, 375–388. [CrossRef]
10. Reisch, C.R.; Moran, M.A.; Whitman, W.B. Bacterial catabolism of dimethylsulfoniopropionate (DMSP). *Front. Microbiol.* **2011**, *2*, 172. [CrossRef]
11. Brinkmeyer, R.; Knittel, K.; Jürgens, J.; Weyland, H.; Amann, R.; Helmke, E. Diversity and structure of bacterial communities in Arctic versus Antarctic pack ice. *Appl. Environ. Microbiol.* **2003**, *69*, 6610–6619. [CrossRef]
12. Zeng, Y.X.; Yu, Y.; Qiao, Z.Y.; Jin, H.Y.; Li, H.R. Diversity of bacterioplankton in coastal seawaters of Fildes Peninsula, King George Island, Antarctica. *Arch. Microbiol.* **2014**, *196*, 137–147. [CrossRef]

13. Zeng, Y.X.; Luo, W.; Li, H.R.; Yu, Y. High diversity of planktonic prokaryotes in Arctic Kongsfjorden seawaters in summer 2015. *Polar Biol.* **2021**, *44*, 195–208. [CrossRef]

14. Esterman, E.S.; Wolf, Y.I.; Kogay, R.; Koonin, E.V.; Zhaxybayeva, O. Evolution DNA packaging in gene transfer agents. *Virus Evol.* **2021**, *7*, veab015. [CrossRef]

15. Hynes, A.P.; Mercer, R.G.; Watton, D.E.; Buckley, C.B.; Lang, A.S. DNA packaging bias and differential expression of gene transfer agent genes within a population during production and release of the *Rhodobacter capsulatus* gene transfer agent, RcGTA. *Mol. Microbiol.* **2012**, *85*, 314–325. [CrossRef]

16. Brinkhoff, T.; Giebel, H.A.; Simon, M. Diversity, ecology, and genomics of the *Roseobacter* clade: A short overview. *Arch. Microbiol.* **2008**, *189*, 531–539. [CrossRef]

17. Lang, A.S.; Beatty, J.T. Importance of widespread gene transfer agent genes in alpha-proteobacteria. *Trends Microbiol.* **2007**, *15*, 54–62. [CrossRef]

18. Zhao, Y.; Wang, K.; Budinoff, C.; Buchan, A.; Lang, A.; Jiao, N.; Chen, F. Gene transfer agent (GTA) genes reveal diverse and dynamic *Roseobacter* and *Rhodobacter* populations in the Chesapeake Bay. *ISME J.* **2009**, *3*, 364–373. [CrossRef]

19. Zeng, Y.X.; Dong, P.Y.; Qiao, Z.Y.; Zheng, T.L. Diversity of the aerobic anoxygenic phototrophy gene *pufM* in Arctic and Antarctic coastal seawaters. *Acta Oceanol. Sin.* **2016**, *35*, 68–77. [CrossRef]

20. Zeng, Y.X.; Qiao, Z.Y. Diversity of dimethylsulfoniopropionate degradation genes reveals the significance of marine *Roseobacter* clade in sulfur metabolism in coastal areas of Antarctic Maxwell Bay. *Curr. Microbiol.* **2019**, *76*, 967–974. [CrossRef]

21. Zeng, Y.X.; Zhang, F.; He, J.F.; Lee, S.H.; Qiao, Z.Y.; Yu, Y.; Li, H.R. Bacterioplankton community structure in the Arctic waters as revealed by pyrosequencing of 16S rRNA genes. *Antonie Leeuwenhoek* **2013**, *103*, 1309–1319. [CrossRef]

22. Bosshard, P.P.; Santini, Y.; Grüter, D.G.; Stettler, R.; Bachofen, R. Bacterial diversity and community composition in the chemocline of the meromictic alpine Lake Cadagno as revealed by 16S rRNA gene analysis. *FEMS Microbiol. Ecol.* **2000**, *31*, 173–182. [CrossRef]

23. Bano, N.; Hollibaugh, J.T. Diversity and distribution of DNA sequences with affinity to ammonia-oxidizing bacteria of the β subdivision of the class Proteobacteria in the Arctic Ocean. *Appl. Environ. Microbiol.* **2000**, *66*, 1960–1969. [CrossRef]

24. Zeng, Y.; Zheng, T.; Li, H. Community composition of the marine bacterioplankton in Kongsfjorden (Spitsbergen) as revealed by 16S rRNA gene analysis. *Polar Biol.* **2009**, *32*, 1447–1460. [CrossRef]

25. Huber, T.; Faulkner, G.; Hugenholtz, P. Bellerophon: A program to detect chimeric sequences in multiple sequence alignments. *Bioinformatics* **2004**, *20*, 2317–2319. [CrossRef]

26. Kurahashi, M.; Yokota, A. *Tateyamaria omphalii* gen. nov., sp. nov., an α-Proteobacterium isolated from a top shell *Omphalius pfeifferi pfeifferi*. *Syst. Appl. Microbiol.* **2007**, *30*, 371–375. [CrossRef]

27. Kim, Y.O.; Park, S.; Nam, B.H.; Lee, C.; Park, J.M.; Kim, D.G.; Yoon, J.H. *Ascidiaceihabitans donghaensis* gen. nov., sp. nov., isolated from the golden sea squirt *Halocynthia aurantium*. *Int. J. Syst. Evol. Microbiol.* **2014**, *64*, 3970–3975. [CrossRef]

28. Sinha, R.K.; Krishnan, K.P.; Kerkar, S.; Thresyamma, D.D. Influence of glacial melt and Atlantic water on bacterioplankton community of Kongsfjorden, an Arctic fjord. *Ecol. Indic.* **2017**, *82*, 143–151. [CrossRef]

29. Zeng, Y.X.; Zhang, Y.H.; Li, H.R.; Luo, W. Complete genome of *Sulfitobacter* sp. BSw21498 isolated from seawater of Arctic Kongsfjorden. *Mar. Genom.* **2020**, *53*, 100769. [CrossRef]

30. Barak-Gavish, N.; Frada, M.J.; Ku, C.; Lee, P.A.; DiTullio, G.R.; Malitsky, S.; Aharoni, A.; Green, S.J.; Rotkopf, R.; Kartvelishvily, E.; et al. Bacterial virulence against an oceanic bloom-forming phytoplankter is mediated by algal DMSP. *Sci. Adv.* **2018**, *4*, eaau5716. [CrossRef]

31. Amin, S.A.; Hmelo, L.R.; van Tol, H.M.; Durham, B.P.; Carlson, L.T.; Heal, K.R.; Morales, R.L.; Berthiaume, C.T.; Parker, M.S.; Djunaedi, B.; et al. Interaction and signalling between a cosmopolitan phytoplankton and associated bacteria. *Nature* **2015**, *522*, 98–101. [CrossRef]

32. Zeng, Y.X.; Qiao, Z.Y.; Yu, Y.; Li, H.R.; Luo, W. Diversity of bacterial dimethylsulfoniopropionate degradation genes in surface seawater of Arctic Kongsfjorden. *Sci. Rep.* **2016**, *6*, 33031. [CrossRef] [PubMed]

33. Wirth, J.S.; Whitman, W.B. Phylogenomic analyses of a clade within the roseobacter group suggest taxonomic reassignments of species of the genera *Aestuariivita*, *Citreicella*, *Loktanella*, *Nautella*, *Pelagibaca*, *Ruegeria*, *Thalassobius*, *Thiobacimonas* and *Tropicibacter*, and the proposal of six novel genera. *Int. J. Syst. Evol. Microbiol.* **2018**, *68*, 2393–2411.

34. Chen, Y.H.; Kuo, J.; Sung, P.J.; Chang, Y.C.; Lu, M.C.; Wong, T.Y.; Liu, J.K.; Weng, C.F.; Twan, W.H.; Kuo, F.W. Isolation of marine bacteria with antimicrobial activities from cultured and field-collected soft corals. *World J. Microbiol. Biotechnol.* **2012**, *28*, 3269–3279. [CrossRef] [PubMed]

35. Villasante, A.; Catalán, N.; Rojas, R.; Lohrmann, K.B.; Romero, J. Microbiota of the digestive gland of red abalone (*Haliotis rufescens*) is affected by withering syndrome. *Microorganisms* **2020**, *8*, 1411. [CrossRef]

36. Wang, Y.; Liu, M.; Wang, B.; Jiang, K.; Wang, M.; Wang, L. Response of the *Litopenaeus vananmei* intestinal bacteria and antioxidant system to rearing density and exposure to *Vibrio paraheamolyticus* E1. *J. Invertebr. Pathol.* **2020**, *170*, 107326. [CrossRef] [PubMed]

37. Zheng, S.; Zhou, S.; Lukwambe, B.; Nicholaus, R.; Yang, W.; Zheng, Z. Bacterioplankton community assembly in migratory fish habitat: A case study of the southern East China Sea. *Environ. Sci. Pollut. Res. Int.* **2022**, *29*, 33725–33736. [CrossRef]

38. Ivanova, E.P.; Pham, D.K.; Wright, J.P.; Nicolau, D.V. Detection of coccoid forms of *Sulfitobacter mediterraneus* using atomic force microscopy. *FEMS Microbiol Lett.* **2002**, *214*, 177–181. [CrossRef]

39. Ivanova, E.P.; Mitik-Dineva, N.; Wang, J.; Pham, D.K.; Wright, J.P.; Nicolau, D.V.; Mocanasu, R.C.; Crawford, R.J. *Staleya guttiformis* attachment on poly(tert-butylmethacrylate) polymeric surfaces. *Micron* **2008**, *39*, 1197–1204. [CrossRef]
40. Labrenz, M.; Tindall, B.J.; Lawson, P.A.; Collins, M.D.; Schumann, P.; Hirsch, P. *Staleya guttiformis* gen. nov., sp. nov. and *Sulfitobacter brevis* sp. nov., alpha-3-Proteobacteria from hypersaline, heliothermal and meromictic antarctic Ekho Lake. *Int. J. Syst. Evol. Microbiol.* **2000**, *50*, 303–313. [CrossRef]
41. Van Trappen, S.; Mergaert, J.; Swings, J. *Loktanella salsilacus* gen. nov., sp. nov., *Loktanella fryxellensis* sp. nov. and *Loktanella vestfoldensis* sp. nov., new members of the Rhodobacter group, isolated from microbial mats in Antarctic lakes. *Int. J. Syst. Evol. Microbiol.* **2004**, *54*, 1263–1269. [CrossRef]
42. Louca, S. The rates of global bacterial and archaeal dispersal. *ISME J.* **2022**, *16*, 159–167. [CrossRef]
43. Staley, J.T.; Gosink, J.J. Poles apart: Biodiversity and biogeography of sea ice bacteria. *Ann. Rev. Microbiol.* **1999**, *53*, 198–215. [CrossRef]
44. Jung, Y.T.; Park, S.; Park, J.M.; Yoon, J.H. *Loktanella ponticola* sp. nov., isolated from seawater. *Int. J. Syst. Evol. Microbiol.* **2014**, *64*, 3717–3723. [CrossRef]
45. Sun, F.L.; Wang, Y.S.; Wu, M.L.; Jiang, Z.Y.; Sun, C.C.; Cheng, H. Genetic diversity of bacterial communities and gene transfer agents in Northern South China Sea. *PLoS ONE* **2014**, *9*, e111892. [CrossRef]
46. Dang, H.; Li, T.; Chen, M.; Huang, G. Cross-ocean distribution of Rhodobacterales bacteria as primary surface colonizers in temperate coastal marine waters. *Appl. Environ. Microbiol.* **2008**, *74*, 52–60. [CrossRef]
47. Roux, S.; Brum, J.R.; Dutilh, B.E.; Sunagawa, S.; Duhaime, M.B.; Loy, A.; Poulos, B.T.; Solonenko, N.; Lara, E.; Poulain, J.; et al. Ecogenomics and potential biogeochemical impacts of globally abundant ocean viruses. *Nature* **2016**, *537*, 689–693. [CrossRef]
48. Bárdy, P.; Füzik, T.; Hrebík, D.; Pantůček, R.; Thomas Beatty, J.; Plevka, P. Structure and mechanism of DNA delivery of a gene transfer agent. *Nat. Commun.* **2020**, *11*, 3034. [CrossRef]

 *diversity*

Article

Mesoscale Spatial Patterns of Gulf of Maine Rocky Intertidal Communities

Thomas J. Trott

Maine Coastal Program, Department of Marine Resources, West Boothbay Harbor, ME 04575, USA; tom.trott@maine.gov; Tel.: +1-207-633-9578

Abstract: Community similarity among macroinvertebrate species assemblages from 12 exposed rocky headlands surveyed in 2004, 2007, and 2012 was examined to resolve mesoscale patterns along an east–west linear distance of 366 km in the coastal Gulf of Maine. The goals were: (1) detect latitudinal patterns of species assemblage similarity and (2) relate species assemblage similarities to environmental factors. Assemblage similarities were correlated with latitude. There was a distinguishable grouping of sampling sites fitting two Gulf regions that separate at mid-coast Maine. This pattern was uniquely intertidal and not shown by subtidal species assemblages. β diversity was high, did not differ between regions, and species turnover accounted for 91% of it. Molluscs and crustaceans, major components of surveyed communities, contributed most of the dissimilarity between regions. Satellite-derived shore and sea surface temperatures explained a significant amount of the variation responsible for producing regional patterns. The regions corresponded with the two principal branches of the Gulf of Maine Coastal Current. These hydrographic features and associated environmental conditions are hypothesized to influence community dynamics and shape the dissimilarity between Gulf regions. The predicted warming of the Gulf of Maine portend change in species turnover from species invasions and range shifts potentially altering rocky intertidal community patterns.

Keywords: nearshore biodiversity; benthic marine organisms; marine benthic ecology; species similarity; biogeography; sea surface temperature; thermogeography

Citation: Trott, T.J. Mesoscale Spatial Patterns of Gulf of Maine Rocky Intertidal Communities. *Diversity* **2022**, *14*, 557. https://doi.org/10.3390/d14070557

Academic Editors: Bert W. Hoeksema and Fernando Tuya

Received: 14 May 2022
Accepted: 5 July 2022
Published: 11 July 2022

Publisher's Note: MDPI stays neutral with regard to jurisdictional claims in published maps and institutional affiliations.

1. Introduction

The delineation of broad scale spatial biodiversity patterns is valuable for detecting, gauging, and predicting the response of communities to environmental change. If variation in community structure produces detectable new patterns, such change can signal modification of community composition with altered or novel species interactions as a consequence [1]. The extent that communities are buffered against change depends on the stability of their populations to recover from perturbations stemming from both environmental and biological factors [2], which in turn will determine the degree of local species extinctions and long-term consequences for community dynamics [3–5]. The number and types of potential species interactions in novel communities that emerge can change ecosystem function and linked ecosystem services [6]. Outcomes can have direct economic, demographic, and social consequences for coastal communities especially when commercially valuable species are lost.

The diversity and structure of intertidal communities living at the land–sea interface are shaped by the aggregative effects of local and broad scale marine, terrestrial, and atmospheric processes. What shapes broad scale diversity and complexity of intertidal community structure is intimately tied to coastal circulation and oceanic processes [7,8] and the biogeographic patterns that result are strongly associated with these features [9,10]. The dispersal and delivery of nutrients, food, and propagules are steered by ocean currents, which also set physical limits that intertidal species tolerate. Intertidal communities are also vulnerable to the effects of aerial exposure during periods of low tide. Air temperature, humidity, and precipitation, influenced by broad scale meteorological systems [11], exert selective influences on community composition according to the physiological requirements of species [12]. In addition, many intertidal communities are subject to the near- and far-field effects of freshwater outflow from associated watersheds [13–15]. Overall, the shoreline frames a habitat subject to extremes where the effects of climate change from altered terrestrial, atmospheric, and marine environments are concentrated.

Global warming will directly affect thermally sensitive processes, and temperature is a pervasive force on all biological phenomena [16]. Global scale distribution patterns of rocky shore intertidal communities are strongly related to temperature [17]. Latitudinal species distributions are limited by the effects of temperature [18,19], which constrain rates of reproduction and pelagic larval development [20,21]. The thermal challenges encountered by species dispersing into the extremes of their geographic ranges tests their physiological adaptations to the intertidal environment [22,23]. Shifts in species distributional ranges from global warming are limited by the genetic capacity for evolving thermal and phenological adaptations [6,24]. Thus, range shifting may not be continuous progressions in space and time but instead can be punctuated over short time and spatial scales. Local extinctions can happen when intertidal thermal environments surpass the capacity of species to acclimate [25,26] due to limited phenotypic plasticity [12]. Understanding the outcomes of warming is not straightforward since the degree to which changes in temperature effect species interactions is not well understood [25,26]. Species establishment is dependent on a variety of abiotic and biotic factors, such as oceanographic conditions, food limitation, competition, and predation. Given the complexity of abiotic and biotic effects, determining regional patterns in community structure will help to facilitate the prediction of changes from global warming.

The association of coastal circulation and temperature was explored to develop testable predictions concerning the role of these features in structuring intertidal communities within the Gulf of Maine (GoM) (Figure 1). The GoM is described by the expanse of water between Cape Cod, Massachusetts, and southwestern Nova Scotia. Isolated by Browns Bank and Georges Bank from the open Northwest Atlantic, the GoM is a semi-enclosed marginal sea with distinct oceanographic and meteorological features [27–29]. The Gulf of Maine Coastal Current (GMCC) is one such feature and is a major influence on the Gulf's biological productivity [30]. The GMCC receives water from the Scotian Shelf as it flows cyclonically near the 100 m isobath from the Grand Banks to Massachusetts Bay south-west. The GMCC has two principal branches, the Eastern Maine Coastal Current (EMCC), where there is an offshore component, and Western Maine Coastal Current (WMCC). The EMCC extends along the eastern Gulf and flows southwest towards Penobscot Bay and the WMCC originates immediately south of the bay and flows into the southern Gulf [31]. Among other physical characteristics, sea water temperature distinguishes these two currents, the EMCC being colder [32,33]. The general cyclonic flow pattern of the GMCC changes seasonally. During spring, summer, and fall the EMCC flows uninterrupted to mid-coast where upon encountering the Penobscot Bay region a portion flows cyclonically away from the coastline. A portion continues past the mouth of the bay to join the WMCC [31,33]. Flow to the WMCC is regulated by complex hydrographic processes and can range from continuous to complete disruption during different years. Keafer et al. [34] described another GoM hydrographic feature, the low salinity Gulf of Maine Coastal Plume (GOMCP), which lies sandwiched between the coast and GMCC and receives water from major rivers along its southwest flowing course extending from eastern Maine. Winter circulation in the GoM is less well known. However, the cyclonic pattern is less organized and slows down [35–37].

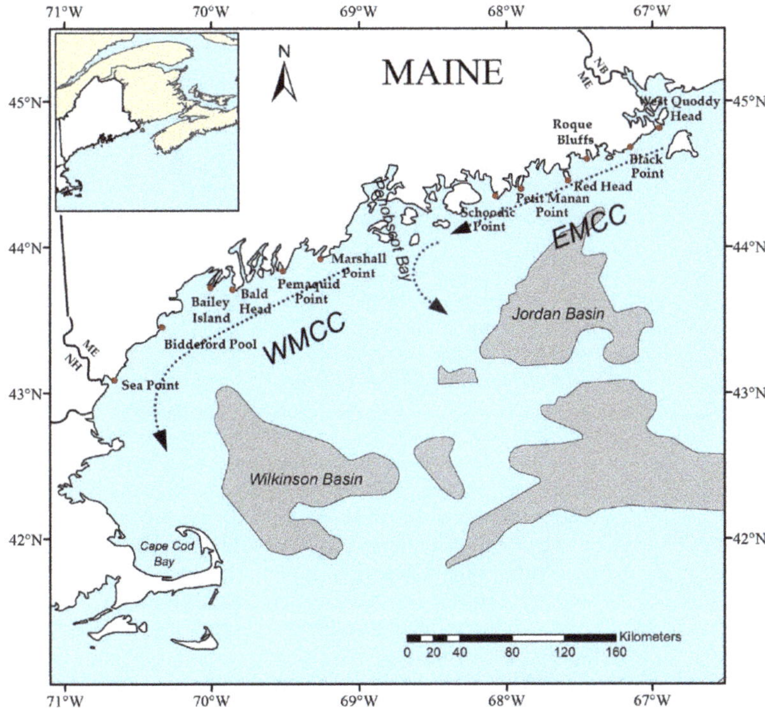

Figure 1. The Gulf of Maine with survey locations, geographic features, 200 m bathymetric contour and generalized flow of the principal segments of the Gulf of Maine Coastal Current, the Eastern Maine Coastal Current (EMCC), and Western Maine Coastal Current (WMCC). Inset for context with the northeastern US (white) and Canada. Abbreviations: NH, New Hampshire; ME, Maine; NB, New Brunswick.

The main purpose of this investigation was to examine the similarity of species assemblages among rocky intertidal communities in the GoM to reveal mesoscale spatial patterns and their persistence in time. The strong thermal gradient established by summer coastal circulation was predicted to influence the similarity among species assemblages on northern and southern Gulf shores. This was evaluated using a combination of multivariate and nonparametric approaches to compare patterns in temperatures and community variability across space. On the basis of these analyses, it was determined whether intertidal communities were similar throughout the GoM or if there were regional differences. This led to comparisons with subtidal communities across the same spatial extent. By doing so, differences found between these habitats were used to delineate where community similarities were found.

2. Materials and Methods

2.1. Study Sites and Rocky Intertidal Surveys

The study area spanned approximately 2 degrees of latitude, a distance of 336 km. A total of 12 exposed headlands were surveyed in 2004, 2005, and 2012. Rocky, exposed headlands were selected as study locations to have some degree of habitat similarity for comparisons and give a mesoscale geographic representation of the GoM coast. They were distributed from Sea Point, near the New Hampshire–Maine state border, to West Quoddy Head, near the Maine–New Brunswick, Canada border (Figure 1). In order to keep some degree of congruence among habitats, Central Maine, primarily occupied by Penobscot Bay, was not surveyed because the majority of the exposed rocky shores in the Penobscot

Bay estuary are on islands and not the mainland. Estuarine mud covers most of bottom of this island bay complex which receives freshwater from the Penobscot River and its watershed [38]. Intertidal communities were sampled at low tide in summer (June–August) during 2004 with line-transects, and in 2005 and 2012 with walk-about surveys. Time of low water and tidal amplitudes relative to mean lower low water (MLLW) were taken from the WWW Tide and Current Predictor [39]. The presence of macroinvertebrates ($\geq$1 mm) was recorded, identified in the field to the lowest taxon possible, usually species, or if unknown, collected, and identified in the laboratory the same day.

Sampling Methods

Headlands were surveyed in 2004 with line transects extending the full intertidal range from low water (chart datum) occurring at the predicted time to the high water line marked prior to low tide. There were three line transects of equal length for each location and the positions of endpoints recorded with WAAS GPS. Reconnaissance surveys were conducted as part of a pre-selection process for positioning sample transects that best avoided tidepools, large boulders, and upturned bedrock benches. Tide pools were not sampled and when encountered, the meter interval free of standing water closest to the immersed transect sample was selected instead. All macroinvertebrates in every meter interval which contacted a transect line were recorded.

During 2004 surveys, substrate types and dominant algae were assessed in four, non-random 1 m^2 quadrats positioned along one transect randomly selected from the three line transects. One quadrat was located about 2 m above the lowest exposed point on the shore, another approximately 2 m below high water, and two situated at quarter marks between these stations so that adjacent pairs were equidistant from each other. Within each quadrat, the primary substrate type was identified by visual estimate after dividing each 1 m^2 quadrat into 0.25 m^2 subsamples. The substrate class that covered > 50% of the surface was classified as primary. Substrate classes were gravel, cobble, boulder, and rock as defined by Brown [40].

Headlands were surveyed in 2007 and 2012 using walk-about surveys. The area surveyed varied among locations because of differences in slope, shoreline contour, and topography which determined the amount of exposed shore in addition to tidal amplitude. In 2007, each location was sampled over the course of 2 or 3 days during one low tide for 4 h each day. In 2012, each location was sampled in one day during one low tide. The difference in times for completing surveys reflects a time gaining experience with each intertidal site and funding objectives. The procedure for walk about surveys was as follows. Intertidal macroinvertebrates were sampled at randomly selected points with 10 $\times$ 10 cm quadrats. These sample points were at the terminus of path segments of random length. Path segments were oriented in randomly chosen compass bearings from a sample point. In this way, each quadrat was an independent, randomly selected sample. Sampling began towards a seaward horizon away from the high tide mark. Upon reaching the water's edge, the general heading switched to a landward horizon until reaching the high water mark, when the general heading switched back to seaward. Sampling continued until species accumulation curves reached an asymptote. Start and end points of sample paths were recorded with WAAS GPS, landmarks, and photographically. Boundaries to most survey areas were taken from maps accessed from the Critical Areas files in the Maine State Archives Library, otherwise they were defined using ArcMap™.

2.2. Exposure Index

Exposure was estimated for each survey location using an index that combines wind energy and effective fetch [41]. For this index, wind energy (W) depends on the duration and average speed (knots) the wind blows in each compass direction defined by 22.5° sectors. It was calculated using the equation:

$$W = \left(\frac{\text{percentage of time the wind blows in a 22.5° compass sec tor}}{100} \right) \times (\text{mean wind speed})^2$$

Effective fetch introduces a bathymetric component to the exposure index and is the quotient of actual fetch (F) divided by the sum of the extent (nautical miles) of shallow water <6 m deep joining the shoreline (CS) plus shallow water <6 m deep beyond that margin (DS). Fetch has a 100 NM maximum, i.e., distances greater than 100 NM are recorded as 100 NM. In summary, the exposure index is the sum of wind energy and effective fetch within each 22.5° compass sector of shoreline calculated using the Equation E1 of Thomas [41]:

$$\sum \log W \times \log[1 + F/(CS + 0.1\ DS)]$$

The measurement of each variable was achieved using the following method. Wind roses with 22.5° sectors were generated with WRPLOT View^{TM} (version 8.0.2) using wind velocity data recorded over a 5-year period between 2004 and 2012 at nearby coastal weather stations. From these, wind duration and mean speed were used to calculate wind energy for each sector and subsequently summed to calculate W. Using Google Earth Pro, wind roses were digitally centered on top of survey locations so that the first compass sector of the rose aligned with true north. After adjusting the transparency of the wind rose, the maximum extents of shallow water (CS and DS) within each sector were measured from NOAA Office of Coast Survey raster navigational charts overlaid on Google Earth imagery.

2.3. Subtidal Species Assemblages

Subtidal species assemblages among GoM benthic communities were explored for patterns of species assemblage similarity to compare with those of intertidal assemblages and the biogeographic analysis by Hale [42]. To carry this out, Environmental Protection Agency National Coastal Condition Assessment (NCCA) [43] data collected from subtidal stations July–September during 2000 to 2004 were selected from the same set of data analyzed by Hale [42] for western Atlantic biogeographic patterns. Station data includes benthic macroinvertebrate species abundance from 0.05 m^2 grabs, one grab sample per station, with no resampling, and water quality and temperature measurements. Sample mean depth, after removing rivers and ponds, was 19 m (max = 77.9 m, min = 1.1 m, mode = 18 m). Proxy stations were selected for nearness to intertidal survey locations (within 1km) and, when possible, shallow (<10 m) depths. Species presence data were used to characterize subtidal species assemblages. Comparisons with rocky subtidal epifauna were not possible because no data were available for the complete set of intertidal locations and the data that were accessible were collected outside of the time frame of intertidal surveys.

2.4. Coastal Temperature Data Acquisition and Analysis

Coastal land and sea surface temperatures during intertidal surveys were estimated using Copernicus Climate Change Service (C3S) Climate Data Store data. Land temperatures (2 m temperature, i.e., air temperature 2 m above the ground) were downloaded from *ERA5—Land monthly data from 1950 to present* [44]. The data are monthly averages calculated from average daily temperatures and are gridded with a horizontal resolution of 0.1° × 0.1°. Sea surface temperatures were downloaded from *Sea surface temperature daily data from 1981 to present*, derived from satellite observations [45]. The chosen Level 4 processing (Version 2.1) yielded temperatures resulting from a combination of measurements made by multiple sensor types (AVHRR, ATSR, SLSTR, and MetOp) and satellites (NOAA, ERS, Envisat, and Sentinel). Gridded SST data have a horizontal resolution of 0.05° × 0.05°. Preliminary examination showed that for all intertidal survey years, the warmest temperatures occurred during July–September and coldest during December–February. Therefore, mean temperatures for the two periods, called summer and winter from hereon, were calculated using contiguous months. For winter, the December of the year preceding January and February was used. For example, 2004 winter temperatures were assembled from 2003 December and 2004 January and February temperatures. Radiometer-based SST temperatures were ground truthed with temperatures measured during surveys and recordings by buoys of the Northeastern Regional Association of Coastal and Ocean Observing System (NERACOOS).

A three-way ANOVA (Sigma Plot 14.5) was used to explore differences in temperatures among survey years, the region where surveys were conducted, and where the temperature was estimated (land versus seas surface) as a factor. Data passed the Shapiro-Wilk and Brown-Forsythe tests for normality and equal variance, respectively. When significance was detected, the Holm-Sidak test was used for multiple comparisons among means to find which were statistically different.

2.5. Statistical Analysis of Species Assemblages

The similarity of species assemblages was compared among survey locations using Plymouth Routines in Multivariate Ecological Research (PRIMER 7) [46,47], PER-MANOVA+ [48], and their various subroutines. Only species incidence data were analyzed. Spatial analysis of species distributions within and among locations was not explored. Datasets collected using line transects, walk-about surveys, and NCCA benthic grabs were analyzed separately to accommodate differences in temperatures, surveyed locations, sampling protocols, and year sampled. Species accumulation curves for 2004, 2007, and 2012 surveys showed that all assemblages were adequately sampled with species richness reaching an asymptote (Supplementary Materials, Table S1, Figure S1). Samples were pooled for species presence at each intertidal survey location prior to analysis. Species presence was compiled from species abundance for subtidal NCCA grab samples. Patterns of species assemblage similarity among these four sets of data were investigated using hierarchical cluster analysis, canonical analysis of principle components (CAP), nonparametric multidimensional scaling (nMDS), and tests of mesoscale differences between species assemblages north and south of the mid-coast Penobscot Bay region (ANOSIM). Species similarity within regions and dissimilarity between regions was computed and compared (SIMPER). β diversity and its components was assessed for GoM species assemblages using R. Statistical significance for all tests was defined by p values less than 0.05.

Regional patterns in species assemblage similarity were investigated with hierarchal cluster analysis using the group average as the cluster mode on Bray-Curtis similarity matrices of species presence data. Evidence of statistically distinct clusters was explored with the similarity of profiles test (SIMPROF). An association of species assemblage similarity with latitude was evaluated using the canonical analysis of principle components (CAP). This method was used to visualize the distances between centroids of survey location similarity using latitude as the predictor variable. CAP also assessed the strength of correlation (δ_x^2) of the constrained ordination of samples with latitude. Patterns in similarity among assemblages were visualized with nonparametric multidimensional scaling (nMDS) on Bray-Curtis similarities. A spatial relationship of assemblage similarity at a coarser scale than latitude was explored by grouping survey locations by region and performing a one-way analysis of similarity (ANOSIM) test, with regions as unordered groups, to evaluate a statistical difference between regions north and south of mid-coast Maine. There were only three replicates, i.e., locations, per group for the 2007 walk-about surveyed headlands, too few to give meaningful ANOSIM results [47]. Instead, a one-way PERMANOVA was used to test for difference with Bray-Curtis similarities and region as a fixed factor in the model, followed by a pair-wise test to resolve statistical differences between north and south regions using Monte Carlo p values ($p_{(MC)}$). Unlike ANOSIM, PERMANOVA permutes similarity values rather than ranks, and evaluates the difference between centroids. The problem of a small number of replicates was surmounted by using Monte Carlo p values. Average similarity within and dissimilarity between regions were measured using the similarity percentages routine (SIMPER). This test also named species that contributed most (up to 70%) to the within-region similarity and differences between regions.

β diversity was evaluated for each survey year and for species incidence pooled among years at the Gulf scale using the Sørensen dissimilarity coefficient. Sørensen was chosen since the Bray-Curtis coefficient, used throughout analyses, is identical when calculated on presence/absence data [46]. The contributions of nestedness and turnover to structuring β diversity was assessed by partitioning β diversity into the components of species richness difference and species replacement. These computations were performed using the function *beta.multi* in the betapart package [49] in R (Version 4.2.0). Next, differences in β diversity between the north and south Gulf regions were explored using the *betadisper* function in vegan (version 2.6-2) [50]. This analysis used PERMDISP [51] to test if β diversity differed significantly between regions.

2.6. Temperature and Species Assemblage Similarities

The relationships of summer land temperature, summer SST, and exposure with species similarities among surveyed locations were examined. Environmental variables were not strongly collinear (Pearson $|r| \geq 0.95$) and were normalized prior to analyses to place them on a common scale. Summer land and sea surface temperatures and exposure were fitted with Bray–Curtis similarity matrixes using the distance-based linear models (DISTLM) routine in PERMANOVA + [48]. This procedure modelled the relationship between species assemblage similarities using the environmental variables as predictor variables. In general, DISTLM partitions the variation in multivariate data described by a resemblance matrix, and predictor variables are fit individually or sequentially to the model. Thus, the proportion of variation explained by each variable alone (marginal tests) and the proportion explained by each variable when added sequentially to a specified set of variables (conditional tests) are calculated with associated *p*-values acquired by permutation methods. The conditional sequential tests can disentangle the proportion of variation explained by each variable when added after ones previously fitted to the model. Finally, fitted models were visualized using the distance-based redundancy analysis (dbRDA) routine in PERMANOVA + and the patterns of sample ordination seen on plots examined.

3. Results

3.1. Coastal Temperatures

The thermogeography of the region features temperatures, which vary according to season and latitude. Summer coastal land temperatures were warmer than sea surface temperatures and cooler in the northern Gulf compared to the south (Figure 2). Sea surface temperatures followed this same latitudinal trend. Winter featured coastal land temperatures colder than sea surface temperatures (Figure 3). The summer trend with latitude was not present. Year, region, and temperature type (land versus SST) were statistically significant by the three-way ANOVA test, with a significant interaction of region and temperature type (Table 1). Multiple pairwise comparisons found statistical significance: (1) among all survey years, (2) northern and southern regions, and (3) land versus sea surface temperatures. Pair-wise comparisons exploring the interaction between region and temperature type found significance in all combinations of these two factors. In other words, land and sea surface temperatures differed significantly in the north and south, as the north and south regions differed in land temperature and SST.

3.2. Intertidal Description

All of the 12 exposed headlands were bedrock, and many were covered with boulder and cobble in varying degrees (Table 2). The exposure index ranged most often from 20–32, although the full range was 75.26–18.48 (median = 26.49). Generally, the dominant macroalgae present were: *Fucus vesiculosus*, *Chondrus crispus*, and *Mastocarpus stellatus*, with *Fucus distichus*, *Saccharina latissima*, *Laminaria digitata*, and *Alaria esculenta* lowest intertidally. Rocky surfaces were coated in patches with *Hildenbrandia* sp. and *Ceramium* sp. mid-intertidally. *Cladophora* sp. and *Corallina officinalis* were part of this understory.

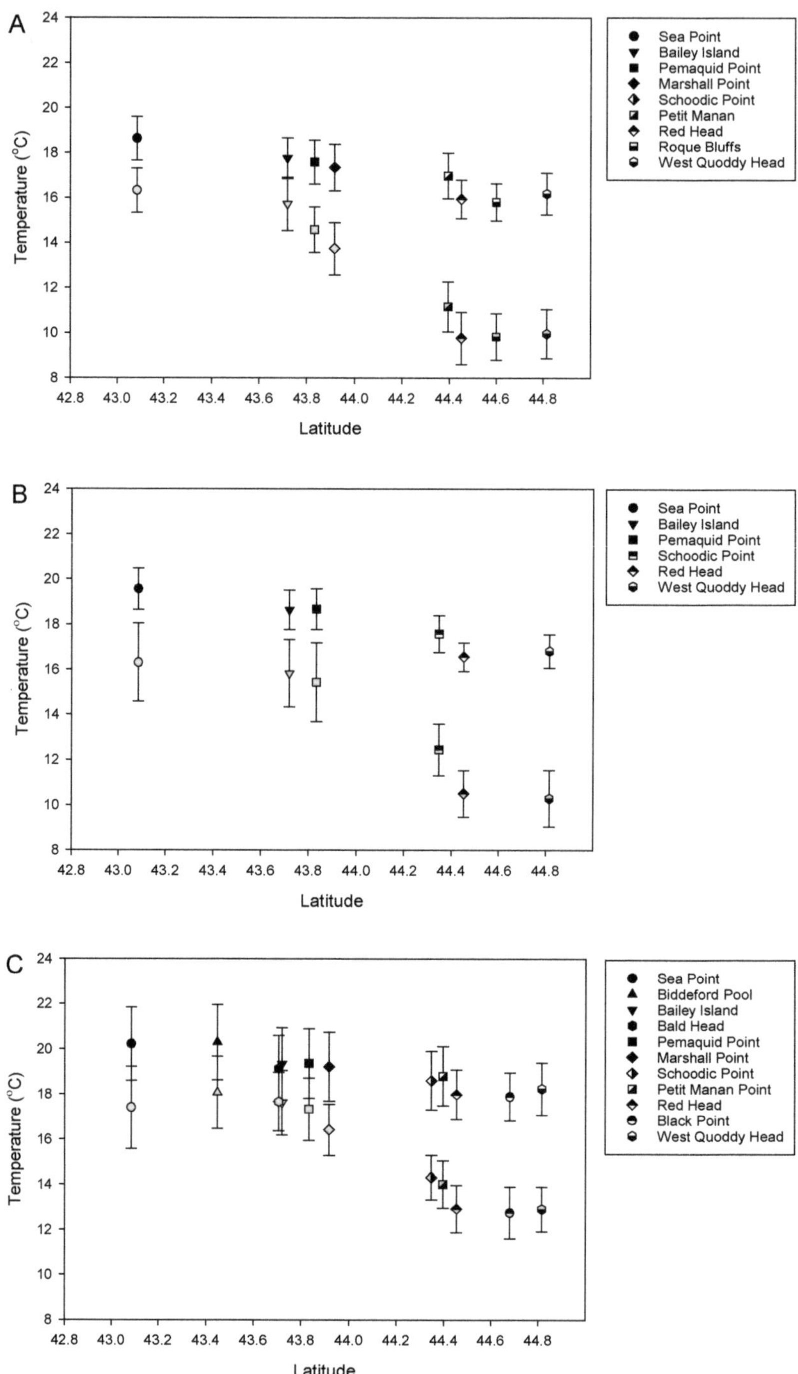

Figure 2. Summer temperature ($\bar{x} \pm$ SE) profiles for the GoM study area. (**A**), 2004; (**B**), 2007; (**C**), 2012. Symbology: Coastal land 2 m air temperature, black symbols; sea surface temperature, green symbols; filled symbols correspond to southern GoM region; half-filled symbols correspond to the northern GoM region.

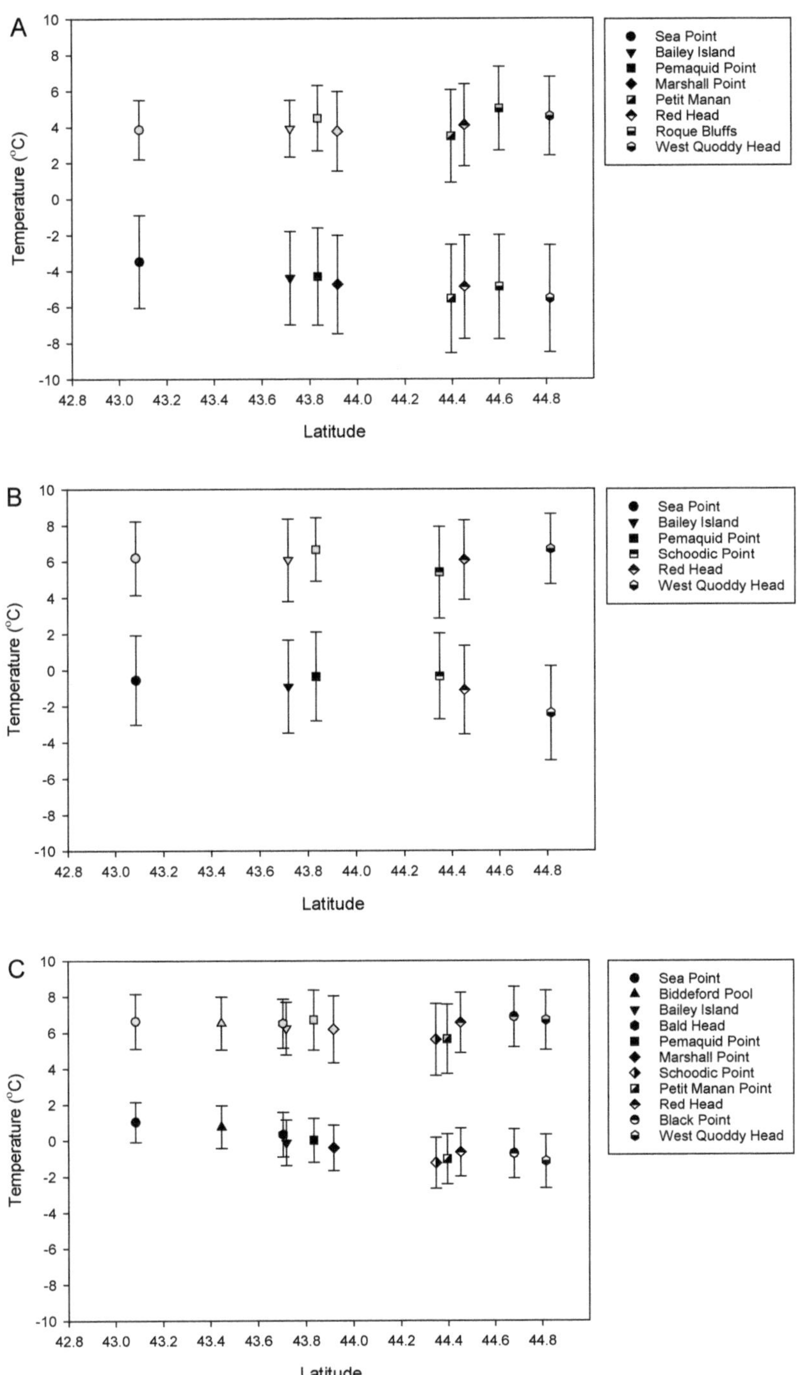

Figure 3. Winter temperature ($\bar{x} \pm$ SE) profiles for the GoM study area. (**A**), 2004; (**B**), 2007; (**C**), 2012. Symbology: Coastal land 2 m air temperature, black symbols; sea surface temperature, green symbols; filled symbols correspond to southern GoM region; half-filled symbols correspond to the northern GoM region.

Table 1. Summary table for three-way ANOVA test of summer land temperatures (2 m air) and SST during intertidal survey years. Region refers to northern versus southern GoM, defined by survey locations. Temperature Type refers to land (2m air) temperature versus sea surface temperature (SST).

Source of Variation	DF	SS	MS	F	p
Year	2	61.256	30.628	40.958	<0.001
Location	1	110.079	110.079	147.207	<0.001
Position	1	187.953	187.953	251.346	<0.001
Year × Region	2	1.41	0.705	0.943	0.398
Year × Temperature Type	2	3.342	1.671	2.234	0.12
Region × Temperature Type	1	20.165	20.165	26.966	<0.001
Year × Region × Temperature Type	2	0.0917	0.0459	0.0613	0.941
Residual	41	30.659	0.748		
Total	52	446.112	8.579		

Holm–Sidak Pairwise Multiple Comparison Tests

Comparison	Diff of Means	t	p
2012 vs. 2004	2.431	8.888	<0.001
2012 vs. 2007	1.65	5.268	<0.001
2007 vs. 2004	0.78	2.452	0.019
South vs. North	2.995	12.133	<0.001
Land Temperature vs. SST	3.913	15.854	<0.001

Region × Temperature Type Comparison Tests

Comparison	Diff of Means	t	p
Temperature Type within South			
Land Temperature vs. SST	2.632	7.294	<0.001
Temperature Type within North			
Land Temperature vs. SST	5.195	15.417	<0.001
Region within Land Temperature			
South vs. North	1.713	4.88	<0.001
Region within SST			
South vs. North	4.277	12.321	<0.001

Table 2. Aspect, exposure, and primary substrate for exposed headland intertidal locations surveyed. Abbreviations for substrate: B, bedrock; Bo, boulder; C, cobble. Symbols show year of survey: *, 2004; [†], 2007; [§], 2012.

Survey Site	Location	Aspect	Exposure	Substrate
Sea Point *,[†],[§]	43.09°–70.66°	120° SE	18.48	B/Bo/C
Biddeford Pool [§]	43.45°–70.33°	116° ESE	75.26	B/Bo
Bailey Island *,[†],[§]	43.72°–70.00°	218° SW	27.98	B
Bald Head [§]	43.70°–69.85°	245° WSW	62.61	B
Pemaquid Point *,[†],[§]	43.83°–69.51°	190° S	23.69	B
Marshall Point *,[§]	43.92°–69.26°	122° SE	24.58	B/Bo
Schoodic Point [†],[§]	44.35°–68.08°	230° SW	20.26	B/C
Petit Manan Point *,[§]	44.40°–67.90°	256° WSW	31.78	B/Bo
Red Head *,[†],[§]	44.45°–67.58°	231° SW	25.18	B
Roque Bluffs *	44.68°–67.15°	198° SSW	24.79	B/Bo
Black Point [§]	44.68°–67.15°	63° ENE	27.79	B
West Quoddy Head *,[†],[§]	44.81°–66.95°	90° E	32.17	B/Bo

3.3. Species Diversity

A pooled total of 117 taxa (Supplementary Materials, Table S2) was dominated by molluscs (29%) and crustaceans (17%). β diversity of species assemblages was moderate among survey years (Table 3). There was no statistical difference in β diversity between south and north regions. Species turnover accounted for 78% to 88% of the β diversity. This indicates that variation among species assemblages results from species replacement along the longitudinal gradient of the GoM shore and not because locations are nested subsets. These trends in β diversity and its components were consistent when species incidence was pooled among years except β diversity was high and species turnover was greater.

Table 3. β diversity and contribution of its components for GoM exposed headland rocky intertidal species assemblages for each survey year and all years pooled.

Year	β diversity	Turnover	Proportion of β	Nestedness	Proportion of β
2004	0.63	0.56	0.89	0.07	0.11
2007	0.49	0.38	0.78	0.11	0.22
2012	0.67	0.59	0.88	0.08	0.09
Pooled	0.86	0.78	0.91	0.08	0.09

3.4. Regional Comparison of Similarity among Exposed Rocky Headland Species Assemblages

Exposed headland species assemblages differed in similarity on a regional scale. For all survey years, assemblages clustered into two statistically distinct groups corresponding to regions north and south of mid-coastal Penobscot Bay. However, there were no significant differences among assemblages within each region (Figure 4). Survey locations based on assemblage similarity clustered by region and latitude in constrained CAP ordinations of species assemblages (Figure 5). The relationship was strong (δ_1^2, range 0.79–0.94) and canonical correlations were highly significant (Table 4), except for the 2007 survey due to the low number of locations. Regional dissimilarity was clear for all survey years in unconstrained two-dimensional nMDS ordinations, each with low stress value and resembling CAP ordinations (Figure 6). Southern locations grouped together and separate from northern ones that grouped together. Southern and northern regions differed significantly when assemblage similarities were compared (Table 4). Among all years, the average Bray-Curtis similarity of species present within southern and northern regions ranged among years from 67.21 to 77.3 (Table 5). Average Bray-Curtis dissimilarity between regions ranged from 32.97 to 41.28. Overall, most species which contributed up to 70% of the average dissimilarity were arthropods and molluscs. However, when only the species found exclusively in one or the other region were considered, the dominant taxa changed to a mixed group. In general, more species were found only in the southern region with invasive species appearing in 2007 and 2012. Interestingly, subtidal species assemblages did not cluster according to similarity by region, were not significantly correlated in CAP analysis with latitude ($\delta_1^2 = 0.287$, $p = 0.219$), and there was no statistical difference between regional groupings (ANOSIM R = 0.136, $p = 0.12$). nMDS ordination showed no clear pattern of separation among species assemblages according to where grab samples were taken in respect to regions north and south (Figure 7).

Table 4. Summary of results from the statistical comparison of southern versus northern GoM species assemblage similarities (ANOSIM) and the relationship of species assemblage similarity with latitude (CAP).

Year	ANOSIM	*p*-Value	CAP (δ_1^2)	*p*-Value
2004	0.839	0.003	0.93879	0.002
2007	2.12 [†]	0.03 [††]	0.79364	0.1099
2012	0.768	0.002	0.79643	0.002

[†] Test statistic, *t*, for 2007 was calculated from PERMANOVA, not ANOSIM. [††] Monte Carlo adjusted *p*.

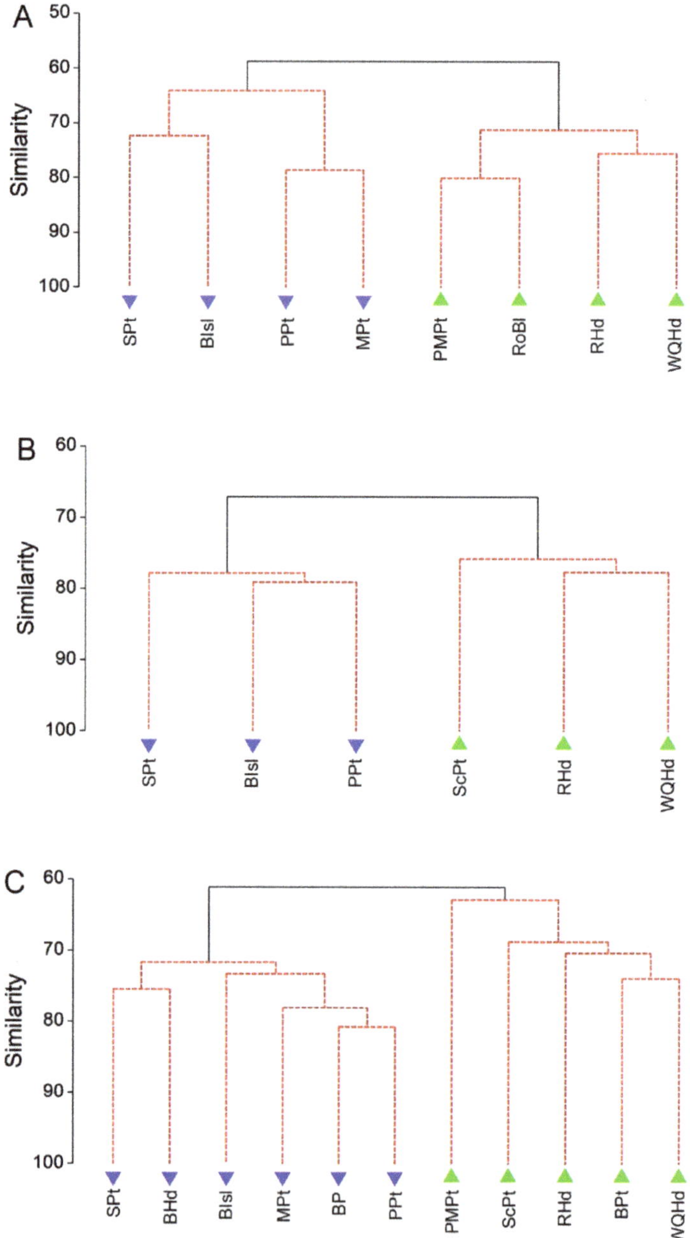

Figure 4. Hierarchical cluster analysis of Bray-Curtis similarities with group average as the cluster mode for rocky intertidal species assemblages. (**A**) 2004; (**B**) 2007; (**C**) 2012. Solid black lines connect samples that differ significantly. Red dashed lines connect samples not significantly different (SIMPROF). Symbology: Blue triangles, southern GoM region; green triangles, northern GoM region. Abbreviations: BIsl, Bailey Island; BHd, Bald Head; BP, Biddeford Pool; BPt, Black Point; MPt, Marshall Point; PPt, Pemaquid Point; ScPt, Schoodic Point; SPt, Sea Point; PMPt, Petit Manan Point; RHd, Red Head; RoBl, Roque Bluffs; WQHd, West Quoddy Head.

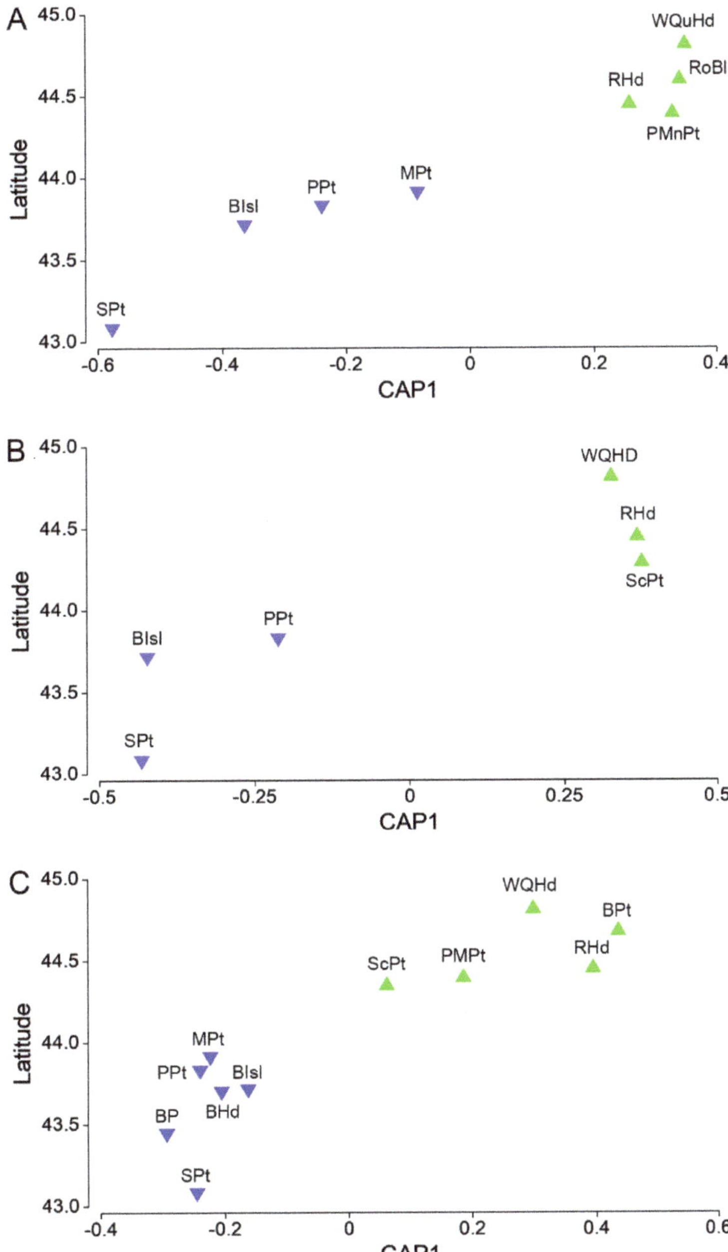

Figure 5. Canonical ordinations of rocky intertidal species assemblage similarity with latitude. (**A**) 2004; (**B**) 2007; (**C**) 2012. Symbology: Blue triangles, southern GoM region; green triangles, northern GoM region. Abbreviations: BIsl, Bailey Island; BHd, Bald Head; BP, Biddeford Pool; BPt, Black Point; MPt, Marshall Point; PPt, Pemaquid Point; ScPt, Schoodic Point; SPt, Sea Point; PMPt, Petit Manan Point; RHd, Red Head; RoBl, Roque Bluffs; WQHd, West Quoddy Head.

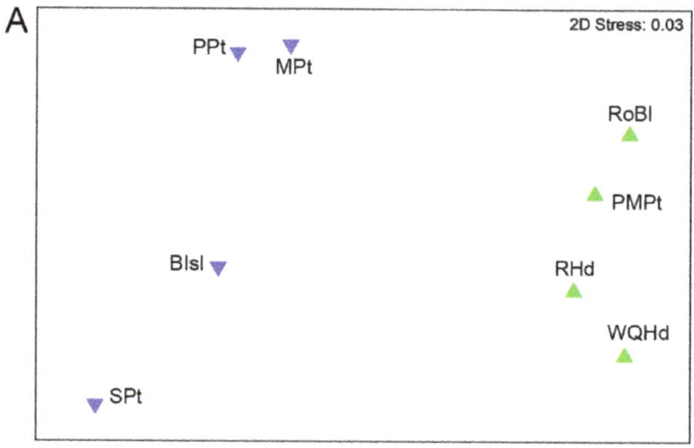

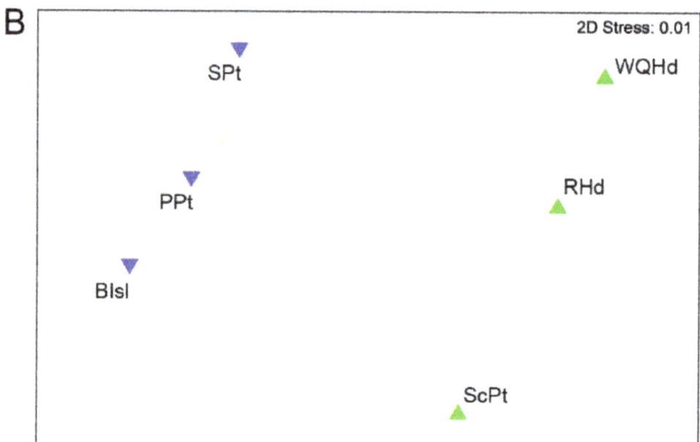

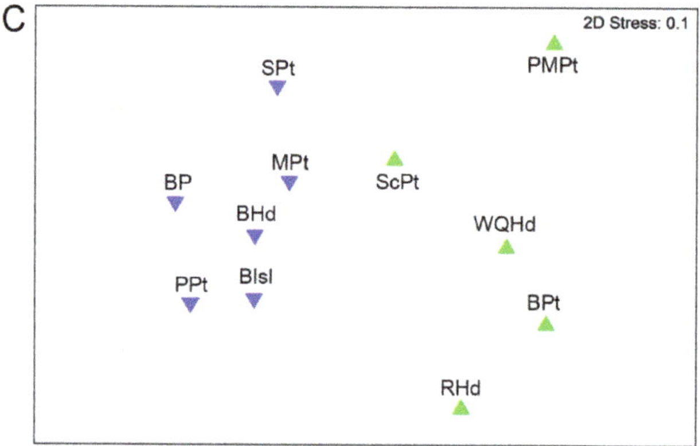

Figure 6. Ordination of rocky intertidal species assemblages by nMDS of Bray-Curtis similarities. (**A**) 2004; (**B**) 2007; (**C**) 2012. Symbology: Blue triangles, southern GoM region; green triangles, northern GoM region. Abbreviations: BIsl, Bailey Island; BHd, Bald Head; BP, Biddeford Pool; BPt, Black Point; MPt, Marshall Point; PPt, Pemaquid Point; ScPt, Schoodic Point; SPt, Sea Point; PMPt, Petit Manan Point; RHd, Red Head; RoBl, Roque Bluffs; WQHd, West Quoddy Head.

Table 5. Species that contributed most to dissimilarity (up to 70%) between regions determined by SIMPER. Average Bray-Curtis dissimilarity between regions in parentheses. Average Bray-Curtis similarity within regions in braces.

2004		2007		2012	
(41.28)		(32.97)		(38.91)	
South	**North**	**South**	**North**	**South**	**North**
{67.85}	{77.33}	{78.21}	{76.4}	{73.71}	{67.21}
Halichondia panacea	*Lacuna vincta*	*Botryloides violaceus*	*Scycon ciliarus*	*Botryloides violaceus*	*Cucumaria frondosa*
Halisarca sp.	*Metridium senile*	*Harmothoe imbricata*	*Leptasterias littoralis*	*Botryllus schlosseri*	*Buccinum undatum*
Pectinaria gouldii	*Stenosemus albus*	*Homarus americanus*	*Aulactinia stella*	*Didemnum vexillum*	
Lineus viridis	*Urticina felina*	*Idotea phosphorea*		*Hemigrapsus sanguineus*	
	Jaera albifronds	*Nereis pelagica*		*Crepidula fornicata*	
	Halocynthia pyriformis	*Clava multicoirnis*		*Diplosoma listerium*	
		Tubularia indivisa		*Tubularia indivisa*	
		Cadlina laevis		*Disporella hispida*	
		Crepidula fornicata			
		Hemigrapsus sanguineus			
		Onchidoris bilamellata			
		Phoxichilidium fematorum			

3.5. Spatial Relationship of Communities with Temperature and Exposure

Summer sea surface temperature and land temperature explained a large, statistically significant amount of the variation in intertidal species assemble similarity. When each variable was considered individually in marginal tests, SST explained 37–53% of the variation in species assemblage similarity among all survey years. Land temperatures explained 31–44%. These relationships were statistically significant in all cases except 2007 where only SST was significant (Table 6). Exposure did not explain a significant amount of variation in marginal tests ($\leq$11%, $p > 0.05$). In sequential conditional tests where the ordering of SST and land temperature was switched, the variable explaining the largest proportion of variation was the first one in the sequence evaluated (Table 7). When SST was first, it explained a statistically significant proportion (37–53%) of variation. Land temperatures explained 5–18% more, statistically insignificant amounts. When land temperature was first, it explained more of the variation (31–45%) in species assemblage similarity than SST, statistically significant amounts except for 2007. Sea surface temperature contributed an additional 10–27%, insignificant amounts except for 2012. Placing exposure first in the testing sequence did not change the outcomes for temperatures, and the amount of variation it explained was always the smallest and insignificant. In summary, land temperatures and SST together explained a significant amount of the variability in assemblage similarity but not exposure. The models performed well and captured most of the variation in species assemblage variation as shown by their associated plots produced by distance-based redundancy analysis (Figure 8). Among all years, the first two dbRDA axes explained 90–100% of the fitted variation, which was about 48–78% of the total variation in species assemblage similarities. The separation of species assemblages into northern and southern groups by dbRDA was clear in all plots and consistent among all survey years.

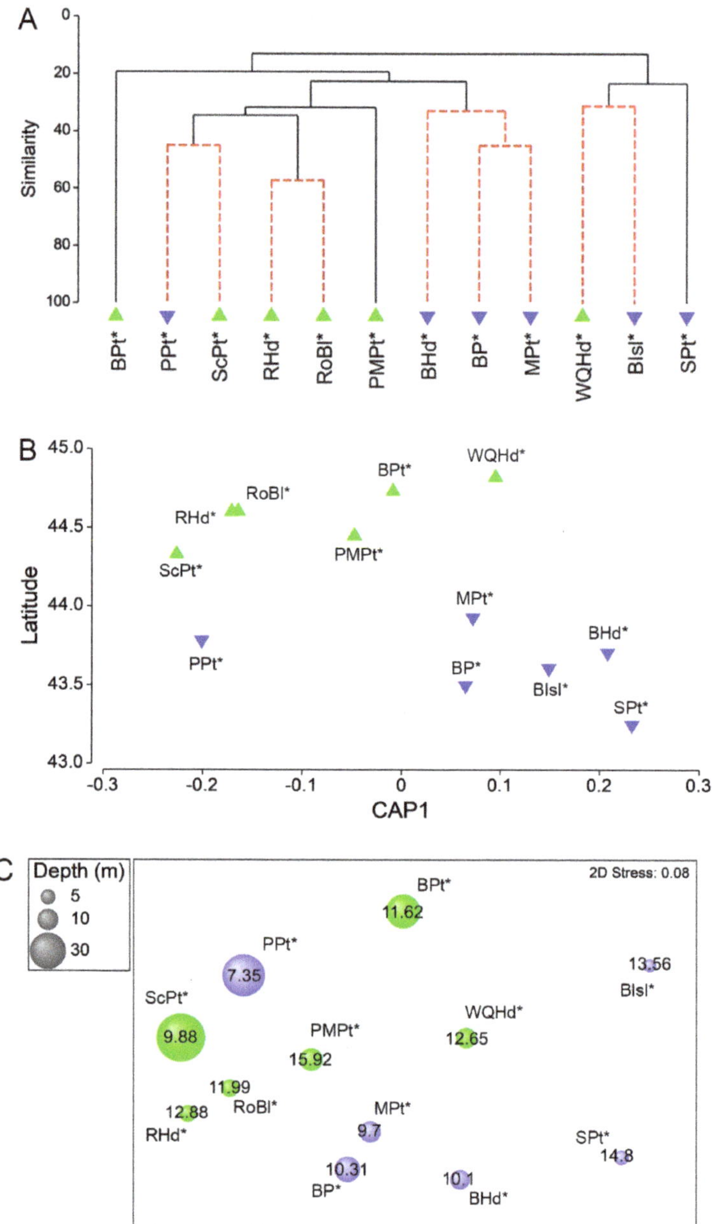

Figure 7. Similarity among subtidal species assemblages at locations proximal to rocky intertidal survey locations. (**A**) Hierarchical cluster analysis. Solid black lines connect assemblages that differ significantly; red dashed lines connect assemblages that are not significantly different (SIMPROF). (**B**) Canonical ordination of species assemblage similarity with latitude. (**C**) nMDS of assemblage similarities displayed with bottom sea water temperatures (°C) and depth. Symbology: Green bubbles, southern GoM region; blue bubbles, northern GoM region; *, indicates proxy station. Abbreviations: BIsl, Bailey Island; BHd, Bald Head; BP, Biddeford Pool; BPt, Black Point; MPt, Marshall Point; PPt, Pemaquid Point; ScPt, Schoodic Point; SPt, Sea Point; PMPt, Petit Manan Point; RHd, Red Head; RoBl, Roque Bluffs; WQHd, West Quoddy Head.

Table 6. Marginal test results of distance-based linear modelling of species similarity with temperatures and exposure for survey years. Results show the proportion of variability (Prop.) explained for each variable and its level of significance.

Year	Variable	Pseudo-*F*	*p*	Prop.
2004	Summer Mean SST	4.873	0.003	0.45
	Summer Mean Land Temp	4.147	0.004	0.41
	Exposure	0.757	0.590	0.11
2007	Summer Mean SST	4.564	0.020	0.53
	Summer Mean Land Temp	3.217	0.062	0.45
	Exposure	0.188	0.922	0.04
2012	Summer Mean SST	5.266	0.001	0.37
	Summer Mean Land Temp	4.076	0.001	0.31
	Exposure	0.783	0.614	0.08

Table 7. Sequential conditional test results of distance-based linear modelling of species similarity with temperatures and exposure for survey years. Variables are fit to models as covariables and evaluated as sets. The sequence of SST and land temperature was switched in tests for each year. The proportion of variation (Prop.) of an added variable is its contribution to the amount already explained by the preceding variable (Res. df, residual degrees of freedom).

Year	Variable Test Order	Pseudo-*F*	*p*	Prop.	Res. df
2004	+Summer Mean SST	4.873	0.003	0.45	6
	+Summer Mean Land Temp	0.608	0.778	0.06	5
	+Exposure	0.735	0.624	0.08	4
	+Summer Mean Land Temp	4.147	0.001	0.41	6
	+Summer Mean SST	1.010	0.453	0.10	5
	+Exposure	0.735	0.614	0.08	4
2007	+Summer Mean SST	4.564	0.018	0.53	4
	+Summer Mean Land Temp	1.894	0.217	0.18	3
	+Exposure	0.492	0.683	0.06	2
	+Summer Mean Land Temp	3.217	0.070	0.45	4
	+Summer Mean SST	2.808	0.068	0.27	3
	+Exposure	0.492	0.662	0.06	2
2012	+Summer Mean SST	5.266	0.001	0.37	9
	+Summer Mean Land Temp	1.635	0.113	0.11	8
	+Exposure	0.461	0.886	0.03	7
	+Summer Mean Land Temp	4.076	0.001	0.31	9
	+Summer Mean SST	2.512	0.012	0.16	8
	+Exposure	0.461	0.873	0.03	7

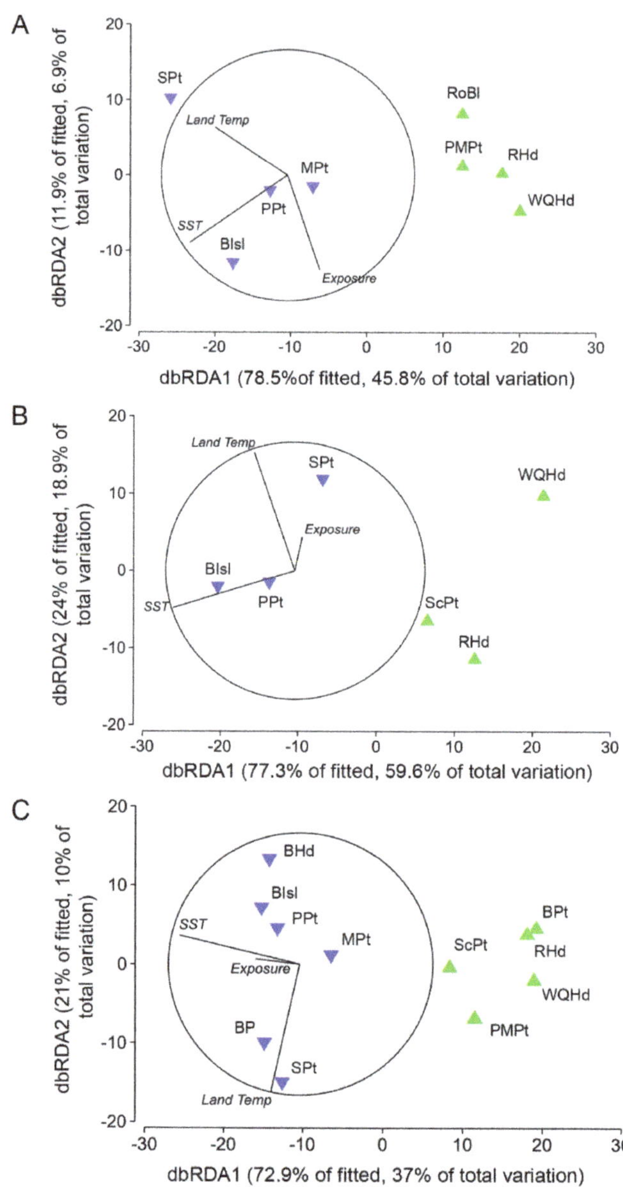

Figure 8. dbRDA of the modelled species assemblage similarity data with the predictor variables land temperature, SST, and exposure. Vectors show the strength and direction of the effect each variable had in relation to the others in constructing the ordinations. (**A**) 2004; (**B**) 2007; (**C**) 2012. Symbology: Blue triangles, southern GoM region; green triangles, northern GoM region. Abbrevi-ations: BIsl, Bailey Island; BHd, Bald Head; BP, Biddeford Pool; BPt, Black Point; MPt, Marshall Point; PPt, Pemaquid Point; ScPt, Schoodic Point; SPt, Sea Point; PMPt, Petit Manan Point; RHd, Red Head; RoBl, Roque Bluffs; WQHd, West Quoddy Head.

4. Discussion

Exposed rocky headland intertidal species assemblages of northern and southern GoM shores were dissimilar. Penobscot Bay located at mid-coast Maine, the largest estuary in Maine and the second largest on the US east coast [52], marked the division between the two regions. This pattern was persistent among three sets of data, which varied in collection methods and spanned a total of eight years. Variation in species assemblage similarity was correlated with latitude and persistent among surveys separated in time by 8 years. The regional differences were consistent with prior exposed rocky shore studies conducted in similar if not the same locations [53,54]. The dissimilarity was confined to the intertidal since it was not evident among subtidal species assemblages. In general, the northern and southern Gulf regions matched the northern and southern faunal zones proposed by Bousfield and Thomas [55]. Their biogeographic scheme was based on temperature and divides the GoM coastline into three zones: a < 12 °C subarctic zone in northern Maine, a 12–15 °C boreal region in central Maine, and a 15–18 °C cool temperate zone that extends to Massachusetts Bay. Central Maine, primarily occupied by Penobscot Bay, was not surveyed by this study. Adey and Heyek [56] documented differences between northern rocky intertidal communities from those in the southern GoM, work that contrasted the pre-existing idea that the Gulf was part of a single biogeographic unit extending from Cape Cod to Labrador [57]. The pattern of dissimilarity between Gulf regions is not limited to exposed rocky shores. It was found among macroinvertebrate species assemblages of sand beaches [58] and low energy intertidal areas dominated by *Ascophyllum nodosum* [59]. The pattern of dissimilarity was a key feature of GoM intertidal communities with a signal strong enough to be detected by species presence data.

Species were not evenly distributed among all surveyed locations, which lead to the dissimilarity between north and south regions. Species turnover contributed most to this pattern. Analysis of these dissimilarities with SIMPER showed trends that coincided with GoM invasive species histories, each with sea water temperature identified as a common factor underlying population dynamics. Some species were found exclusively in one region, a situation that occurred more often in the southern Gulf, particularly in 2007 and 2012 when invasive species accounted for some of them. These included the colonial tunicates *Botryloides violaceus*, *Didemnum vexillum*, and *Diplosoma listereum*, and *Botryllus schlosseri* currently understood to be cryptogenic [60]. Around the 2007, the southern region was subject to invasion by those species and their community dynamics, competition for space in particular, were shown to be correlated with seasonal changes in sea water temperatures [61]. Similarly identified among the southern Gulf exclusives was the invasive Asian shore crab *Hemigrapsus sanguineus* that appeared in the region around 2001 [62]. Based on 2002 to 2005 coastal Maine surveys that over-lapped the range in latitude of the current study, Stephenson et al. [62] suggested that sea water temperatures might limit range expansion to the warmer coasts south of Penobscot Bay. Since then, *H. sanguineus* has increased in density in the southern region where established populations remain confined [63].

Shore and sea surface temperatures explained a large proportion of the variation of similarity among species assemblages. When evaluated individually or as covariates, the degree of variation explained was statistically significant except when the number of surveyed locations was small, i.e., the 2007 surveys. The influence of shore temperatures reinforced the idea that regional community dissimilarity is a feature of the shore and not the subtidal. Their contribution was significant despite the lower resolution of the gridded data compared to SST. The different thermogeography of northern and southern Gulf regions is largely influenced by ocean circulation, and the flow of major coastal currents match the pattern of dissimilarity among species assemblages of the two regions. Coastal water temperatures do not appear to drive land temperatures [64], but they do act as a buffer, cooling coastal land temperatures in summer and warming them in winter [26,27]. Across-shore thermal gradients that are discontinuous in summer could sort community composition according to species thermal tolerance to produce the patterns in species similarity. Elsewhere, sea surface temperatures were found to play key roles in the regional distribution patterns of species assemblages [9,10,65–67]. Likewise, shore temperatures influenced species intertidal distributions [1,12,23,68]. Temperature can affect major ecological patterns and community assembly by driving metabolism, resulting in modifications of longevity, population growth, and consequently competition [16] among other species interactions [25,69]. The effects of temperature are pervasive throughout biological processes, and temperature zonation and biogeographic regions are some macroscale manifestations [18,70].

The lack of a significant effect from wave exposure was unexpected since it is known to influence species richness and diversity among intertidal communities in the same biogeographic region [57,61–73] and elsewhere [74]. The method for calculating exposure was not an issue since it was supported by survey data [41,75]. This suggests that the effects of exposure are best explored using abundance data and vertically stratified sampling methods to include shore height. In addition, the range in exposure indices was not very large especially when the three extreme measures were ignored (range: 20.26–32.17; SD = 3.86). Therefore, the variation in exposure among survey locations may be too slight to evaluate a relationship with assemblage similarity. Future studies examining an effect of wave exposure on intertidal communities might include sheltered and partially exposed shores [74] to increase the range of variation of exposure among survey locations.

Rocky intertidal communities on northern and southern shores are dissimilar and the mid-coast Penobscot Bay region marks the area where the two regions diverge. However, other research indicates that this feature is not expected. Within the GoM, population genetics of some of the same species found in surveys show no discontinuities [76,77]. Instead, a review of population genetics data [77] showed a discontinuity displaced to the south of Cape Cod, not within the GoM. However, since that review, two species of rocky intertidal gastropods with non-planktonic development, *Nucella lapillus* [78] and *Littorina saxatilis* [79], were shown to split into northern and southern clades within the same two regions defined by the current study. Models integrating ocean currents and species with high-dispersal larvae did not predict a peak of range boundaries within the GoM [80]. That said, oceanographic features of the northern Gulf, the EMCC in particular, appear to set the southern range boundary for the mussel *Mytilus trossulus* [81]. Large-scale analyses of West Atlantic species distribution patterns support conclusions from population genetics and modelling [82]. Furthermore, Hale [43] did not find a transition area for subtidal benthic invertebrates at mid-coast Maine, a finding corroborated by the present study, but instead found one to the south of Casco Bay. That embayment is located approximately 90 km south of the mid-coast. In summary, there is much evidence to the contrary of a mid-coast Maine rocky intertidal discontinuity.

How might the differences between predicted boundaries and the results of the present study be reconciled? Firstly, the transition area for benthic species assemblages was identified for species that are subtidal soft bottom inhabitants, not rocky intertidal ones. Additionally, the species assemblages were different. Subtidal assemblages were dominated by polychaetes, with crustaceans and molluscs the next most abundant taxa [43]. Rocky intertidal assemblages were dominated by molluscs and crustaceans. Next, the incongruity stemming from population genetics has value in the sense that these studies rule out the possibility that hydrodynamic barriers to gene flow via larval dispersal shape regional divergence. Genetic differences among populations are not a prerequisite for community dissimilarity. Species interactions can influence community assembly [83]. Likewise, the meta-analysis of species distributions [82] did not consider community dynamics and its consequences for predicting discontinuities. Finally, high-dispersing larvae are not characteristic of all rocky intertidal species so the hydrodynamic modelling of discontinuities [80] is limited. These models also did not consider how community dynamics might influence their predictions. While species interactions were not specifically examined by the present study, they are implicit in structuring the similarities among the surveyed species assemblages. The dissimilarities between northern and southern shores are likely a signal of the interactions particular to the sets of species present in those regions, an idea supported by their differences in predation [84], recovery from disturbance [54], and possibly recruitment [81,85].

The GoM is currently warming faster than most other bodies of waters globally [86,87]. The survey conducted in 2012 occurred during an ocean heat wave when warming was especially pronounced in the GoM [86]. In the decade since, there have been profound consequences for fisheries [86,88,89], kelp forests [90], and phenologies [91–93]. Given the predicted conditions in the GoM [94], change in species turnover from species invasions [95] and range shifts [96] portend novel species interactions with consequences for rocky intertidal community patterns.

5. Conclusions

GoM rocky intertidal communities were similar within regions, but the regions were distinct and located south and north of the Penobscot Bay estuary. This discontinuity did not extend into the subtidal; it was a uniquely intertidal feature. Both land and SST explained a significant amount of the variation which gave rise to regional dissimilarity, but they did not explain all of it. Species interactions and community dynamics are predicted to play important roles.

Supplementary Materials: The following supporting information can be downloaded at: https://www.mdpi.com/article/10.3390/d14070557/s1, Figure S1: Species accumulation curves for all survey years. Table S1: Survey metadata for exposed rocky intertidal locations; Table S2: List of species present among surveyed GoM rocky intertidal exposed headlands.

Funding: Surveys conducted in 2004 were financially supported by The Nature Conservancy, Maine Chapter and a Program Development Grant from Maine Sea Grant. Surveys conducted in 2007 were supported by the Census of Marine Life programs Natural Geography in Shore Areas (NaGISA), History of Marine Animal Populations (HMAP), and the GoM Area Program (GoMA). Surveys conducted in 2012 received no financial support.

Institutional Review Board Statement: Not applicable.

Informed Consent Statement: Not applicable.

Data Availability Statement: Available upon request within a reasonable amount of time.

Acknowledgments: Statistical advice from M.L. Anderson, Massey University, Auckland, New Zealand, and P. Somerfield and K.R. Clarke, Plymouth Marine Laboratory, Plymouth, United Kingdom, proved invaluable and is greatly appreciated. The results contain modified Copernicus Climate Change Service information 2020. Neither the European Commission nor ECMWF is responsible for any use that may be made of the Copernicus information or data it contains. Much appreciation expressed to the field crews and others who collected and processed the NCCA data and made it available. The U.S. National Park System and U.S. Fish and Wildlife approved licenses for work on their properties. The Maine Department of Marine Resources gave in-kind support throughout many phases of this work. The R.S. Friedman Field Station of Suffolk University, Boston, USA, graciously provided facilities. Adrianna Zito-Livingston, The Nature Conservancy, Delmont, New Jersey, USA, assisted with the 2004 field surveys and data transcription. Many thanks expressed to the Mateo family for access to their property. Gerhard Pohle, Huntsman Marine Science Centre, St. Andrews, New Brunswick, provided many helpful suggestions which improved an earlier draft.

Conflicts of Interest: The funders had no role in the design of the study; in the collection, analyses, or interpretation of data; in the writing of the manuscript, or in the decision to publish the results.

Abbreviations

ANOSIM	Analysis of Similarities
CAP	Canonical Analysis of Principal Components
dbRDA	Distance-based Redundancy Analysis
DISTL	Distance-based Linear Analysis
EMCC	Eastern Maine Costal Current
GoM	Gulf of Maine
NM	Nautical Mile
nMDS	Nonparametric Multidimensional Scaling
SIMPER	Similarity of Percentages

References

1. Barry, J.P.; Baxter, C.H.; Sagarin, R.D.; Gilman, S.E. Climate-related, long-term faunal changes in a California rocky intertidal community. *Science* **1995**, *267*, 672–675. [CrossRef] [PubMed]
2. Montoya, J.M.; Raffaelli, D. Climate change, biotic interactions and ecosystem services. *Phil. Trans. R. Soc. B* **2010**, *365*, 2013–2018. [CrossRef] [PubMed]
3. Hughes, T.; Bellwood, D.; Folke, C.; Steneck, R.; Wilson, J. New paradigms for supporting the resilience of marine ecosystems. *Trends Ecol. Evol.* **2005**, *20*, 380–386. [CrossRef] [PubMed]
4. Montoya, J.M.; Prim, S.L.; Solé, R.V. Ecological networks and their fragility. *Nature* **2006**, *442*, 259–264. [CrossRef]
5. Doney, S.C.; Ruckelshaus, M.; Duffy, J.E.; Barry, J.P.; Chan, F.; English, C.A.; Galindo, H.M.; Grebmeier, J.M.; Hollowed, A.B.; Knowlton, N.; et al. Climate change impacts on marine ecosystems. *Annu. Rev. Mar. Sci.* **2012**, *4*, 11–37. [CrossRef]
6. Bernhardt, J.R.; Leslie, H.M. Resilience to climate change in coastal ecosystems. *Annu. Rev. Mar. Sci.* **2013**, *5*, 371–392. [CrossRef]
7. Connolly, S.R.; Roughgarden, J.A. Latitudinal gradient in northeast Pacific intertidal community structure: Evidence for an oceanographically based synthesis of marine community theory. *Am. Nat.* **1998**, *151*, 311–326. [CrossRef]
8. Menge, B.A.; Lubchenco, J.; Bracken, M.E.S.; Chan, F.; Foley, M.M.; Freidenburg, T.L.; Gaines, D.S.; Hudson, G.; Krenz, C.; Leslie, H.; et al. Coastal oceanography sets the pace of rocky intertidal community dynamics. *Proc. Natl. Acad. Sci. USA* **2003**, *100*, 12229–12234. [CrossRef]
9. Broitman, B.R.; Navarrete, S.A.; Smith, F.; Gaines, S.D. Geographic variation of southeastern Pacific intertidal communities. *Mar. Ecol. Prog. Ser.* **2001**, *224*, 21–34. [CrossRef]
10. Blanchette, C.A.; Miner, C.M.; Raimondi, P.T.; Lohse, D.; Heady, K.E.; Broitman, B.R. Biogeographical patterns of rocky intertidal communities along the Pacific coast of North America. *J. Biogeogr.* **2008**, *35*, 1593–1607. [CrossRef]
11. Hurrell, J.; Kushner, Y.; Ottersen, G.; Visbeck, M. (Eds). The North Atlantic Oscillation: Climatic significance and environmental impact. *Geoph. Monogr. Ser.* **2003**, *134*, 1–279.
12. Somero, G.N. Thermal physiology and vertical zonation of intertidal animals: Optima, limits, and costs of living. *Integr. Comp. Biol.* **2002**, *42*, 780–789. [CrossRef] [PubMed]
13. Skreslet, S. *The Role of Freshwater Outflow in Coastal Ecosystems*; NATO AS1 Series G; Springer: Berlin, Germany, 1986; Volume 7.
14. Strayer, D.L.; Smith, L.C. Macroinvertebrates of a rocky shore in the freshwater tidal Hudson River. *Estuar. Coast* **2000**, *23*, 359–366. [CrossRef]
15. Gillanders, B.M.; Kingsford, M.J. Impact of changes in flow of freshwater on estuarine and open coastal habitats and the associated organisms. *Oceanogr. Mar. Biol.* **2002**, *40*, 233–309.
16. Saito, V.S.; Perkins, D.M.; Kratina, P. A metabolic perspective of stochastic community assembly. *Trends Ecol. Evol.* **2021**, *36*, 280–283. [CrossRef]
17. Cruz-Motta, J.J.; Miloslavich, P.; Palomo, G.; Iken, K.; Konar, B.; Pohle, G.; Trott, T.; Benedetti-Cecchi, L.; Herrera, C.; Hernández, A.; et al. Patterns of spatial variation of assemblages associated with intertidal rocky shores: A global perspective. *PLoS ONE* **2010**, *5*, e14354. [CrossRef]
18. Hutchins, L.W. The basis for temperature zonation in geographical distribution. *Ecol. Monogr.* **1947**, *17*, 325–335. [CrossRef]
19. Southward, A.J. Note on the temperature tolerance of some intertidal animals in relation to environmental temperature and geographic distribution. *J. Mar. Biol. Assoc. UK* **1958**, *37*, 49–66. [CrossRef]
20. Orton, J.H. Sea temperature, breeding and the distribution of marine animals. *J. Mar. Biol. Assoc. UK* **1920**, *12*, 339–366. [CrossRef]
21. O'Connor, M.L.; Bruno, J.F.; Gaines, S.D.; Halpern, B.S.; Lester, S.E.; Kinlan, B.P.; Weiss, J.M. Temperature control of larval dispersal and the implications for marine ecology, evolution, and conservation. *Proc. Natl. Acad. Sci. USA* **2007**, *104*, 1266–1271. [CrossRef]
22. Somero, G.N. Linking biogeography to physiology: Evolutionary and acclimatory adjustments of thermal limits. *Front. Zool.* **2005**, *2*, 1. [CrossRef] [PubMed]
23. Somero, G.N. The physiology of climate change: Linking patterns to mechanisms. *Annu. Rev. Mar. Sci.* **2012**, *4*, 39–61. [CrossRef] [PubMed]
24. Helmuth, B.; Harey, C.D.G.; Halpin, P.M.; O'Donnell, M.; Hoffmann, G.E.; Blanchette, C.A. Climate change and latitudinal patterns of intertidal stress. *Science* **2002**, *298*, 1015–1017. [CrossRef] [PubMed]
25. Harley, C.D.G.; Randall, H.A.; Hultgren, K.M.; Miner, B.G.; Sorte, C.J.B.; Thornber, C.S.; Rodriguez, L.F.; Tomanek, L.; Williams, S.L. The impacts of climate change in coastal marine systems. *Ecol. Lett.* **2006**, *9*, 228–241. [CrossRef] [PubMed]
26. Kordas, R.L.; Haley, C.D.G.; O'Connor, M.I. Community ecology in a warming world: The influence of temperature on interspecific interactions in marine systems. *J. Exp. Mar. Ecol. Biol.* **2011**, *400*, 218–226. [CrossRef]
27. Hertzman, O. Meteorology of the Gulf of Maine. In Proceedings of the Gulf of Maine Scientific Workshop, Woods Hole, MA, USA, 8–10 January 1991.
28. Angevine, W.M.; Trainer, M.; McKeen, S.A.; Berkowitz, C.M. Mesoscale meteorology of the New England coast, Gulf of Maine, and Nova Scotia: Overview. *J. Geophys. Res.* **1996**, *101*, 28893–28901. [CrossRef]
29. Townsend, D.W.; Thomas, A.C.; Mayer, L.M.; Thomas, M.; Quinlan, J.A. Oceanography of the Northwest Atlantic continental shelf. In *The Sea*; Robinson, A.R., Brink, K.H., Eds.; Harvard University Press: Cambridge, MA, USA, 2006; Volume 14A, pp. 119–168.
30. Townsend, D.W. Influences of oceanographic processes on the biological productivity of the Gulf of Maine. *Rev. Aquat. Sci.* **1991**, *5*, 211–230.

31. Pettigrew, N.R.; Churchill, J.H.; Janzen, C.D.; Mangum, L.J.; Signell, R.P.; Thomas, A.C.; Townsend, D.W.; Wallinga, J.P.; Xue, H. The kinematic and hydrographic structure of the Gulf of Maine Coastal Current. *Deep-Sea Res. II* **2005**, *52*, 2369–2391. [CrossRef]

32. Bisagni, J.J.; Gifford, D.J.; Ruhsam, C.M. The spatial and temporal distribution of the Maine Coastal Current during 1982. *Cont. Shelf Res.* **1996**, *16*, 1–24. [CrossRef]

33. Pettigrew, N.R.; Townsend, D.W.; Xue, H.; Wallinga, J.P.; Brickley, P.J.; Hetland, R.D. Observations of the Eastern Maine Coastal Current and its offshore extensions. *J. Geophy. Res.* **1998**, *103*, 30623–30639. [CrossRef]

34. Kaefer, B.A.; Churchill, J.H.; McGillicuddy, D.J., Jr.; Anderson, D.M. Bloom development and transport of toxic *Alexandrium fundyense* populations within a coastal plume in the Gulf of Maine. *Deep Sea Res. II* **2005**, *52*, 2674–2697. [CrossRef]

35. Bumpus, D.F.; Lauzier, L.M. *Surface Circulation on the Continental Shelf of the East Coast of the United States between Newfoundland and Florida*; American Gergraphy Society: New York, NY, USA, 1965; pp. 1–4.

36. Vermersch, J.A.; Beardsley, R.C.; Brown, W.S. Winter circulation in the Western Gulf of Maine: Part 2. Current and pressure observation. *J. Phys. Oceanogr.* **1979**, *9*, 768–784. [CrossRef]

37. Xue, H.; Chai, F.; Pettigrew, N.R. A model study of the seasonal circulation in the Gulf of Maine. *J. Phys. Oceanogr.* **2000**, *30*, 1111–1135. [CrossRef]

38. Kelley, J.T.; Barnhardt, W.A.; Belknap, D.F.; Dickson, S.M.; Kelley, A.R. *The Seafloor Revealed: The Geology of the Northwestern Gulf of Maine Inner Continental Shelf*; Maine Geological Survey, Open-File Report 96-6; Maine Geological Survey Publications: Augusta, ME, USA, 1998; Volume 119, pp. 1–55.

39. WWW Current Predictor. Available online: Tbone.biol.sc.edu/tide (accessed on 5 May 2012).

40. Brown, B. *A Classification System of Marine and Estuarine Habitats in Maine: Part 1. Benthic Habitats*; Maine Natural Areas Program; Department of Community Development: Augusta, ME, USA, 1993; pp. 1–51.

41. Thomas, M.L.H. A physically derived exposure index for marine shorelines. *Ophelia* **1986**, *25*, 1–13. [CrossRef]

42. Hale, S.S. Biogeographical patterns of marine benthic macroinvertebrates along the Atlantic coast of the Northeastern USA. *Estuaries Coasts* **2010**, *33*, 1039–1053. [CrossRef]

43. NCCA [National Coastal Condition Assessment]. Available online: www.epa.gov/national-aquatic-resource-surveys/ncca (accessed on 1 January 2022).

44. Muñoz Sabater, J. ERA5-Land Monthly Averaged Data from 1981 to Present. *Copernic. Clim. Change Serv. (C3S) Clim. Data Store (CDS)* **2019**, *10*. [CrossRef]

45. Merchant, C.J.; Embury, O.; Bulgin, C.E.; Block, T.; Corlett, G.K.; Fiedler, E.; Good, S.A.; Mittaz, J.; Rayner, N.A.; Berry, D.; et al. Satellite-based time-series of sea-surface temperature since 1981 for climate applications. *Sci. Data* **2019**, *6*, 223. [CrossRef]

46. Clarke, K.R.; Warwick, R.M. *Change in Marine Communities: An Approach to Statistical Analysis and Interpretation*, 2nd ed.; Plymouth Marine Laboratory: Plymouth, UK, 2001.

47. Clarke, K.R.; Gorley, R.N. *PRIMER v7: User Manual/Tutorial*; PRIMER-E Ltd., Plymouth Marine Laboratory: Plymouth, UK, 2006; pp. 1–296.

48. Anderson, M.J.; Gorley, R.N.; Clarke, K.R. *PERMANOVA+ for PRIMER: Guide to Software and Statistical Methods*; PRIMER-E Ltd., Plymouth Marine Laboratory: Plymouth, UK, 2008; pp. 1–214.

49. Baselga, A.; Orme, D.; Villeger, S.; De Bortoli, J.; Leprieur, F. Package 'Betapart': Partitioning Beta Diversity into Turnover and Nestedness Components. 2022. Available online: https://CRAN.R-project.org/package=betapart (accessed on 16 June 2022).

50. Oksanen, J.; Simpson, G.L.; Blanchet, F.G.; Kindt, R.; Legendre, P.; Minchin, P.R.; O'Hara, R.B.; Solymos, P.; Stevens, M.H.H.; Szoecs, E.; et al. Vegan: Community Ecology Package, R Package Version 2.6-2. 2018. Available online: https://CRAN.R-project.org/package=vegan (accessed on 16 June 2022).

51. Anderson, M.J. Distance-based tests for homogeneity of multivariate dispersions. *Biometrics* **2006**, *62*, 245–253. [CrossRef]

52. Xue, H.; Xu, Y.; Brooks, D.; Pettigrew, N.; Wallinga, J. Modeling the circulation in Penobscot Bay, Maine. In *Estuarine and Coastal Modeling, Proceedings of the Sixth International Conference of the American Society of Civil Engineers, New Orleans, LA, USA, 3–5 November 2000*; Spaulding, M.A., Butler, H.L., Eds.; American Society of Civil Engineers: Reston, VA, USA, 2000.

53. Trott, T.J. Zoogeography and changes in macroinvertebrate community diversity of rocky intertidal habitats in the Maine coast. In *Challenges in Environmental Management in the Bay of Fundy-Gulf of Maine*; Pohle, G.W., Wells, P.G., Rolston, S.J., Eds.; Bay of Fundy Ecosystem Partnership Report; Bay of Fundy Ecosystem Partnership: Wolfville, NS, Canada, 2007; pp. 54–73.

54. Bryson, E.S.; Trussell, G.C.; Ewanchuk, P. Broad scale geographic variation in the organization of rocky intertidal communities in the Gulf of Maine. *Ecol. Monogr.* **2014**, *84*, 579–597. [CrossRef]

55. Bousfield, E.L.; Thomas, M.L.H. Postglacial distribution of littoral marine invertebrates in the Canadian Atlantic region. In *Environmental Change in the Maritimes*; Ogden, J.G., III, Harvey, M.J., Eds.; Nova Scotian Institute of Science: Halifax, UK, 1975; pp. 47–60.

56. Adey, W.H.; Heyek, L.-A. The biogeographic structure of the western North Atlantic rocky intertidal. *Cryptogam. Algol.* **2005**, *26*, 35–66.

57. Mathieson, A.C.; Penniman, C.A.; Harris, L.G. Chapter 7. Northwest Atlantic Rocky Shore Ecology. In *Intertidal and Littoral Ecosystems*; Mathieson, A.C., Nienhuis, P.H., Eds.; Ecosystems of the World; Elsevier: New York, NY, USA, 1991; Volume 24, pp. 109–191.

58. Larsen, P.F.; Doggett, L.F. Sand beach macrofauna of the Gulf of Maine with inference on the role of oceanic fronts in determining community composition. *J. Coast. Res.* **1990**, *6*, 913–926.

59. Larsen, P.F. The macroinvertebrate fauna of rockweed (*Ascophyllum nodosum*)-dominated low-energy rocky shores of the Northern Gulf of Maine. *J. Coast. Res.* **2012**, *28*, 36–42. [CrossRef]

60. Yund, P.O.; Collins, C.; Johnson, S.L. Evidence of a native Northwest Atlantic COI haplotype clade in the cryptogenic colonial ascidian *Botryllus schlosseri*. *Biol. Bull.* **2015**, *228*, 201–216. [CrossRef] [PubMed]

61. Dijkstra, J.; Harris, L.G.; Westermann, E. Distribution and long-term temporal patterns of four invasive colonial ascidians in the Gulf of Maine. *J. Exp. Mar. Biol. Ecol.* **2007**, *342*, 61–68. [CrossRef]

62. Stephenson, E.H.; Steneck, R.S.; Seeley, R.H. Possible temperature limits to range expansion of non-native Asian shore crabs in Maine. *J. Exp. Mar. Biol. Ecol.* **2009**, *375*, 21–31. [CrossRef]

63. Lord, J.P.; Williams, L.M. Increase in density of genetically diverse invasive Asian shore crab (*Hemigrapsus sanguineus*) populations in the Gulf of Maine. *Biol. Invasions* **2017**, *19*, 1153–1168. [CrossRef]

64. Wannamaker, A.D.; Jr Kruetz, K.J.; Shöne, B.R.; Pettigrew, N.; Borns, H.W.; Introne, D.S.; Belknap, D.; Masch, K.A.; Feindel, S. Coupled North Atlantic slope water forcing on Gulf of Maine temperatures over the past millennium. *Clim. Dyn.* **2008**, *31*, 183–194. [CrossRef]

65. Zacharias, M.A.; Rolff, J.C. Explanations of patterns of intertidal diversity at regional scales. *J. Biogeogr.* **2001**, *28*, 471–483. [CrossRef]

66. Sagarin, R.D.; Gaines, S.D. Geographic abundance distributions of coastal invertebrates: Using one-dimensional ranges to test biogeographic hypotheses. *J. Biogeogr.* **2002**, *29*, 985–997. [CrossRef]

67. Kotta, J.; Orav-Kotta, H.; Jänes, H.; Hummel, H.; Arvanitidis, C.; Van Avesaath, P.; Bachelet, G.; Benedetti-Cecchi, L.; Bojanić, N.; Como, S.; et al. Essence of the patterns of cover and richness of intertidal hard bottom communities: A pan-European study. *J. Mar. Biol. Assoc. UK* **2017**, *97*, 525–538. [CrossRef]

68. Jones, S.J.; Lima, F.P.; Wethey, D.S. Rising environmental temperatures and biogeography: Poleward range contraction of the blue mussel, *Mytilus edulis* L., in the western Atlantic. *J. Biogeogr.* **2010**, *37*, 2243–2259. [CrossRef]

69. Leonard, G.H. Latitudinal variation in species interactions: A test in the New England rocky intertidal zone. *Ecology* **2000**, *81*, 1015–1030. [CrossRef]

70. Adey, W.H.; Steneck, R. Thermogeography over time creates biogeographic regions: A temperature/space/time-integrated model and an abundance-weighted test for benthic marine algae. *J. Phycol.* **2001**, *37*, 677–698. [CrossRef]

71. Arribas, L.P.; Donnarumma, L.; Gabriela Palomo, G.; Scrosati, R.A. Intertidal mussels as ecosystem engineers: Their associated invertebrate biodiversity under contrasting wave exposures. *Mar. Biodiv.* **2014**, *44*, 203–211. [CrossRef]

72. Scrosati, R.; Heaven, C. Spatial trends in community richness, diversity, and evenness across rocky intertidal environmental stress gradients in eastern Canada. *Mar. Ecol. Prog. Ser.* **2007**, *342*, 1–14. [CrossRef]

73. Heaven, C.S.; Scrosati, R.A. Benthic community composition across gradients of intertidal elevation, wave exposure, and ice scour in Atlantic Canada. *Mar. Ecol. Prog. Ser.* **2008**, *369*, 13–23. [CrossRef]

74. Hummel, H.; Van Avesaath, P.; Wijnhoven, S.; Kleine-Schaars, L.; Degraer, S.; Kerckhof, F.; Bojanic, N.; Skejic, S.; Vidjak, O.; Rousou, M.; et al. Geographic patterns of biodiversity in European coastal marine benthos. *J. Mar. Biol. Assoc. UK* **2017**, *97*, 507–523. [CrossRef]

75. Thomas, M.L.H. Littoral community structure and zonation in the rocky shores of Bermuda. *Bull. Mar. Sci.* **1984**, *37*, 857–870.

76. Wares, J.P.; Cunningham, C.W. Phylogeography and historical ecology of the North Atlantic intertidal. *Evolution* **2001**, *55*, 2455–2469. [CrossRef]

77. Wares, J.P. Community genetics in the Northwestern Atlantic intertidal. *Mol. Ecol.* **2002**, *11*, 1131–1144. [CrossRef]

78. Doelman, M.M.; Trussel, G.C.; Grahame, J.W.; Vollmer, S.V. Phylogenetic analysis reveals a deep lineage spilt within North Atlantic *Littorina saxatilis*. *Proc. R. Soc. B* **2011**, *278*, 3175–3183. [CrossRef]

79. Chu, N.D.; Kaluziak, S.T.; Trussell, C.; Vollmer, S.V. Phylogenetic analyses reveal latitudinal population structure and polymorphisms in heat stress genes in the North Atlantic snail *Nucella lapillus*. *Mol. Ecol.* **2014**, *23*, 1863–1873. [CrossRef]

80. Pringle, J.M.; Byers, J.E.; Ruoying, H.; Pappalardo, P.; Wares, J. Ocean currents and competitive strength interact to cluster benthic species range boundaries in the coastal ocean. *Mar. Ecol. Prog. Ser.* **2017**, *567*, 29–40. [CrossRef]

81. Yund, P.O.; Tilburg, C.E.; McCartney, M.A. Across-shelf distribution of blue mussel larvae in the northern Gulf of Maine: Consequences for population connectivity and a species range boundary. *R. Soc. Open Sci.* **2015**, *2*, 150513. [CrossRef]

82. Pappalardo, P.; Pringle, J.M.; Wares, J.P.; Byers, J.E. The location, strength, and mechanisms behind marine biogeographic boundaries of the east coast of North America. *Ecography* **2015**, *38*, 722–731. [CrossRef]

83. Gross, C.P.; Duffy, E.; Hovel, K.A.; Kardish, M.R.; Reynolds, P.L.; Bortröm, C.; Boyer, K.E.; Cuson, M.; Eklöf, J.; Engelen, A.H.; et al. The biogeography of community assembly: Latitude and predation drive variation in community trait distribution in a guild of epifaunal crustaceans. *Proc. R. Soc. B* **2022**, *289*, 20211762. [CrossRef]

84. Large, S.I.; Smee, D.L. Biogeographic variation in behavioral and morphological responses to predation risk. *Oecologia* **2013**, *171*, 961–969. [CrossRef]

85. Tilburg, C.E.; McCartney, M.A.; Yund, P.O. Across-shelf transport of bivalve larvae: Can the interface between a coastal current and inshore waters act as an ecological barrier to larval dispersal? *PLoS ONE* **2012**, *7*, e48960. [CrossRef]

86. Pershing, A.J.; Alexander, M.A.; Hernandez, C.M.; Kerr, L.A.; Bris, A.L.; Mills, K.E.; Nye, J.A.; Record, N.R.; Scannell, H.A.; Scott, J.D.; et al. Slow adaptation in the face of rapid warming leads to collapse of the Gulf of Maine cod fishery. *Science* **2015**, *350*, 809–812. [CrossRef]

87. Saba, V.S.; Griffies, S.M.; Anderson, W.G.; Winton, M.; Alexander, M.A.; Delworth, T.L.; Hare, J.A.; Harrison, M.J.; Riosati, A.; Vecchi, G.A.; et al. Enhanced warming of the Northwest Atlantic Ocean under climate change. *J. Geophys. Res. Oceans* **2016**, *121*, 118–132. [CrossRef]

88. Goode, A.G.; Brady, D.C.; Steneck, R.S.; Wahle, R.A. The brighter side of climate change: How local oceanography amplified a lobster boom in the Gulf of Maine. *Glob. Change Biol.* **2019**, *25*, 3906–3917. [CrossRef]

89. Mills, K.E.; Pershing, A.J.; Brown, C.J.; Chen, Y.; Chiang, F.-S.; Holland, D.S.; Lehuta, S.; Nye, J.A.; Sun, J.C.; Thomas, A.C.; et al. Fisheries management in a changing climate: Lessons from the 2012 ocean heat wave in the Northwest Atlantic. *Oceanography* **2013**, *26*, 191–195. [CrossRef]

90. Witman, J.D.; Lamb, R.W. Persistent differences between coastal and offshore kelp forest communities in a warming Gulf of Maine. *PLoS ONE* **2018**, *13*, e0189388. [CrossRef]

91. Thomas, A.C.; Pershing, A.J.; Friedland, K.D.; Nye, J.A.; Mills, K.E.; Alexander, M.A.; Record, N.R.; Weatherbee, R.; Henderson, E.M. Seasonal trends and phenology shifts in sea surface temperature on the North American northeastern continental shelf. *Elem. Sci. Anth.* **2017**, *5*, 48. [CrossRef]

92. Record, N.R.; Balch, W.M.; Stamieszkin, K. Century-scale changes in phytoplankton phenology in the Gulf of Maine. *PeerJ* **2019**, *7*, e6735. [CrossRef]

93. Staudinger, M.D.; Mills, K.E.; Stamieszkin, K.; Record, N.R.; Hudak, C.A.; Allyn, A.; Diamond, A.; Friedland, K.D.; Golet, W.; Henderson, M.E.; et al. It's about time: A synthesis of changing phenology in the Gulf of Maine ecosystem. *Fisheries Oceanogr.* **2019**, *28*, 532–566. [CrossRef]

94. Pershing, A.J.; Alexander, M.; Brady, D.C.; Brickman, D.; Curchitser, E.N.; Diamond, A.W.; McClenachan, L.; Mills, K.E.; Nichols, O.C.; Pendleton, D.E.; et al. Climate impacts on the Gulf of Maine ecosystem: A review of observed and expected changes in 2050 from rising temperatures. *Elem. Sci. Anth.* **2021**, *9*, 76. [CrossRef]

95. Dijkstra, J.A.; Harris, L.G.; Mello, K.; Litterer, A.; Wells, C.; Ware, C. Invasive seaweeds transform habitat structure and increase biodiversity of associated species. *J. Ecol.* **2017**, *105*, 1668–1678. [CrossRef]

96. Sorte, C.J.B.; Williams, S.L.; Carlton, J.T. Marine range shifts and species introductions: Comparative spread rates and community impacts. *Glob. Ecol. Biogeogr.* **2010**, *19*, 303–316. [CrossRef]

 *diversity*

Article

Roving Diver Survey as a Rapid and Cost-Effective Methodology to Register Species Richness in Sub-Antarctic Kelp Forests

Gonzalo Bravo [1,2,*,†], Julieta Kaminsky [3,*,†], María Bagur [3,*,†], Cecilia Paula Alonso [4], Mariano Rodríguez [4], Cintia Fraysse [3], Gustavo Lovrich [3] and Gregorio Bigatti [1,2,5]

1   Instituto de Biología de Organismos Marinos (IBIOMAR-CCTCONICET-CENPAT), Bvd. Brown 2825, Puerto Madryn U9120ACF, Argentina
2   Facultad de Ciencias Naturales, Universidad Nacional de la Patagonia San Juan Bosco (UNPSJB), Bvd. Brown 3051, Puerto Madryn U9120ACE, Argentina
3   Centro Austral de Investigaciones Científicas (CADIC-CONICET), Bernardo Houssay 200, Ushuaia V9410CAB, Argentina
4   Instituto de Ciencias Polares, Ambiente y Recursos Naturales, Universidad Nacional de Tierra del Fuego (ICPA-UNTDF), Fuegia Basket 251, Ushuaia V9410BXE, Argentina
5   Escuela de Ciencias Ambientales, Universidad de Especialidades Espíritu Santo, Av. Samborondón, Guayaquil 09230, Ecuador
*   Correspondence: gonzalobravoargentina@gmail.com (G.B.); kaminsky.julieta@gmail.com (J.K.); mariabagur@conicet.gov.ar (M.B.)
†   These authors contributed equally to this work.

Abstract: Underwater sampling needs to strike a balance between time-efficient and standardized data that allow comparison with different areas and times. The roving diver survey involves divers meandering and actively searching for species and has been useful for producing fish species lists but has seldom been implemented for benthic taxa. In this study, we used this non-destructive technique to register species associated with kelp forests at the sub-Antarctic Bécasses Island (Beagle Channel, Argentina), detecting numerous species while providing the first multi-taxa inventory for the area, including macroalgae, invertebrates, and fish, with supporting photographs of each observation hosted on the citizen science platform iNaturalist. This research established a timely and cost-effective methodology for surveys with scuba diving in cold waters, promoting the obtention of new records, data sharing, and transparency of the taxonomic curation. Overall, 160 taxa were found, including 41 not reported previously for this area and three records of southernmost distribution. Other studies in nearby areas with extensive sampling efforts arrived at similar richness estimations. Our findings reveal that the roving diver survey using photographs is a good approach for creating inventories of marine species, which will serve for a better understanding of underwater biodiversity and future long-term monitoring to assess the health of kelp environments.

Keywords: benthic species; scuba diving; Bécasses Islands; iNaturalist; Patagonia; underwater photography; biodiversity; rocky reefs

Citation: Bravo, G.; Kaminsky, J.; Bagur, M.; Alonso, C.P.; Rodríguez, M.; Fraysse, C.; Lovrich, G.; Bigatti, G. Roving Diver Survey as a Rapid and Cost-Effective Methodology to Register Species Richness in Sub-Antarctic Kelp Forests. *Diversity* **2023**, *15*, 354. https://doi.org/10.3390/d15030354

Academic Editors: Thomas J. Trott and Michael Wink

Received: 3 February 2023
Revised: 23 February 2023
Accepted: 26 February 2023
Published: 1 March 2023

1. Introduction

Making reliable and effective biodiversity surveys is crucial to evaluate the status of the marine environment and for conservation planning. Species richness is a key parameter used as basic information for community ecology and is considered among the biological and ecological essential ocean variables (EOVs, [1]). Monitoring the presence of marine species in space and time at local and global scales is necessary to reduce the knowledge gap in biodiversity, particularly in subtidal habitats. Furthermore, global platforms with open accessibility for uploading species occurrences, such as the Ocean Biogeographic Information System (OBIS) or the Global Biodiversity Information Facility (GBIF), provide databases to test ecological and biogeographic hypotheses. Recently, websites like iNaturalist.org enable shortcuts for adding observations to GBIF, and researchers have started integrating iNaturalist data in their studies [2–5].

Although there is an increase in underwater biodiversity studies, there are still gaps in the knowledge of benthic communities in the Southwest Atlantic [6,7], especially at shallow (<30 m) rocky shores in Atlantic Patagonia, Argentina. The Beagle Channel and the sub-Antarctic region, recognized as a conservation priority site for coastal biodiversity, houses the most southern *Macrocystis pyrifera* kelp forests globally [8]. These structurally complex and highly productive giant kelp forests provide habitat and food for marine mammals, seabirds, invertebrates, fish, and macroalgae, e.g., [9–13].

Previous studies in nearby areas have examined *M. pyrifera* kelp forest communities by using traditional underwater samplings such as transects, in situ quadrats, photoquadrats, or extractive samples, e.g., [14–22], that are difficult to perform due to weather conditions in these cold environments. Notwithstanding, many areas in this spatial and temporally heterogeneous region remain to be explored. With this in mind, we performed an active search photographic survey with the roving diver technique to investigate species richness associated with kelp forests at Bécasses Island. The roving diver survey involves divers meandering and actively searching for species [23] and has been useful for producing fish species lists in tropical seas, e.g., [24,25], but only seldom implemented for benthic taxa, e.g., [26,27]. This study constitutes a good example to establish a timely and cost-effective methodology for surveys with scuba diving in this area, characterized by strong winds and low water temperatures (average 6.8 °C [28]), especially during winter (minimum 5.1 °C, [28]). We created a species list and field photographic record of invertebrates, macroalgae, and fishes that occurred in kelp forests (2–30 m depth) at Bécasses Island, Beagle Channel, that will serve as a baseline of biodiversity and future monitoring to determine the health of sub-Antarctic kelp forests. Then, we discuss and compare our results to other approaches developed to study kelp forest communities in nearby areas.

2. Materials and Methods

2.1. Study Site

The Bécasses Islands are located at the eastern end of the Beagle Channel (Figure 1). They constitute a group of two main islands and a few islets, the larger one "Bécasses Island", also known as Septentrional Island, is approximately 750 m in length from north to south (Figure 1). Geomorphology in the Beagle Channel has been modeled during the Last Glacial Maximum previous to ca. 11,000 Ka [29], whereas the present fluvial and marine processes mainly have modeled coastal landscapes, e.g., cliffs, capes, and bays [30–32]. This natural Channel connects the Pacific and Atlantic Oceans with sub-Antarctic waters, particularly the Cape Horn Current, determining its hydrodynamics [33]. Oceanographic and meteorological conditions present a seasonal pattern in water and air temperatures and light and nutrient availability [28,34]. The Beagle Channel has subpolar wet weather, strong exposure to prevailing southwest winds, and a mixed semidiurnal tide regime with an average amplitude of 1.15 m [35,36]. During warmer months, freshwater inputs from glacial melting and river runoff reduce surface salinities, driving water column stratification and reducing light availability in the water column [33,37]. The Bécasses Islands are a nesting site for seabirds (*Phalacrocorax atriceps* and *P. magellanicus*, [38]) and marine mammals (*Otaria flavescens*, [39]).

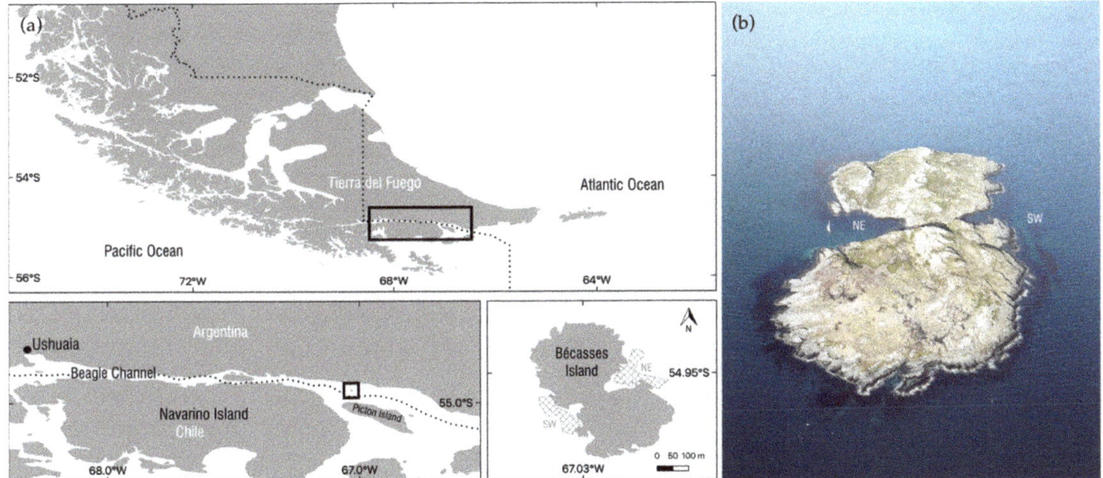

Figure 1. Bécasses Island location in the Beagle Channel, Tierra del Fuego, Argentina. (**a**) A map of the study area, the surveyed zones are indicated with grid lines (SW = Southwest and NE = Northeast). (**b**) Bécasses Island, photo acquired by a drone, survey zones marked with letters (SW and NE); and the sailboat "Kostat" is observed in the NE bay.

2.2. Survey Method

Bécasses Island was explored during a research cruise conducted in August 2021 (winter in the Southern Hemisphere) on board the sailboat "Kostat." An underwater roving diver survey [23] was conducted on the NE and SW subtidal areas of Bécasses Island (Figure 1) to create a benthic species inventory. The SW area is exposed to the dominant SW winds in the area. In both sampling areas, three to four dives were performed by 4–5 divers swimming freely for ~45 min and taking photos of each species they encountered, covering in total more than 700 min of diving and ~20,000 m². The diving range was 2–33 m in depth, and in each dive, more than 100 lineal meters were covered. Two-night dives were performed to record species with nocturnal behavior. Special attention was paid to taking good-quality photos of each specimen. Highly mobile and small-sized species (<1 cm) were not photographed. All divers participating in the surveys had marine biology backgrounds and knowledge of local species, which was relevant for finding rare or cryptic species. The cameras used for the survey were Olympus TG6 (Olympus corporation, Vietnam), Cannon SL1 (Canon, Taiwan, Republic of China), Sony Alpha 7S2 (Sony Corporation, Thailand), and Nikon Coolpix W300 (Nikon Corporation, Indonesia) with external lights or flashes. Only some individuals of specific groups (some macroalgae and sea stars) were collected by hand in order to confirm the species identification under a stereoscopic microscope.

2.3. Data Preparation and Quality Control

All the photos (n = 672) were uploaded to an iNaturalist project [40] that only includes observations of divers in this expedition. The open platform iNaturalist (launched in 2008) allows users to submit species observations along with images and GPS coordinates. Once submitted, the observations were identified by the community and vetted by specialists (curators). In our case, we request local taxonomists to review the observations of this project in order to improve the taxonomic resolution (see Acknowledgments). Higher taxonomic levels as genus or family were only used when the photograph was not clear enough, or the specimen did not show the taxonomic features needed for specific identification. All these observations identified to species level and accepted by two or more iNaturalist users ("Research Grade") were automatically uploaded to GBIF by the platform.

Our sampling method was compared with previous studies involving diving surveys and reporting marine species in the nearby areas (50° S to 56° S) with similar subtidal environments, hence we created a list combining all species (Supplementary Material Table S1). Only studies involving communities sampled through transects or quadrats were selected instead of detailed extractive samplings, e.g., analysis of the macrofauna inhabiting kelp holdfasts [10,22] was not selected because of species sizes and sampling effort differences. All the taxonomic names recovered by this list were verified using the Taxon Match tool [41] of the World Register of Marine Species (WoRMS) in December 2022, to prevent the inflation of taxa richness by synonyms, unaccepted or non-updated names. The websites AlgaeBase [42] and FishBase [43] were also used to check the accepted names of macroalgae and fishes, respectively. Due to some invalid taxonomic names that drive inconsistencies (e.g., Porifera sp. 1, Porifera sp. 2, Porifera sp. 3, etc.), the total number of species reported by each study was calculated using this list instead of using the numbers presented in the original articles. In order to compare the richness variation among the different studies and for each group of taxa, the coefficient of variation (CV = standard deviation divided by the mean) was calculated. Additionally, a presence–absence table was constructed to find "unique" species in each of the studies, i.e., species only present in one of the studies and absent in all others (Supplementary Material Table S2).

3. Results

3.1. Environment

Kelps widely colonized subtidal environments in Bécasses Island where the water temperature was 7 °C during the samplings. In the upper subtidal, down to 2 m of depth, kelp forests were mainly dominated by *Lessonia flavicans*, whereas *Macrocystis pyrifera* and *Lessonia searlesiana* formed dense, mixed forests between 2 and 20 m depth. *Lessonia searlesiana* was also found down to 30 m depth. The underwater landscape presented visual differences between NE and SW coasts: the former, the windward side, with a stepped topography with flat bedrock reaching more than 30 m deep, whereas the SW coast was shallower (13 m maximum deep) and presented bedrock with boulders surrounded by sand patches.

3.2. Bécasses Checklist

A total of 160 taxa were recorded by the roving diver survey at the two zones sampled (see Figure 1) at Bécasses Island, including 121 invertebrates, 7 fishes, and 32 macroalgae (Table 1). Invertebrates were dominated by Mollusca (40 taxa), followed by Echinodermata (27), Cnidaria (14), Arthropoda (13), Tunicata (10), Annelida (7), Bryozoa (5), and Porifera (4). Most species of fish belonged to the Nototheniidae family. Regarding macroalgae, 19 Rhodophyta, 10 Ochrophyta, and 3 Chlorophyta species were found. All observations are publicly available at iNaturalist (see Section 2).

3.3. Comparison with Other Studies

In order to compare the species richness obtained with our photographic survey method with traditional methods performed in nearby areas (such as transects and photo-quadrats), seven articles were selected (Table 2). One paper studies only the understory macroalgae community in nearby kelp forests [9], another studies only the fishes community [11], and the rest investigate the invertebrate community only [18,20], or together with fish [17,19,21]. Careful data control was taken to avoid species name artifacts (see Section 2 and Supplementary Material Table S1) as inputs for Tables 1 and 2.

Diversity 2023, 15, 354

Table 1. List of marine species found during this study with indication of the number of observations (N) with associated photograph and those taxa recorded by other studies in the region (*). Bold letter indicates "unique" records (family, genus, or species) found in our study. N for taxa higher than species corresponds to specimens not possible to be determined to specific level.

Phylum	Class	Order	Family	Genus	Species	N (this study)	Beaton et al. 2020 [18]	Cárdenas and Montiel 2015 [20]	Friedlander et al. 2018 [17]	Friedlander et al. 2020 [19]	Friedlander et al. 2021 [21]	Santelices and Ojeda 1984 [9]	Vanella et al. 2007 [11]
Porifera	Calcarea	Leucosolenida	Syconidae	Sycon		2			*	*	*		
	Demospongiae					2							
		Haplosclerida	Chalinidae	Haliclona		1		*	*	*	*		
		Poecilosclerida	Hymedesmiidae	Phorbas	*Phorbas ferrugineus*	9				*	*		
Cnidaria	Anthozoa	Actiniaria				2		*	*				
			Actiniidae	Bunodactis	*Bunodactis octoradiata*	1	*	*	*	*	*		
			Actinostolidae	Actinostola		3							
			Actinostolidae	Antholoba	*Antholoba achates*	4	*			*	*		
			Halcuriidae	**Halcurias**		4							
			Isanthidae	Isoparactis	***Isoparactis fionae***	1							
			Metridiidae	Metridium	*Metridium senile*	24			*		*		
			Preactiniidae	Dactylanthus	***Dactylanthus antarcticus***	4							
		Alcyonacea	Alcyoniidae	Alcyonium	***Alcyonium haddoni***	6							
			Clavulariidae	Incrustatus		1							
			Primnoidae	Primnoella		4							
			Primnoidae	Primnoella	*Primnoella chilensis*	1		*			*		
	Hydrozoa	Leptothecata	Campanulariidae	Obelia	*Obelia geniculata*	2			*	*	*		
	Staurozoa	Stauromedusae	Haliclystidae	Haliclystus	*Haliclystus antarcticus*	1	*						
Annelida	Polychaeta		Chaetopteridae	Chaetopterus		1							
			Chaetopteridae	Chaetopterus	*Chaetopterus variopedatus*	1	*	*	*	*	*		
		Phyllodocida	Nereididae			1							
			Phyllodocidae	Eulalia		1					*		
			Phyllodocidae			3							
			Polynoidae			1							
		Terebellida	**Cirratulidae**			1							
Mollusca	Bivalvia	Mytilida	Mytilidae	Aulacomya	*Aulacomya atra*	3		*			*		
		Pectinida	Pectinidae	Austrochlamys	***Austrochlamys natans***	2							
	Cephalopoda	Myopsida	Loliginidae	Doryteuthis	***Doryteuthis gahi***	10							
		Octopoda	Enteroctopodidae	Enteroctopus	***Enteroctopus megalocyathus***	4							
	Gastropoda		Limapontiidae	Placida	***Placida sudamericana***	1							
			Nacellidae	Nacella		2			*				
			Nacellidae	Nacella	***Nacella deaurata***	2							
			Nacellidae	Nacella	*Nacella mytilina*	3	*		*	*	*		
			Plakobranchidae	Elysia	***Elysia patagonica***	2							
		Lepetellida	Fissurellidae	Fissurella		1			*	*			
			Fissurellidae	Fissurella	*Fissurella oriens*	4	*						
			Fissurellidae	Fissurella	***Fissurella picta***	1							
		Littorinimorpha	Calyptraeidae	**Crepidula**		1							
			Calyptraeidae	Crepipatella	*Crepipatella dilatata*	2				*	*		
			Cymatiidae	Fusitriton	*Fusitriton magellanicus*	9			*	*	*		
			Velutinidae	Lamellaria		1				*	*		
		Neogastropoda	Cominellidae	Pareuthria	*Pareuthria fuscata*	1	*		*	*	*		
			Muricidae			1							
			Muricidae	Acanthina	*Acanthina monodon*	1	*			*	*		
			Muricidae	Trophon	*Trophon geversianus*	1	*		*	*	*		
			Volutidae	Adelomelon	*Adelomelon ancilla*	6			*	*	*		
			Volutidae	Odontocymbiola	*Odontocymbiola magellanica*	6	*						
		Nudibranchia	Chromodorididae	Tyrinna	*Tyrinna delicata*	3			*		*		

Table 1. *Cont.*

Phylum	Class	Order	Family	Genus	Species	N (this study)	Beaton et al. 2020 [18]	Cárdenas and Montiel 2015 [20]	Friedlander et al. 2018 [17]	Friedlander et al. 2020 [19]	Friedlander et al. 2021 [21]	Santelices and Ojeda 1984 [9]	Vanella et al. 2007 [11]
			Coryphellidae	*Coryphella*	*Coryphella falklandica*	5					*		
			Discodorididae	*Diaulula*	*Diaulula hispida*	1			*	*	*		
			Discodorididae	*Diaulula*	*Diaulula punctuolata*	2			*	*	*		
			Dorididae	*Doris*	*Doris fontainii*	3			*	*	*		
			Polyceridae	*Thecacera*	*Thecacera darwini*	4			*	*	*		
			Tritoniidae	*Tritonia*	*Tritonia challengeriana*	10			*		*		
			Tritoniidae	*Tritonia*	***Tritonia odhneri***	3							
			Tritoniidae	*Tritonia*	***Tritonia vorax***	1							
		Pleurobranchida	Pleurobranchidae	*Berthella*	*Berthella platei*	4					*		
		Trochida				1							
			Calliostomatidae	***Calliostoma***		1							
			Calliostomatidae	*Margarella*	*Margarella violacea*	3	*		*	*	*		
	Polyplacophora	Chitonida	Chitonidae	*Tonicia*		8				*	*		
			Chitonidae	*Chiton*	*Chiton magnificus*	3			*	*	*		
			Chitonidae	*Tonicia*	*Tonicia chilensis*	1			*				
			Chitonidae	*Tonicia*	*Tonicia disjuncta*	5	*						
			Mopaliidae	*Plaxiphora*	*Plaxiphora aurata*	2	*		*	*	*		
Arthropoda	Hexanauplia					2	*	*					
		Balanomorpha	Balanidae	*Austromegabalanus*	*Austromegabalanus psittacus*	4	*		*	*	*		
	Malacostraca	Amphipoda	**Ampithoidae**			1							
			Gammarellidae	**cf. Austroregia**		2							
			cf. Gammarellidae			2							
		Decapoda	Campylonotidae	*Campylonotus*	*Campylonotus vagans*	11			*	*	*		
			Hippolytidae	*Nauticaris*	*Nauticaris magellanica*	5	*		*	*	*		
			Hymenosomatidae	*Halicarcinus*	*Halicarcinus planatus*	2	*		*	*	*		
			Inachidae	*Eurypodius*	***Eurypodius longirostris***	7							
			Lithodidae	*Paralomis*	*Paralomis granulosa*	10	*		*	*	*		
			Munididae	*Grimothea*	*Grimothea gregaria*	7			*	*	*		
			Paguridae	*Pagurus*	*Pagurus comptus*	2	*		*	*	*		
			Trichopeltariidae	*Peltarion*	*Peltarion spinulosum*	1			*	*	*		
Brachiopoda	Rhynchonellata	Terebratulida	Terebratellidae	*Magellania*	*Magellania venosa*	3	*		*	*	*		
Bryozoa						4	*	*	*	*	*		
	Gymnolaemata	Cheilostomatida	Beaniidae	*Beania*	*Beania magellanica*	1		*	*	*	*		
			Cellariidae	*Cellaria*	*Cellaria malvinensis*	2	*	*	*	*	*		
			Alcyonidiidae	*Alcyonidium*	*Alcyonidium australe*	1		*					
	Stenolaemata	Cyclostomatida	Crisiidae	*Crisia*		3			*		*		
Echinodermata	Asteroidea	Forcipulatida	Asteriidae			1							
			Asteriidae	*Anasterias*	*Anasterias antarctica*	4	*		*	*	*		
			Asteriidae	*Diplasterias*	***Diplasterias brandti***	12							
			Heliasteridae	*Labidiaster*	*Labidiaster radiosus*	21	*		*	*	*		
			Stichasteridae	*Allostichaster*	***Allostichaster capensis***	1							
			Stichasteridae	*Cosmasterias*	*Cosmasterias lurida*	16	*		*	*	*		
		Spinulosida	Echinasteridae	*Henricia*	*Henricia obesa*	4			*	*	*		
		Valvatida				1							
			Asterinidae	*Asterina*	*Asterina fimbriata*	1	*		*	*	*		
			Asterinidae	*Cycethra*	*Cycethra verrucosa*	8	*		*	*	*		
			Asterinidae	*Ganeria*	*Ganeria falklandica*	9	*			*	*		
			Odontasteridae			2							

Table 1. *Cont.*

Phylum	Class	Order	Family	Genus	Species	N (this study)	Beaton et al. 2020 [18]	Cárdenas and Montiel 2015 [20]	Friedlander et al. 2018 [17]	Friedlander et al. 2020 [19]	Friedlander et al. 2021 [21]	Santelices and Ojeda 1984 [9]	Vanella et al. 2007 [11]
			Odontasteridae	Diplodontias	*Diplodontias singularis*	3				*	*		
			Odontasteridae	Odontaster	*Odontaster penicillatus*	4	*		*	*	*		
			Poraniidae	Glabraster	*Glabraster antarctica*	25	*		*		*		
			Poraniidae	Poraniopsis	*Poraniopsis echinaster*	6			*		*		
	Echinoidea	Arbacioida	Arbaciidae	Arbacia	*Arbacia dufresnii*	23	*		*	*	*		
		Camarodonta	Parechinidae	Loxechinus	*Loxechinus albus*	24	*		*	*	*		
			Temnopleuridae	Pseudechinus	*Pseudechinus magellanicus*	11	*		*	*	*		
	Holothuroidea	Apodida	Chiridotidae	Chiridota	*Chiridota pisanii*	2				*	*		
		Dendrochirotida	Cucumariidae	Cladodactyla	*Cladodactyla crocea*	4	*						
			Cucumariidae	Trachythyone	**Trachythyone lechleri**	1							
			Psolidae	Psolus	*Psolus patagonicus*	2					*		
	Ophiuroidea	Amphilepidida	Amphiuridae	Ophiophragmus	**Ophiophragmus chilensis**	1							
			Ophiactidae	Ophiactis	*Ophiactis asperula*	3	*		*	*	*		
		Euryalida	Gorgonocephalidae	Gorgonocephalus	**Gorgonocephalus chilensis**	16							
		Ophiacanthida	Ophiomyxidae	Ophiomyxa	*Ophiomyxa vivipara*	1	*		*	*	*		
Chordata	Actinopterygii	Perciformes	Bovichtidae	Cottoperca	*Cottoperca trigloides*	2			*	*	*		
			Harpagiferidae	Harpagifer	*Harpagifer bispinis*	1				*	*		
			Liparidae	Careproctus	*Careproctus pallidus*	1					*		*
			Nototheniidae	Patagonotothen		9			*	*	*		
			Nototheniidae	Paranotothenia	*Paranotothenia magellanica*	2			*	*	*		*
			Nototheniidae	Patagonotothen	*Patagonotothen tessellata*	1			*	*	*		*
			Zoarcidae	Dadyanos	*Dadyanos insignis*	3					*		
	Ascidiacea					3	*				*		
		Aplousobranchia	Didemnidae			3	*						
			Holozoidae	Sycozoa	*Sycozoa gaimardi*	2	*		*	*	*		
			Polyclinidae	Aplidium		3	*	*	*	*			
			Polyclinidae	Aplidium	*Aplidium fuegiense*	5		*	*	*	*		
		Stolidobranchia	Molgulidae	Paramolgula	*Paramolgula gregaria*	1	*						
			Pyuridae	Pyura	*Pyura legumen*	3			*	*	*		
			Styelidae	Cnemidocarpa		1			*	*	*		
			Styelidae	Cnemidocarpa	*Cnemidocarpa verrucosa*	1			*	*	*		
			Styelidae	Polyzoa	*Polyzoa opuntia*	6			*	*	*		
Rhodophyta	Florideophyceae	Balliales	Balliaceae	Ballia	*Ballia callitricha*	3		*				*	
		Bonnemaisoniales	Bonnemaisoniaceae	Ptilonia	*Ptilonia magellanica*	9		*					
		Ceramiales	Delesseriaceae	**Paraglossum**		1							
			Delesseriaceae	Cladodonta	**Cladodonta lyallii**	1							
			Delesseriaceae	Hymenena	**Hymenena falklandica**	2							
			Delesseriaceae	Pseudophycodrys	**Pseudophycodrys phyllophora**	6							
			Rhodomelaceae	Lophurella	**Lophurella hookeriana**	1							
			Rhodomelaceae	Picconiella	**Picconiella pectinata**	4							
			Wrangeliaceae	Griffithsia		1						*	
		Corallinales				3							
			Corallinaceae	Ellisolandia	**Ellisolandia elongata**	1							
		Gigartinales				1							

Table 1. *Cont.*

Phylum	Class	Order	Family	Genus	Species	N (this study)	Beaton et al. 2020 [18]	Cárdenas and Montiel 2015 [20]	Friedlander et al. 2018 [17]	Friedlander et al. 2020 [19]	Friedlander et al. 2021 [21]	Santelices and Ojeda 1984 [9]	Vanella et al. 2007 [11]
			Gigartinaceae	Sarcopeltis	*Sarcopeltis skottsbergii*	11		*				*	
			Kallymeniaceae	Callophyllis	***Callophyllis atrosanguinea***	5							
			Kallymeniaceae	Callophyllis	*Callophyllis variegata*	2						*	
		Hildenbrandiales	Hildenbrandiaceae	**Hildenbrandia**		3							
		Plocamiales	Plocamiaceae	Plocamium	*Plocamium secundatum*	1						*	
		Rhodymeniales	Rhodymeniaceae	Rhodymenia	***Rhodymenia coccocarpa***	2							
			Rhodymeniaceae	Rhodymenia	*Rhodymenia falklandica*	1						*	
Ochrophyta	Phaeophyceae	Desmarestiales	Desmarestiaceae	Desmarestia		5							
		Dictyotales	Dictyotaceae	Dictyota	***Dictyota falklandica***	5							
		Ectocarpales	Adenocystaceae	Adenocystis	*Adenocystis utricularis*	1		*				*	
			Adenocystaceae	Caepidium	*Caepidium antarcticum*	2		*					
			Scytosiphonaceae	**Colpomenia**		1							
		Laminariales	Laminariaceae	Macrocystis	*Macrocystis pyrifera*	9	*	*	*	*	*	*	
			Lessoniaceae	Lessonia	*Lessonia flavicans*	9				*		*	
			Lessoniaceae	Lessonia	***Lessonia searlesiana***	4							
		Sphacelariales	Stypocaulaceae	Halopteris		1							
		Syringodermatales	Syringodermataceae	Microzonia	***Microzonia velutina***	3							
Chlorophyta	Ulvophyceae	Bryopsidales	Bryopsidaceae	**Bryopsis**		1							
			Codiaceae	Codium	***Codium subantarcticum***	8							
		Ulvales	Ulvaceae	Ulva		4		*					

The number of taxa reported in this study reached similar values to previous studies and the overall CV (31%) was low (Table 2). Comparisons of the number of taxa for each taxonomic group revealed that our estimations had similar values to other studies for most of the groups, except for Porifera, Bryozoa, and fishes. The highest number of taxa (n = 196) was found by Friedlander et al. [21] at the Kawésqar Reserve, Chile. The closest area to our study, Peninsula Mitre, and Isla de los Estados, was surveyed by Friedlander et al. [19], where they recorded 162 taxa in a broader area. Santelices and Ojeda [9] for macroalgae and Vanella et al. [11] for fish, using extractive sampling (no other option available for comparison), found similar species richness estimations compared with our photo surveys (Table 2). *Bunodactis octoradiata*, *Chaetopterus variopedatus*, *Cellaria malvinensis*, *Macrocystis pyrifera*, and *Lessonia* spp. constitute common species recorded by all the studies (Table 1).

From the overall number of taxa found in this work, 41 (30 species, eight genera, and three families) were not reported in previous studies and therefore here considered as "unique" species (Table 1, see names in bold). This number represented the highest as compared to other studies (between 2 and 28 species, see Supplementary Material Table S2). Three of these "unique" species represented the southernmost record of the species (checked in GBIF and local references): the seastar *Allostichaster capensis*, and the molluscs *Elysia patagonica* and *Placida sudamericana* (Figure 2a,c,d).

Figure 2. Field photos of those species that represented first photographic record for the area or southernmost record: (**a**) *Allostichaster capensis* (southernmost record); (**b**) *Diplasterias brandti* (first photo record for Beagle Channel); (**c**) *Elysia patagonica* (southernmost record), arrow indicates specimens; (**d**) *Placida Sudamericana* (southernmost record).

Table 2. List of studies in the nearby areas with information of the type of methodology, the depth range, and the number of taxa found by taxonomic groups. SD = standard deviation and CV = coefficient of variation.

Study	Geographic Area	Method	Depth Range (m)	Porifera	Cnidaria	Annelida	Mollusca	Arthropoda	Bryozoa	Echinodermata	Tunicates	others invertebrates	Fish	Macroalgae	Total
Santelices and Ojeda 1984 [9]	Puerto Toro, Navarino Island, Chile (55°)	Extractive quadrat in transects	4–10											39	39 *
Vanella et al. 2007 [11]	Despard Island, Beagle Channel, Argentina (54°)	Trammel nets and holdfast removal	6										11		11 *
Cárdenas and Montiel 2015 [20]	Santa Ana, Magellan Strait, Chile (53°)	Photoquadrats in vertical walls	0–30	5	5	3	3	2	9	2	6	1		31	67
Friedlander et al. 2018 [17]	Francisco Coloane Reserve, Cape Horn, Diego Ramirez, Chile (53°–56°)	Visual transect survey	7–15	13	12	2	31	15	8	20	11	2	14	2 *	130
Friedlander et al. 2020 [19]	Península Mitre and Isla de los Estados, Argentina (54°)	Visual transect survey	3.7–17	20	14	4	33	14	10	28	14	1	21	3 *	162
Friedlander et al. 2021 [21]	Kawésqar Reserve, Chile (50°–54°)	Visual transect survey	3.5–10	19	20	6	43	18	13	32	19	4	19	2 *	195
Beaton et al. 2020 [18]	Malvinas Islands, Argentina (51°)	Photoquadrats along transects	5–20	19	11	8	31	9	2	21	13	2		2 *	118
This study	Bécasses Island, Argentina (54°)	Roving diver survey	2–33	4	14	7	40	13	5	27	10	1	7	32	160
		SD		7.28	4.89	2.37	14.2	5.64	3.87	10.63	4.36	1.17	5.73	4.36	44.3
		CV		54.64	38.57	47.33	47.09	47.63	49.39	49.08	35.8	63.77	39.77	12.82	31.95

* Not considered for SD and CV calculation. Only those studies that sampled the overall community were taken into account.

4. Discussion

The results of this study provide an updated checklist of marine taxa for Bécasses Island, a location on the eastern Beagle Channel, including several new records for nearby areas. We listed 160 taxa, this study being the first to compile with photographic support invertebrates, fish, and macroalgae species for the Beagle Channel. We stored the photos with geographic positions on the iNaturalist platform. The most powerful benefits of using a citizen science platform as iNaturalist were: (a) the photos of the taxa remain with public access, (b) verified observations were uploaded to GBIF, and (c) the digital collection could serve as an identification guide for other studies, whereas some observations already had additional scientific importance. For example, our observations of *Metridium senile* were used as input on a scientific note aiming to track the movement of this invasive anemone in the last ten years [44]. We also registered the southernmost occurrence of three species (*Allostichaster capensis*, *Elysia patagonica*, and *Placida sudamericana*) and the first record with an in situ field photo of the seastar *Diplasterias brandti* for the Beagle Channel (Figure 2b). The latter is important since previous records were deeper or closer to the Beagle Channel's eastern entrance with Atlantic waters influence [45].

Most of the studies analyzed in Table 2 showed similar species richness compared to our survey, meaning the roving diver survey succeeded in characterizing the local species richness. Compared to our study, the greater number of species recorded by Friedlander et al. [17,21] could be related to their sampling effort and broader survey areas. However, we also notice a high estimation of sponges, bryozoans, and some sea star species that is too detailed for a visual survey without sample extraction and dissection. Sponges of the same species typically vary in color and shape; therefore, identification requires the study of the morphology and size of spicules [46]. This is similar for bryozoans since microscopical analysis might be needed. Fraysse et al. [45] identified 22 sea star species along the Beagle Channel, but some of them are cryptic species that can only be identified under a stereoscopic microscope. With this in mind, we decided to be conservative in identifying these taxa by photos, resulting in fewer species. Although Friedlander et al. [17,19,21] might have overestimated these groups, we probably underestimated them.

For macroalgae, it is often necessary to collect samples and dissect them under a microscope for proper identification. Moreover, at the Beagle Channel, macroalgae communities commonly show variations in composition and biomass between seasons, spring and summer being the seasons with higher abundances [47]. Notwithstanding, we could identify by field photos (using macro lenses in many cases and collecting small samples in a few others) as many as 32 different macroalgae. This richness is similar to that reported by Santelices and Ojeda [9], see Table 2) in the nearby Puerto Toro by means of extractive sampling. Furthermore, in a one-year seasonal extractive sampling conducted in two different kelp forests of the Beagle Channel, we found around 60 macroalgae species [48], double what we found in winter in Bécasses Island with the roving diver survey. Therefore, we believe this kind of survey is a good method for registering macroalgae as an initial monitoring method, which can be complemented later with extractive samplings for more detailed information. Most of the common macroalgae can be identified through pictures by a trained diver. However, small-sized species or specific groups still need collection and processing in the lab. For example, Mendoza [49] found 17 species of Corallinales for Tierra del Fuego, most of them impossible to identify in the field.

4.1. Limitations of the Roving Diver Survey Methodology

Because richness estimates are dependent on the sample design and sampling effort, the comparisons with other studies found here should be considered only qualitative. However, based on the low number of dives employed and the high number of species reported only by this study, we suggest that the roving diver survey should be considered a good method to complement richness estimates. The weakness of this type of survey is that density and cover cannot be estimated, and it is well-known that this information is important for biodiversity studies [50]. However, species richness data and taxa geographic distribution could serve as input for future studies, biodiversity monitoring, and species distribution modeling [51].

Although some small-sized species could be photographed and added to the list (e.g., polychaetes and small crustaceans), the roving diver survey is not recommended for highly mobile and small species (<1 cm). These kinds of organisms need extracting sampling methods (e.g., drags, nets, etc.), with adequate processing (e.g., sieves) and conservation depending on the taxa, in order to identify the species and count individuals. For example, 36 amphipod species have been found with extractive methods associated with the kelp *Macrocystis pyrifera* at the Beagle Channel [52]. However, we only photographed three of the largest species (Table 1).

4.2. Why Have We Found More "Unique" Species Than in Previous Studies?

Several reasons can explain the presence of a higher number of "unique" species when comparing the roving diver survey with more traditional surveys. This method allows the diver to explore a vast area and "free their eyes of other tasks" (e.g., counting and writing down species numbers), gaining time to search for "unique" species. Particularly the following reasons can be explained by examples from this study:

Deep species: The roving diver survey allows for freely exploring a broader area, whereas traditional sampling methods have been conducted in shallow waters (see Table 2) and were mostly restricted to kelp forests. Below 18 m, we found some species normally present at depth ranges of 15–900 m. Examples are the gorgonian-feeding anemone *Dactylanthus antarcticus* (Figure 3b), the orange deep-water anemone *Actinostola* sp. (Figure 3c), the basket star *Gorgonocephalus chilensis* (Figure 3d) and the nudibranchs *Tritonia vorax* (Figure 3f) and *Tritonia odhneri* (Figure 3e) [53].

Small/cryptic species: Small-sized (2–4 cm) and cryptic species are frequently not included (intentionally or not) in traditional samplings such as bottom transects or quadrats. The roving diver survey allows including these kinds of species, by using macro lenses in cameras (to obtain quality pictures of small species) and fundamentally by carefully exploring different types of habitats, which are normally restricted in traditional samplings (e.g., vertical or overhanging surfaces, crevices, species under rocks, biological habitats such as algae, sponges, or shells). The "good eye" and local biodiversity knowledge of biodiversity by the survey divers are also important factors. In this survey, we can mention as this type of "unique" species some polychaetes and small crustaceans (mainly amphipods and isopods), the octopus *Enteroctopus megalocyathus* (Figure 3g), the heterobranch sea slugs *Elysia patagonica* (Figure 2c), and *Placida sudamericana* (Figure 2d). The octopus was hiding in a crevice and the sea slugs were associated with the green algae *Codium subantarcticum*. These sea slugs were 10–20 mm in size and the same color as the algae (see Figure 2c,d); therefore, a careful look was fundamental to find them. On the other hand, small highly mobile species are still very difficult to detect with the roving diver survey and should not be considered when estimating richness. We could easily photograph small sea slugs because they are slow, but highly mobile species such as shrimps are too difficult to photograph, and not because they are necessarily cryptic but because of their exhaust speed.

Rare species: Infrequent species (because of their low density or infrequent presence in one particular environment) could be challenging to detect with traditional methods such as transects or quadrats. As the roving diver survey commonly explores a broader area, the chances to find rare species increase. The possibility of freely exploring different habitats and not being restricted to swimming following a line increases the chances even more. For example, only a few individuals of the sea stars *Allostichaster capensis* and *Diplasterias brandti* (see Figure 2a,b) were photographed during the roving diver survey. However, these records were scientifically important because they constitute the southernmost record for *A. capensis* [54] and the first record with an in situ field photo of *D. brandti* for the Beagle Channel.

Figure 3. Some interesting species recorded in the field. (**a**) *Doryteuthis gahi*; (**b**) *Dactylanthus antarcticus*; (**c**) *Actinostola* sp.; (**d**) *Gorgonocephalus chilensis*; (**e**) *Tritonia odhneri*; (**f**) *Tritonia vorax*; (**g**) *Enteroctopus megalocyathus*; (**h**) *Dictyota falklandica*. The scale bars correspond to 1 cm.

Pelagic/nocturnal species: In contrast with traditional surveys where the focus is generally on benthic species, the roving diver method allows also registering pelagic species (e.g., jellyfish), occasional visitors (e.g., sea lions), and epibenthic species that can be climbing or attached to the kelp at different depths in the water column. Many of these species can also have nocturnal behavior; therefore, it is important to conduct the survey during the day and night. For example, we have registered the squid *Doryteuthis gahi* (Figure 3a). It is common to find egg masses of this species attached to kelp stipes and blades [55], but squids are generally difficult to see.

New species/taxonomic problematic species: Finally, and in this case not concerning the roving diver survey, we have found more "unique" species in comparison with other published studies, simply because new species have been discovered and described in the last few years. Taxonomy is constantly changing, and new species may have been confused with other known species, especially if samples of individuals were not collected and no field pictures were available. As examples, we can mention three new species found on Bécasses Island that were described in the last four years: the macroalgae *Dictyota falklandica* (Figure 3h) [56], the sea anemone *Isoparactis fionae* [57], and the heterobranch sea slug *Placida sudamericana* [58]. Species with taxonomic problems or that are difficult to identify in the field (name in revision, sibling species, etc.) often lead to misinterpretations. In this last category is the kelp *Lessonia searlesiana*, which has often been confused in the selected studies with *Lessonia vadosa* or *L. flavicans*. The genus *Lessonia* is actually under revision. Following Asensi and de Reviers [59], we detected differences in blades and stipes morphology between *L. searlesiana* and *L. flavicans*. Particularly, for these species, we collected some samples and looked for the presence/absence of lagoons in blades through microscope view: *L. flavicans* presented lagoons, whereas the absence was detected in *L. searlesiana*. As mentioned above, the spatial distribution also differed between these species. *Lessonia flavicans* was found in the upper subtidal, whereas *L. searlesiana* was observed at intermediate and deeper subtidal zones, even at 30 m.

4.3. Recommendations for Applying the Roving Diver Survey

The roving diver survey applied in this study has been useful in obtaining a complete checklist of macroalgae, invertebrates, and fish in a fast and easy way in an extreme subtidal environment. Divers optimize their time under the water by freely swimming to wherever they like and searching for species in special habitats (e.g., searching for cryptic species). This method also avoids spending time and effort in carrying and deploying extra equipment, such as transect lines or quadrats. We recommend the roving diver survey for checklist studies by the presence–absence of species, in places difficult to sample due to extreme conditions, and when human resources and equipment are scarce (e.g., when comparing many sites for a marine baseline study).

Marine biodiversity knowledge is an important factor for the roving diver survey. Local knowledge of the diving sites and their fauna will allow scientific divers to easily obtain data on the common species and to search for rare species in specific habitats. An inexperienced diver could easily misidentify or lose cryptic species, while a trained diver is less likely to do so. To avoid confusion and misinterpretations, we strongly recommend using underwater cameras and external light to back up the species identification. A known-species checklist could be filled while diving, but the photos must accompany the checklist. We found it very useful to upload the photos later to the iNaturalist platform, and we encourage researchers and other divers to do this for data validation transparency and accessibility of the community.

In order to improve the survey, an underwater position system that allows errant swimming of divers between kelp forests, e.g., [60], could be used to record the dive trajectory and estimate the density of species with precision. Another option, which does not require additional technology, is to use the SACFOR scale (Superabundant; Abundant; Common; Frequent; Occasional; Rare) (see [61]), where species are recorded, either in terms of percentage cover or density in six logarithmic steps. This scale is quicker, compared to more time-consuming density estimation methods such as quadrats or transects.

In conclusion, our findings reveal that the roving diver survey using photographs is a good approach for creating inventories of subaquatic species in a timely and cost-effective way. This method is very recommendable for kelp forests, where minimum equipment and trajectory freedom help to avoid frequent entanglements, and optimization of the time when diving in extreme environments such as sub-Antarctic cold waters is especially important. We encourage scientific and recreational divers to try this non-destructive method and enjoy the freedom of exploring in every dive. As it has been proven in other parts of the world, the roving diver survey can be easily adapted for citizen science programs in different environments, e.g., Reef Environmental Education Foundation (REEF) Fish Survey Project, and has provided valuable data for scientific research [62,63]. At the same time, unstructured citizen science data stored on iNaturalist can increase the species richness records, especially in those areas where recreational diving is popular [64]. Comparing the species richness obtained in the same site by different sampling methods (e.g., transects vs. roving diver survey) could be a way to improve and optimize the roving diver survey. We hope this proposed method will serve for a better understanding of underwater biodiversity and be implemented for monitoring programs, aiming at the conservation of marine habitats.

Supplementary Materials: The following supporting information can be downloaded at: https://www.mdpi.com/article/10.3390/d15030354/s1, Table S1: overall list of species from all the studies used for comparison; Table S2: unique species: species only present in one of the studies and absent in all others. Videos S1: https://youtu.be/Uvi083RWEz8, https://youtu.be/ZQUlATCEfkY: sailing and underwater images from the August 2021 Bécasses campaign.

Author Contributions: Conceptualization, G.B. (Gonzalo Bravo), M.B., J.K. and C.P.A.; methodology, G.B. (Gonzalo Bravo), M.R., C.P.A., J.K. and M.B.; validation, G.B. (Gonzalo Bravo), J.K., M.R., M.B. and C.P.A.; formal analysis, G.B. (Gonzalo Bravo); investigation, G.B. (Gonzalo Bravo), M.B., J.K., C.P.A. and M.R.; data curation, G.B. (Gonzalo Bravo), M.B., J.K., C.P.A., M.R. and C.F.; writing—original draft preparation, G.B. (Gonzalo Bravo), M.B. and J.K.; writing—review and editing, G.B. (Gonzalo Bravo), M.B., J.K., C.P.A., C.F., G.L. and G.B. (Gregorio Bigatti); visualization, G.B. (Gonzalo Bravo); supervision, G.B. (Gregorio Bigatti); project administration, G.B. (Gonzalo Bravo), M.B., J.K. and C.P.A.; funding acquisition, G.B. (Gonzalo Bravo), M.B. and G.B. (Gregorio Bigatti). All authors have read and agreed to the published version of the manuscript.

Funding: This research was funded by Agencia Nacional de Promoción Científica y Técnica to M.B. (PICT 2017-2731) and G.B. (Gregorio Bigatti) (PICT 2018-0969). G.B. (Gonzalo Bravo) and J.K. are supported by CONICET fellowships.

Institutional Review Board Statement: Not applicable, since non-invasive research was conducted. No vertebrate sampling was conducted and therefore no approval was required by any Animal Care and Use Committee.

Data Availability Statement: Publicly available datasets were analyzed in this study. This data can be found here: https://www.inaturalist.org/projects/biodiversidad-submarina-islas-becasses, accessed on 10 December 2022.

Acknowledgments: We thank the following researchers for their expertise in species identification: Alicia Boraso, María Paula Raffo, and Erasmo Macaya (macroalgae), Daniel Lauretta (Anthozoa), Mariano Martínez (Holothuroidea), Diego Urteaga (Polyplacophora), Cristian Lagger (Tunicates), Ignacio Chiesa (Amphipoda), Facundo Llompart, and Eloísa Giménez (fish). A special thanks to Miguel Porco Fischer, captain of the vessel "Kostat", for his friendship and support on this adventure. Three anonymous reviewers improved the manuscript with their comments. Fieldwork was conducted with permission from Tierra del Fuego Province (Secretaría de Ambiente, Ministerio de Producción y Ambiente, Res. S.A. N° 345) and from Prefectura Naval Argentina. This is a contribution to the program of GrIETA.

Conflicts of Interest: The authors declare no conflict of interest. The funders had no role in the design of the study; in the collection, analyses, or interpretation of data; in the writing of the manuscript; or in the decision to publish the results.

References

1. Miloslavich, P.; Bax, N.J.; Simmons, S.E.; Klein, E.; Appeltans, W.; Aburto-Oropeza, O.; Andersen Garcia, M.; Batten, S.D.; Benedetti-Cecchi, L.; Checkley, D.M.; et al. Essential ocean variables for global sustained observations of biodiversity and ecosystem changes. *Glob. Chang. Biol.* **2018**, *24*, 2416–2433. [CrossRef]
2. Heberling, J.M.; Isaac, B.L. iNaturalist as a tool to expand the research value of museum specimens. *Appl. Plant Sci.* **2018**, *6*, e01193. [CrossRef]
3. Boone, M.E.; Basille, M. Using iNaturalist to contribute your nature observations to science. *EDIS* **2019**, *2019*, 5. [CrossRef]
4. Hewitt, S.; Picton, B.; DuPont, A.; Zahner, T.; Salvador, R. New records of marine molluscs from Saba, Caribbean Netherlands. *Basteria* **2021**, *85*, 59–72.
5. Rosa, R.M.; Cavallari, D.C.; Salvador, R.B. iNaturalist as a tool in the study of tropical molluscs. *PLoS ONE* **2022**, *17*, e0268048. [CrossRef]
6. Miloslavich, P.; Klein, E.; Díaz, J.M.; Hernández, C.E.; Bigatti, G.; Campos, L.; Artigas, F.; Castillo, J.; Penchaszadeh, P.E.; Neill, P.E.; et al. Marine biodiversity in the Atlantic and Pacific coasts of South America: Knowledge and gaps. *PLoS ONE* **2011**, *6*, e14631. [CrossRef]
7. Bigatti, G.; Signorelli, J. Marine invertebrate biodiversity from the Argentine Sea, South Western Atlantic. *ZooKeys* **2018**, *791*, 47–70. [CrossRef]
8. Mora-Soto, A.; Palacios, M.; Macaya, E.C.; Gómez, I.; Huovinen, P.; Pérez-Matus, A.; Young, M.; Golding, N.; Toro, M.; Yaqub, M. A high-resolution global map of Giant kelp (*Macrocystis pyrifera*) forests and intertidal green algae (*Ulvophyceae*) with Sentinel-2 imagery. *Remote Sens.* **2020**, *12*, 694. [CrossRef]
9. Santelices, B.; Ojeda, F. Effects of canopy removal on the understory algal community structure of coastal forests of *Macrocystis pyrifera* from southern South America. *Mar. Ecol. Prog. Ser.* **1984**, *14*, 165–173. [CrossRef]
10. Adami, M.L.; Gordillo, S. Structure and dynamics of the biota associated with *Macrocystis pyrifera* (Phaeophyta) from the Beagle Channel, Tierra del Fuego. *Sci. Mar.* **1999**, *63*, 183–191. [CrossRef]
11. Vanella, F.A.; Fernández, D.A.; Carolina Romero, M.; Calvo, J. Changes in the fish fauna associated with a sub-Antarctic *Macrocystis pyrifera* kelp forest in response to canopy removal. *Polar Biol.* **2007**, *30*, 449–457. [CrossRef]
12. Hockey, P.A.R. Kelp gulls *Larus dominicanus* as predators in kelp *Macrocystis pyrifera* beds. *Oecologia* **1988**, *76*, 155–157. [CrossRef] [PubMed]
13. Ordoñez, C.; Diez, M.J.; Torres, M.A.; Dellabianca, N.A. Habitat use of Peale's dolphin *Lagenorhynchus australis* in the Beagle Channel, Tierra del Fuego, Argentina. *Mastozool. Neotropical* **2020**, *27*, 247–252. [CrossRef]
14. Ojeda, F.; Santelices, B. Invertebrate communities in holdfasts of the kelp *Macrocystis pyrifera* from southern Chile. *Mar. Ecol. Prog. Ser.* **1984**, *16*, 65–73. [CrossRef]
15. Vásquez, J.; Castilla, J.; Santelices, B. Distributional patterns and diets of four species of sea urchins in giant kelp forest (*Macrocystis pyrifera*) of Puerto Toro, Navarino Island, Chile. *Mar. Ecol. Prog. Ser.* **1984**, *19*, 55–63. [CrossRef]
16. Dayton, P.K. The Structure and Regulation of Some South American Kelp Communities. *Ecol. Monogr.* **1985**, *55*, 447–468. [CrossRef]
17. Friedlander, A.M.; Ballesteros, E.; Bell, T.W.; Giddens, J.; Henning, B.; Hüne, M.; Muñoz, A.; Salinas-de-León, P.; Sala, E. Marine biodiversity at the end of the world: Cape Horn and Diego Ramírez islands. *PLoS ONE* **2018**, *13*, e0189930. [CrossRef]
18. Beaton, E.C.; Küpper, F.C.; van West, P.; Brewin, P.E.; Brickle, P. The influence of depth and season on the benthic communities of a *Macrocystis pyrifera* forest in the Falkland Islands. *Polar Biol.* **2020**, *43*, 573–586. [CrossRef]
19. Friedlander, A.M.; Ballesteros, E.; Bell, T.W.; Caselle, J.E.; Campagna, C.; Goodell, W.; Hüne, M.; Muñoz, A.; Salinas-de-León, P.; Sala, E.; et al. Kelp forests at the end of the earth: 45 years later. *PLoS ONE* **2020**, *15*, e0229259. [CrossRef]
20. Cárdenas, C.A.; Montiel, A. The influence of depth and substrate inclination on sessile assemblages in subantarctic rocky reefs (*Magellan region*). *Polar Biol.* **2015**, *38*, 1631–1644. [CrossRef]
21. Friedlander, A.M.; Ballesteros, E.; Goodell, W.; Hüne, M.; Muñoz, A.; Salinas-de-León, P.; Velasco-Charpentier, C.; Sala, E. Marine communities of the newly created Kawésqar National Reserve, Chile: From glaciers to the Pacific Ocean. *PLoS ONE* **2021**, *16*, e0249413. [CrossRef] [PubMed]
22. Ríos, C.; Arntz, W.E.; Gerdes, D.; Mutschke, E.; Montiel, A. Spatial and temporal variability of the benthic assemblages associated to the holdfasts of the kelp *Macrocystis pyrifera* in the Straits of Magellan, Chile. *Polar Biol.* **2007**, *31*, 89–100. [CrossRef]
23. Wartenberg, R. On the Underwater Visual Census of Western Indian Ocean Coral Reef Fishes. Ph.D. Thesis, Rhodes University, Grahamstown, South Africa, 2012.
24. Rassweiler, A.; Dubel, A.K.; Hernan, G.; Kushner, D.J.; Caselle, J.E.; Sprague, J.L.; Kui, L.; Lamy, T.; Lester, S.E.; Miller, R.J. Roving divers surveying fish in fixed areas capture similar patterns in biogeography but different estimates of density when compared with belt transects. *Front. Mar. Sci.* **2020**, *7*, 272. [CrossRef]
25. Schmitt, E.; Sluka, R.; Sullivan-Sealey, K. Evaluating the use of roving diver and transect surveys to assess the coral reef fish assemblage off southeastern Hispaniola. *Coral Reefs* **2002**, *21*, 216–223. [CrossRef]
26. Leite, T.S.; Haimovici, M.; Mather, J.; Oliveira, J.E.L. Habitat, distribution, and abundance of the commercial octopus (*Octopus insularis*) in a tropical oceanic island, Brazil: Information for management of an artisanal fishery inside a marine protected area. *Fish. Res.* **2009**, *98*, 85–91. [CrossRef]

27. Sabdono, A.; Radjasa, O.K.; Trianto, A.; Sibero, M.T.; Kristiana, R.; Larasati, S.J.H. Comparative assessment of gorgonian abundance and diversity among Islands with different anthropogenic stressors in Karimunjawa Marine National Park, Java Sea. *Int. J. Conserv. Sci.* **2022**, *13*, 341–348.

28. Iachetti, C.M.; Lovrich, G.; Alder, V.A. Temporal variability of the physical and chemical environment, chlorophyll and diatom biomass in the euphotic zone of the Beagle Channel (Argentina): Evidence of nutrient limitation. *Prog. Oceanogr.* **2021**, *195*, 102576. [CrossRef]

29. Rabassa, J.; Coronato, A.; Bujalesky, G.; Salemme, M.; Roig, C.; Meglioli, A.; Heusser, C.; Gordillo, S.; Roig, F.; Borromei, A. Quaternary of Tierra del Fuego, southernmost South America: An updated review. *Quat. Int.* **2000**, *68*, 217–240. [CrossRef]

30. Isla, F.; Bujalesky, G.; Coronato, A. Procesos estuáricos en el Canal Beagle, Tierra del Fuego. *Rev. Asoc. Geol. Argent.* **1999**, *54*, 307–318.

31. Bujalesky, G.G. La Inundación del Valle Beagle (11.000 AÑOS AP), Tierra del Fuego. *An. Inst. Patagon.* **2011**, *39*, 5–21.

32. Ponce, J.F.; Fernández, M. *Climatic and Environmental History of Isla de los Estados, Argentina*; Springer Briefs in Earth System Sciences; Springer: Dordrecht, The Netherlands, 2014; ISBN 978-94-007-4362-5.

33. Giesecke, R.; Martín, J.; Piñones, A.; Höfer, J.; Garcés-Vargas, J.; Flores-Melo, X.; Alarcón, E.; Durrieu de Madron, X.; Bourrin, F.; González, H.E. General Hydrography of the Beagle Channel, a Subantarctic Interoceanic Passage at the Southern Tip of South America. *Front. Mar. Sci.* **2021**, *8*, 621822. [CrossRef]

34. Almandoz, G.O.; Hernando, M.P.; Ferreyra, G.A.; Schloss, I.R.; Ferrario, M.E. Seasonal phytoplankton dynamics in extreme southern South America (*Beagle channel*, Argentina). *J. Sea Res.* **2011**, *66*, 47–57. [CrossRef]

35. Iturraspe, R.; Sottini, R.; Schroeder, C.; Escobar, J. Hidrología y variables climáticas del Territorio de Tierra del Fuego. Información básica. *Contrib. Cient. CADIC* **1989**, *7*, 196.

36. Balestrini, C.; Manzella, G.; Lovrich, G.A. Simulación de corrientes en el Canal Beagle y Bahía Ushuaia, mediante un modelo bidimensional. *Serv. Hidrogr. Nav.* **1998**, *98*, 1–58.

37. Flores Melo, X.; Martín, J.; Kerdel, L.; Bourrin, F.; Colloca, C.B.; Menniti, C.; de Madron, X.D. Particle dynamics in Ushuaia Bay (Tierra del Fuego)-Potential effect on dissolved oxygen depletion. *Water* **2020**, *12*, 324. [CrossRef]

38. Harris, S.; Samaniego, R.A.S.; Rey, A.R. Insights into diet and foraging behavior of imperial shags (*Phalacrocorax atriceps*) breeding at Staten and Becasses Islands, Tierra del Fuego, Argentina. *Wilson J. Ornithol.* **2016**, *128*, 811–820. [CrossRef]

39. Schiavini, A.C.M.; Crespo, E.A.; Szapkievich, V. Status of the population of South American sea lion (*Otaria flavescens* Shaw, 1800) in southern Argentina. *Mamm. Biol.* **2004**, *69*, 108–118. [CrossRef]

40. Bravo, G.; Kaminsky, J.; Bagur, M.; Alonso, C.P.; Rodríguez, M. INaturalist Project: "Biodiversidad Submarina Islas Bécasses". Available online: https://www.inaturalist.org/projects/biodiversidad-submarina-islas-becasses (accessed on 10 December 2022).

41. Ahyong, S.; Boyko, C.B.; Bailly, N.; Bernot, J.; Bieler, R.; Brandão, S.N.; Daly, M.; De Grave, S.; Gofas, S.; Hernandez, F.; et al. World Register of Marine Species (WoRMS). Available online: https://www.marinespecies.org (accessed on 10 December 2022).

42. Guiry, M.D.; Guiry, G.M. AlgaeBase. World-Wide Electronic Publication, National University of Ireland, Galway. Available online: https://www.algaebase.org (accessed on 10 December 2022).

43. Froese, R.; Pauly, D. FishBase. World Wide Web Electronic Publication. Available online: www.fishbase.org (accessed on 10 December 2022).

44. Häussermann, V.; Molinet, C.; Díaz Gómez, M.; Försterra, G.; Henríquez, J.; Espinoza Cea, K.; Matamala Ascencio, T.; Hüne, M.; Cárdenas, C.A.; Glon, H.; et al. Recent massive invasions of the circumboreal sea anemone *Metridium senile* in North and South Patagonia. *Biol. Invasions* **2022**, *24*, 3665–3674. [CrossRef]

45. Fraysse, C.; Boy, C.; Ojeda, M.; Rodriguez, M.; Pérez, A. Distribution and development patterns in Asteroidea of the Southern Atlantic, including marine protected areas, Argentina. 2023; *to be submitted*.

46. Willenz, P.; Hajdu, E.; Desqueyroux-Faúndez, R.; Lôbo-Hajdu, G.; Carvalho, M. Porifera-Sponges. In *Marine Benthic Fauna of Chilean Patagonia*; Nature in Focus: Santiago, Chile, 2009; Volume 1000, pp. 93–94.

47. Ojeda, J.; Marambio, J.; Rosenfeld, S.; Contador, T.; Rozzi, R.; Mansilla, A. Seasonal changes of macroalgae assemblages on the rocky shores of the Cape Horn Biosphere Reserve, Sub-Antarctic Channels, Chile. *Aquat. Bot.* **2019**, *157*, 33–41. [CrossRef]

48. Kaminsky, J.; Bagur, M.; Boraso, A.; Schloss, I.R.; Quartino, M.L. Centro Austral de Investigaciones Científicas (CADIC-CONICET), Ushuaia, Argentina. 2023; *manuscript in preparation*.

49. Mendoza, M.L. State of knowledge of the Corallinales (Rhodophyta) of Tierra del Fuego and the Antarctic Peninsula. *Sci. Mar.* **1999**, *63*, 139–144. [CrossRef]

50. Noss, R.F. Indicators for Monitoring Biodiversity: A Hierarchical Approach. *Conserv. Biol.* **1990**, *4*, 355–364. [CrossRef]

51. Costello, M.J.; Tsai, P.; Wong, P.S.; Cheung, A.K.L.; Basher, Z.; Chaudhary, C. Marine biogeographic realms and species endemicity. *Nat. Commun.* **2017**, *8*, 1057. [CrossRef]

52. Alonso, G.M. Amphipod crustaceans (Corophiidea and Gammaridea) associated with holdfasts of Macrocystis pyrifera from the Beagle Channel (Argentina) and additional records from the Southwestern Atlantic. *J. Nat. Hist.* **2012**, *46*, 1799–1894. [CrossRef]

53. Häussermann, V.; Försterra, G. *Marine Benthic Fauna of Chilean Patagonia*; Nature in Focus: Santiago, Chile, 2009; Volume 1000.

54. GBIF Secretariat; GBIF Backbone Taxonomy. Checklist Dataset. via GBIF.org. Available online: https://doi.org/10.15468/39omei (accessed on 10 July 2022).

55. Rosenfeld, S.; Ojeda, J.; Hüne, M.; Mansilla, A.; Contador, T. Egg masses of the Patagonian squid *Doryteuthis* (*Amerigo*) *gahi* attached to giant kelp (*Macrocystis pyrifera*) in the sub-Antarctic ecoregion. *Polar Res.* **2014**, *33*, 21636. [CrossRef]

56. Küpper, F.C.; Peters, A.F.; Kytinou, E.; Asensi, A.O.; Vieira, C.; Macaya, E.C.; De Clerck, O. *Dictyota falklandica* sp. *nov.* (Dictyotales, Phaeophyceae) from the Falkland Islands and southernmost South America. *Phycologia* **2019**, *58*, 640–647. [CrossRef]

57. Lauretta, D.; Häussermann, V.; Brugler, M.R.; Rodríguez, E. *Isoparactis fionae* sp. nov. (Cnidaria: Anthozoa: Actiniaria) from Southern Patagonia with a discussion of the family Isanthidae. *Org. Divers. Evol.* **2014**, *14*, 31–42. [CrossRef]

58. Cetra, N.; Gregoric, D.E.G.; Roche, A. A New Species of *Placida* (Gastropoda: Sacoglossa) from Southern South America. *Malacologia* **2021**, *64*, 109–120. [CrossRef]

59. Asensi, A.; De Reviers, B. Illustrated catalogue of types of species historically assigned to *Lessonia* (Laminariales, Phaeophyceae) preserved at PC, including a taxonomic study of three South-American species with a description of *L. searlesiana* sp. nov. and a new lectotypification of *L. flavicans*. *Cryptogam. Algol.* **2009**, *30*, 209–249.

60. UWIS Underwater Navigation System. Available online: https://uwis.fi/en (accessed on 15 July 2022).

61. Hiscock, K. *Marine Nature Conservation Review: Rationale and Methods*; Coasts and Seas of the United Kingdom. MNCR Series; Joint Nature Conservation Committee: Peterborough, UK, 1996; ISBN 1-86107-410-7.

62. Semmens, B.X.; Auster, P.J.; Paddack, M.J. Using Ecological Null Models to Assess the Potential for Marine Protected Area Networks to Protect Biodiversity. *PLoS ONE* **2010**, *5*, e8895. [CrossRef] [PubMed]

63. Auster, P.J.; Semmens, B.X.; Barber, K. Pattern in the Co-Occurrence of Fishes Inhabiting the Coral Reefs of Bonaire, Netherlands Antilles. *Environ. Biol. Fishes* **2005**, *74*, 187–194. [CrossRef]

64. Roberts, C.J.; Vergés, A.; Callaghan, C.T.; Poore, A.G.B. Many cameras make light work: Opportunistic photographs of rare species in iNaturalist complement structured surveys of reef fish to better understand species richness. *Biodivers. Conserv.* **2022**, *31*, 1407–1425. [CrossRef]

Communication

Trends in Dominican Republic Coral Reef Biodiversity 2015–2022

Robert S. Steneck [1,*] and Rubén Torres [2]

1   School of Marine Sciences, University of Maine, Orono, ME 04469, USA
2   Reef Check Dominican Republic, Jacinto Mañón #20, Edificio Paraiso, apt 1B, Ensanche Paraiso, Santo Domingo 11005, Dominican Republic
*   Correspondence: steneck@maine.edu

Abstract: In 2015, we initiated a country-wide coral reef ecosystem-monitoring program in the Dominican Republic (DR) to establish biodiversity baselines against which trends in the most important components of coral reef ecosystem's structure and function could be tracked. Replicate transects were set at a 10 m depth at each of the 12 coral reef sites within 6 DR regions in 2015, 2017, 2019, and 2022. We quantified the species-level abundances of adult and juvenile corals, reef fishes, sea urchins, lionfishes, and algal functional groups. Country-wide, coral cover and reef fishes have declined. The steepest declines occurred for reefs that had been among the best in the Caribbean in 2015. However, by 2022, adult and juvenile coral, parrotfish, and other herbivores had declined, and macroalgae had increased. The declines in north-shore coral abundance corresponded with the observed disturbances from coral bleaching, hurricanes, and disease. The capacity of reefs to recover from such disturbances has been compromised by abundant and increasing macroalgae that have likely contributed to north-shore declines in juvenile corals. Country-wide, the abundance of all reef fish species has declined below those of other regions of the Caribbean. Improved management of fishing pressure on coral reefs would likely yield positive results.

Keywords: Caribbean; coral reefs; fisheries management; marine biodiversity; marine ecosystem trends; recovery resilience; tropical marine

Citation: Steneck, R.S.; Torres, R. Trends in Dominican Republic Coral Reef Biodiversity 2015–2022. *Diversity* **2023**, *15*, 389. https://doi.org/10.3390/d15030389

Academic Editors: Thomas J. Trott and Bert W. Hoeksema

Received: 16 January 2023
Revised: 27 February 2023
Accepted: 28 February 2023
Published: 8 March 2023

1. Introduction

Never have scleractinian-dominated coral reefs been as threatened as they are today. Corals die because of hurricanes, ocean warming, disease, and over-growing seaweed. In most cases, factors that kill corals cannot be stopped by current human interventions. As a result, the best hope is to manage coral reefs for efficient and rapid recovery from those events [1–3]. While there may be a general agreement that management for coral reef recovery is necessary, how to do it and how to monitor changes require serious consideration. Specifically, managers and policy makers need to know not only the status of a coral reef and the factors important to them but also how they are trending. With such information, managers can determine how best to use their limited resources available to aid in coral reef recovery.

It is widely recognized that coral reefs are among the world's most endangered ecosystems, and this is especially true in the Caribbean, where trends in declining herbivores, increasing algae, and declining coral effectively lock coral reefs in a degraded algal state [1,3–6]. Until recently, reef recovery from disturbances in the Caribbean was unknown [2–5]. In fact, too little is known about where and why reef condition is declining for most of a country's coral reefs. While studying the trends in coral reefs is necessary, some studies focused on only a few locations within an island or a country, e.g., [4]. It is likely that different factors drive reef condition at different locations within a country or an island. It is possible that different factors in different regions may affect changes to coral reefs that will not be evident at a single site.

What is needed are assessments of the rates of decline or recovery of reefs over time and space. This informs managers where the most urgent action should be taken so they can mitigate the decline or improve the prospects for recovery. Effective action taken early can prevent the coral reef ecosystem from becoming "locked" into a coral-free state [4,5]. It is important to learn from reefs that are healthier than others. For example, recent studies determined that coral reefs in the eastern Caribbean, where fishing had been limited, were healthier and had more juvenile corals than reefs with unconstrained fishing practices [3,6]. Several Caribbean islands have effectively limited fishing pressure—especially on herbivorous parrotfish [7]. One of those islands, Bonaire (Dutch Antilles), thrives today and has highly resilient coral reefs that have fully recovered from a severe coral bleaching event in 2010 [3]. To achieve such recovery resilience requires conditions in which coral larvae can settle and grow to become juvenile corals, and injured corals can recover. If the juvenile corals survive and grow, they improve the recovery of the coral reef ecosystem [3,8,9].

Here, we report on the status and seven-year trends in 12 coral reefs distributed among 6 regions within three sectors along the Dominican Republic (DR) coastline (Figure 1). We monitored the prime drivers and indicators of coral reef health. A driver is a factor that contributes to or causes a change in coral reef structure or function. Heavy emphasis was placed on the drivers that allow a reef to be healthy, thrive, or recover following a disturbance. We measured the trends in key variables among the replicated samples at a 10 m depth in identical locations during each monitoring period. Specifically, we sought to determine the trends in the abundance of reef coral, macroalgae, parrotfish, and juvenile corals. We also censused *Diadema* sea urchins, other herbivorous fishes such as surgeonfishes, and carnivorous fishes. This study provides not only the status of the DR's coral reefs in 2022 but specifically how and where they have changed since our first survey in 2015.

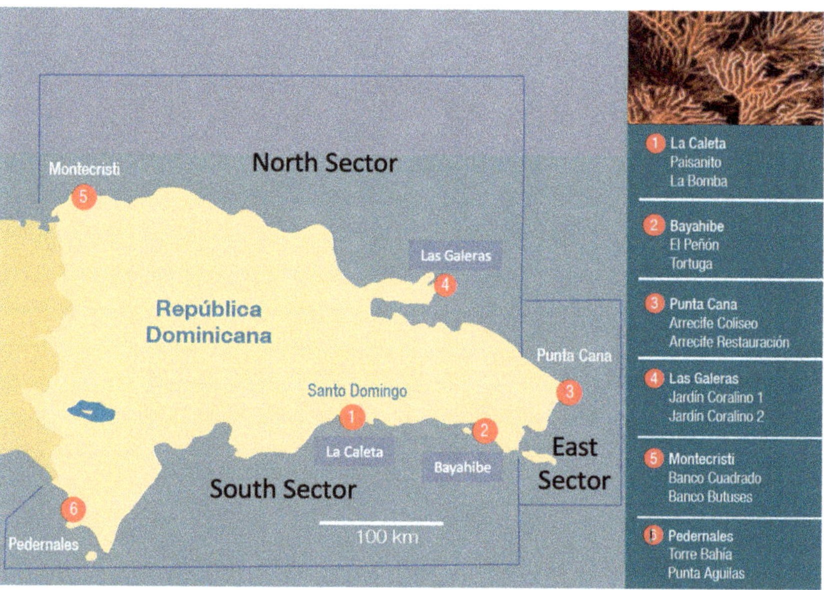

Figure 1. Six monitored regions (numbered in the figure and panel to the right). Each region contained two monitoring sites (listed in panel to the right). Regions were further subdivided into north, east, and south sectors.

2. Methods

We established 12 coral reef monitoring sites, distributed among 6 regions from near the southern border with Haiti to the eastern-most region to near the northern border with Haiti (Figure 1). All sites had well-developed coral reefs that were used for diving and fishing. Most coral reefs were well-developed bank-barrier reefs. Most unique were those at the Montecristi region, where reef development was 10 km from the shore.

For our monitoring, we used a modified Atlantic and Gulf Rapid Reef Assessment (AGRRA) protocol [10] to quantify sessile benthic community structure at each site. At each site, at a 10 m water depth, replicate 10 m transect lead lines were placed on the reef (n = 4 per reef). We recorded the number of centimeters on each transect intercepted by live coral (measured for each coral species), sponges, gorgonians, and benthic algae (measured by functional group such as algal turfs, encrusting coralline algae (Corallinales), and noncoralline (peyssonnelid) crusts, and macroalgae). Benthic macroalgae were further subdivided into erect corticated macrophytes [11] and specific genera *Lobophora* and *Halimeda* because of their damaging effects on reef corals [12,13]. Canopy heights were measured to the nearest millimeter for all nonencrusting algae. The percent cover of each benthic organism (per transect) was determined by summing the centimeters intercepted by the organism and then dividing that total by the length of the transect. Transect sampling was further stratified to only include hard substrates (i.e., sand channels were excluded from our analysis). Macroalgal biomass was inferred by quantifying the algal volume, which is the product of percent cover and canopy height, also called "algal index" [6].

We also used a variety of methods to quantify other reef-dwelling organisms. Juvenile coral densities were quantified by deploying 25 cm × 25 cm quadrats at 5 locations (0, 2.5, 5, 7.5, and 10 m marks) adjacent to each 10 m transect. Quadrats were placed on the reef substrata largely devoid of adult corals (that is, <25% cover of live corals). Operationally, we defined juvenile corals as those less than 40 mm in maximum diameter. Each juvenile coral was identified to the species level and measured to the nearest millimeter. The sea urchin *D. antillarum* was censused by surveying belt transects (2 m wide) centered on the 10 m long transect used for quantifying sessile benthic organisms. Thus, within each belt transect, we surveyed a 20 m² area. We measured the test size of each urchin encountered to the nearest millimeter. Fish population density and body sizes were quantified for all large fishes (that is, excluding small planktivores, such as chromis, and all blennies and gobies) using replicate 30 m × 4 m belt transects [14]. This involved attaching a spool with 30 m of line to dead coral and swimming slowly, recording all large vagile species including most predatory and herbivorous fishes. On the return swim, smaller, less vagile species were recorded. The biomass of each fish species (per transect) was determined using published length–weight relationships [15], and http://fishbase.se (accessed on 1 February 2023).

3. Results and Discussion

For any coral reef, coral abundance is the single most important component; however, the status of corals and coral reefs depends on many other factors. Patterns in coral abundance over time show variability at all levels from single transects, to reef sites, regions, sectors, and the entire country. All error bars reflect the standard error. Below, we summarize the trends in the important components of coral ecosystem structure and function, both geographically and spatially. The details for all monitored variables at each station for each year are provided in the Supplemental Information (Figure S1). Monitoring was conducted March 2015, May 2017 and 2019, and June 2022.

3.1. Trends in Corals, Their Drivers, and the Next Generation

Country-wide coral abundance has declined by nearly 5% since 2015 (Figure 2A). While coral abundances among sites in the South Sector were variable, they showed no net trend (Figure 2B). In contrast, coral cover in most North Sector sites declined (Figure 2C). By averaging all coral cover percentage data among sites in the North, South, and East Sectors

(Figure 2D), we found regionally relevant patterns. Consistently low coral abundance predominated in the East Sector (Punta Cana). The East Sector lacks connectivity to coral larval sources due to the Mona Passage oceanographic barrier [16]. It is uniquely isolated by easterly trade winds that create long-shore currents to the north and south shores. The monitored reefs along the south shore generally averaged more abundant corals over the period but without any distinct temporal trend. In contrast, coral cover along the north shore (especially in the Montecristi region) have consistently declined from their high values in 2015.

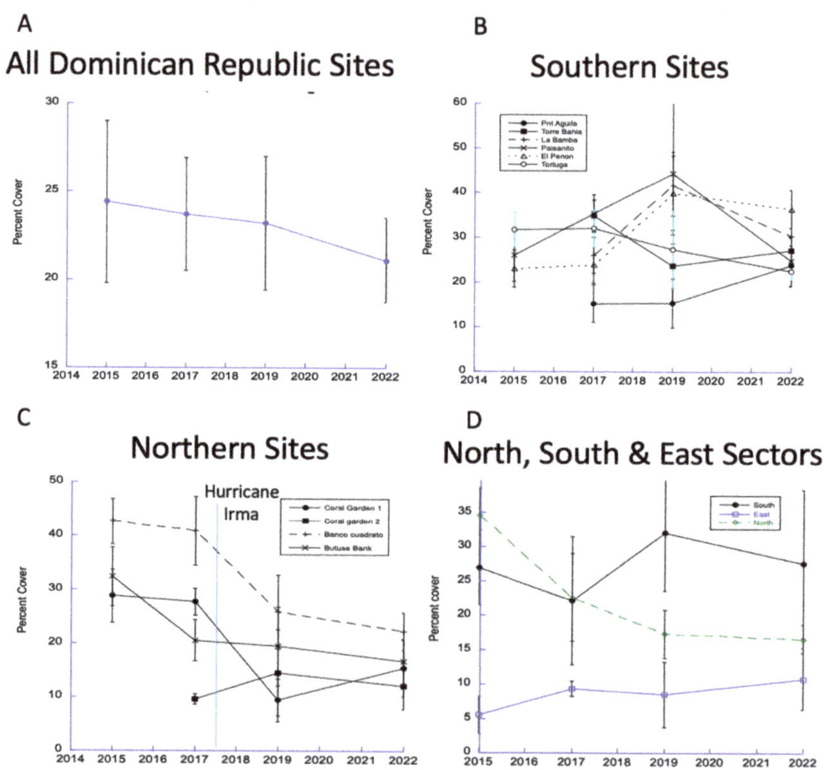

Figure 2. Trends in coral cover from all DR sites (**A**), in the south (**B**), and in the north (**C**). Data from North, South, and East Sectors show strong and protracted declines in coral cover in the North Sector (**D**).

Next, we considered the interactions that may limit the abundance of coral. Recently, disease-related mortality has affected numerous coral species throughout the Caribbean. Most concerning is the recently introduced Stony Coral Rapid Tissue Loss (SCTL) disease [16]. A recent study in the Dominican Republic quantified all coral diseases as well as other agents of mortality in the last few years [17,18]. The coral disease study was held near or on our study sites (Figure 1). We examined the changes in coral abundance between 2019 and 2022 to determine if the disease outbreaks corresponded with changes in coral abundance. We found no relationship country-wide. However, among the southern sites (Figure 2B), Tortuga was observed to have the highest disease percentage in 2019 and continuing in 2022. That site went from highest percent cover of coral along the south shore in 2015 to lowest percent in 2022 (i.e., from 32% to 22%). It is possible that the disease is just starting. By June 2022, we did not detect significant country-wide losses in coral abundance resulting from disease.

Chief among the other agents of mortality is macroalgae, which have been shown to out-compete and poison adult coral [12,13] and to prevent the recruitment of juvenile corals [9,19]. To determine if corals are limited by macroalgae, we pooled all data from all transects for all years to determine the maximum coral cover that can coexist with macroalgae (Figure 3A) and conducted a similar analysis using all juvenile coral quadrat data (Figure 3B). In both cases, there was a significant negative relationship. In a similar analysis, we determined that the maximum abundance of macroalgae was inversely correlated with the abundance of parrotfishes (Figure 3C).

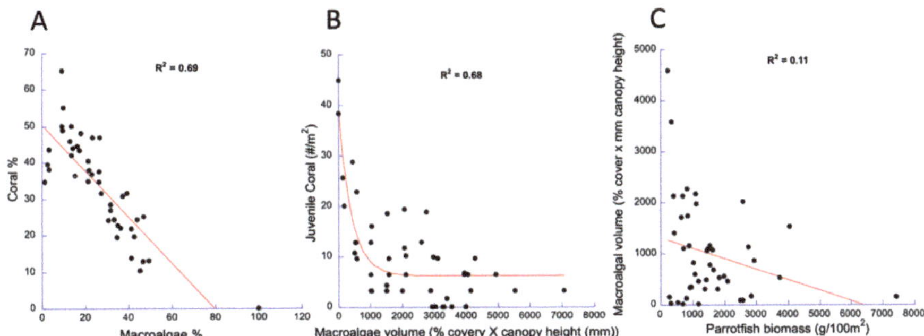

Figure 3. Support for strong drivers that limit the abundances of important components of the DR's coral reef ecosystems. Coral cover (**A**) and juvenile coral density (**B**) were limited by macroalgae cover (N = 180, 10 m transects, N = 895, 625 cm² quadrats, respectively) and macroalgae limited by parrotfish abundance N = 360, 120 m² belt transects (**C**). Data pooled from all surveys from all years.

Given the importance of parrotfishes as herbivores, their trends are most important. Parrotfish (Labridae: Scarinae) abundance was greatest in the North and South Sectors in 2015 and 2017 but declined by 2019 and remained relatively low during 2022 (Figure 4A). A similar pattern was reported for another important family of herbivores: the surgeonfishes (acanthurids) (Figure 4B).

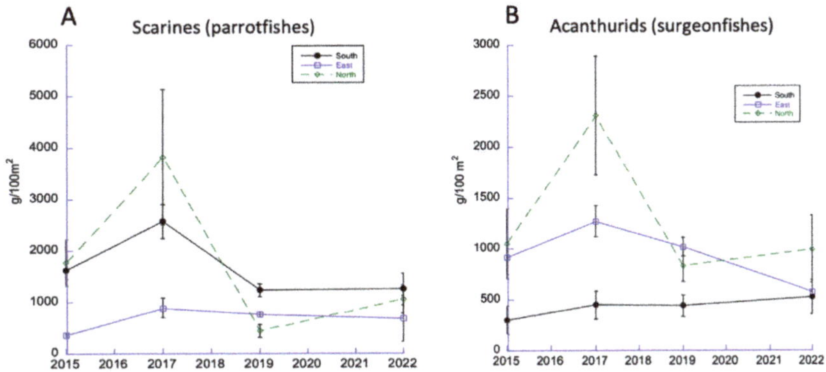

Figure 4. Trends in dominant DR herbivorous fishes (parrotfishes (**A**) and surgeonfishes (**B**)).

Macroalgae have dominated reefs in the East Sector since monitoring began (Figure 5). This has created a hostile environment for corals. The South and North Sectors have less macroalgae; in 2022, algal abundance was increasing in the North and declining in the South Sector (discussed below). We monitored the abundance of crustose coralline algae but found no significant changes among sites in the North or South Sectors.

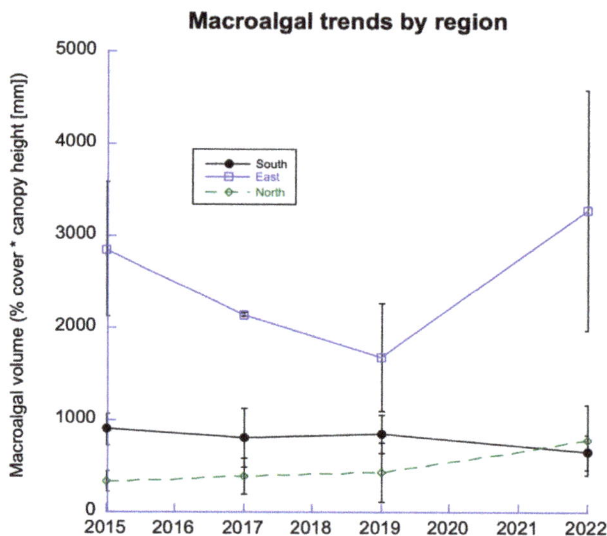

Figure 5. Trends in macroalgae. Data from 12 sites among 6 DR regions from the 3 sectors.

Juvenile corals are important for the recovery of damaged coral reefs. While there has been no net change in the abundance of juvenile corals since 2015 in the East and South Sectors, the North Sector has shown a consistent decline since 2017 (Figure 6).

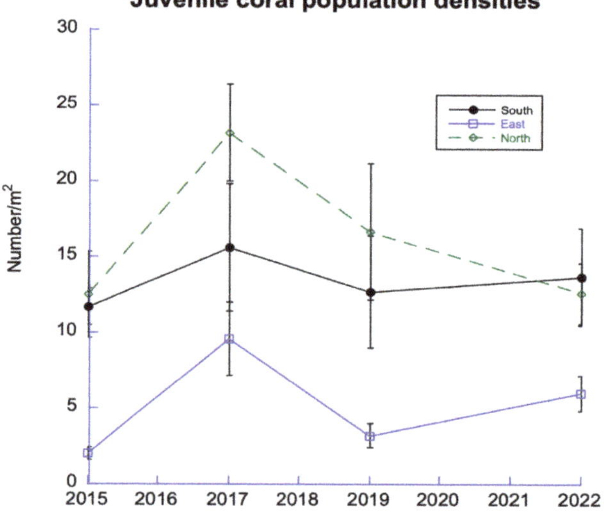

Figure 6. Trends in juvenile corals. Data from 12 sites among 6 DR regions from the 3 sectors.

3.2. Trends in Interactions among Fish Species, Their Prey, and Ecosystem Feedback

Fish assemblages play important roles in the health of coral reefs. All fishes declined in abundance following the 2017 survey and have not recovered (Figure 7A). The steepest decline was recorded in the North Sector. Since 2019, the fish biomass total has been significantly less than published accounts for the eastern Caribbean archipelago [6].

Fishing and hurricanes may have caused the decline in reef fish biomass (Figure 7). Illegal hookah spearfishing occurs in most of the DR, but it is more intense on the north coast (Figure 8A,B), resulting in large quantities and great diversity of reef fish being taken (Figure 8C). After our monitoring in May of 2017, Hurricanes Irma (Figure 9A) and Maria (Figure 9B) damaged coral reefs along the north shore. The observed destruction of branching corals may have flattened the reefs, thereby reducing available nursery habitats for recruiting fishes [20]; see photographs in Figure S2.

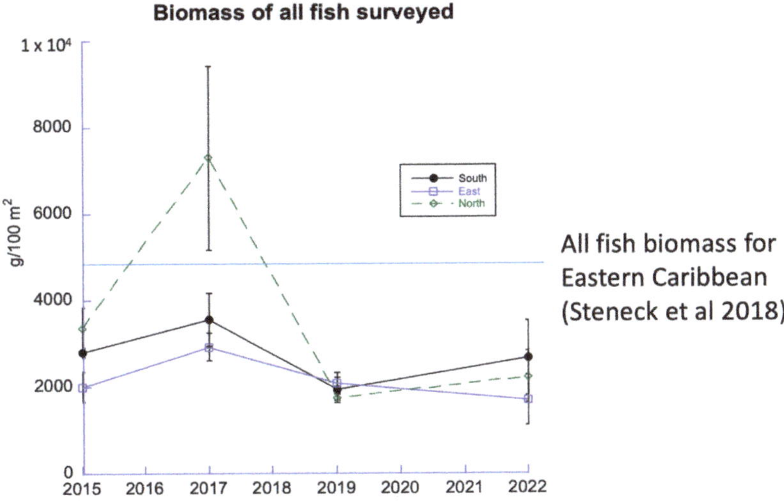

Figure 7. Biomass of all fish from all transects, 2015 to 2022. The horizontal line illustrates the biomass of all fish surveyed for the Eastern Caribbean [6].

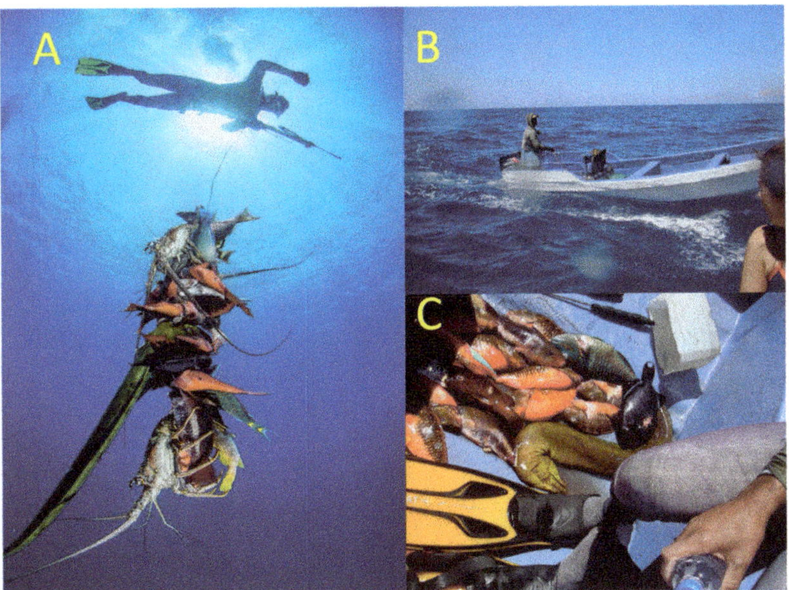

Figure 8. (**A**) Observed spearfishing in the Montecristi Region March 2015. (**B**) Fisher with a hookah rig for supplying air to divers. (**C**) An afternoon's catch of mostly parrotfish. Photographs by Jose Alejandro Alvarez.

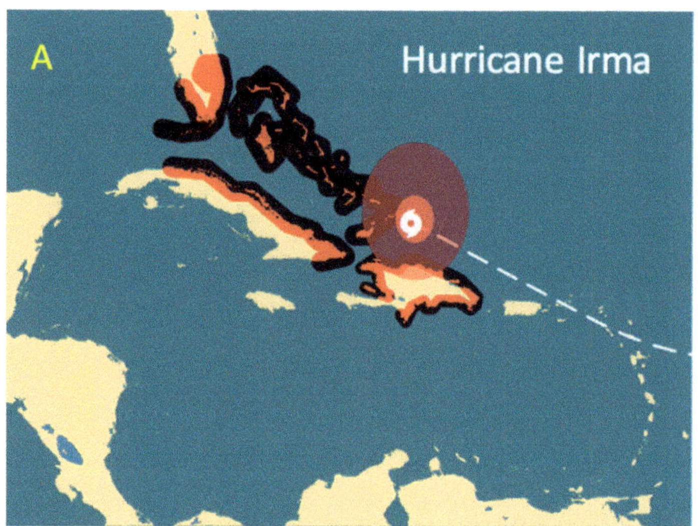

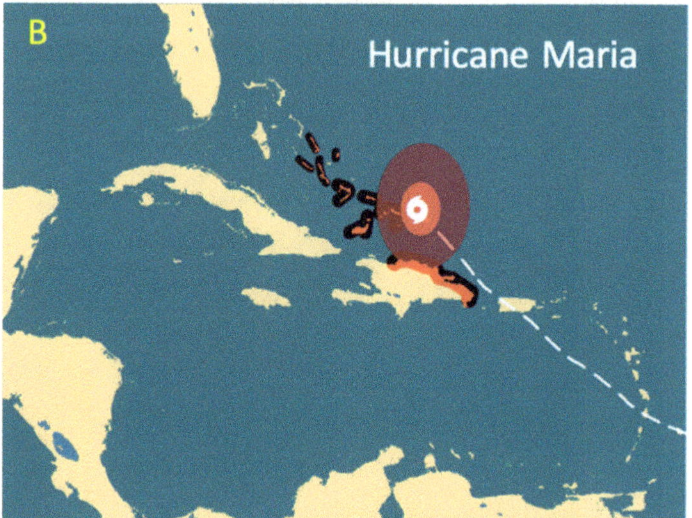

Figure 9. The hurricane tracks for (**A**) Hurricanes Irma (7 September 2017) and (**B**) Maria (21 September 2017).

Snappers, seabass, and jacks were the three most abundant carnivorous reef fish on the DR's coral reefs (Figure 10). Among those three families, the most abundant was the serranid (seabass) family, which includes groupers, graysbys, coney, and hinds. However, since monitoring began in 2015, serranids have declined most steeply every year. According to our 2022 monitoring study, serranids had declined by nearly 80%.

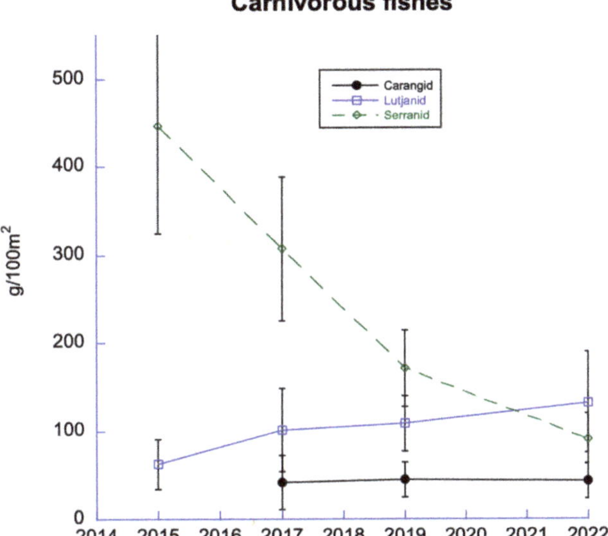

Figure 10. Carnivorous fishes include carangids (jacks), lutjanids (snappers), and serranids (seabass). All were low in abundance, but serranids have declined the most since 2015.

The steep decline among serranids was observed at three of our six sites in the South Sector (Figure 11A). However, even at the time of our 2015 study, the Punta de Aguilas site already had low abundances of predatory fishes. The low abundance of predatory fishes there continued though our 2022 monitoring study; Punta de Aguilas was the only site where we recorded sharp increases in the abundances of sea urchins (*Diadema antillarum,* and *Echinometra viridis*; Figure 11B,C) as well as lionfish (Figure 11D). Several studies have suggested predators control sea urchin abundances, specifically including *Diadema* [21] and *Echinometra* species (reviewed in [22]). Other studies suggested that lionfish abundances in the Caribbean can be limited by serranid groupers [23]. Therefore, it is possible that the low predator abundance at the Punta de Aguilas site facilitated the increases in those prey species. Importantly, *Diadema* functions as an herbivore that depresses macroalgae, but only when its population density exceeds one per square meter ([22,24], i.e., 20 per 20 m² belt transects, e.g., Figure 11B). At the Punta de Aguilas site, *Diadema* densities averaged over 2.75 per square meter. As a result of *Diadema* grazing, this site also had the lowest macroalgal abundance among the southern sites, which likely contributed to the slight decline in macroalgae in the South Sector (Figure 5).

Taken together, the trends among the factors that drive coral reef ecosystems (Figure 12) revealed that the highest proportion of negative trends were along the north coast, with the fewest in the south. We used those trends, in part, to track the drivers of coral reef resilience and to clearly specify to managers and policy makers which components of coral reef ecosystems are in decline. So, operationally, we must pay particular attention to the trends in the abundances of coral, macroalgae, parrotfishes, and juvenile corals (boldface in Figure 12). Whereas we saw both positive and negative trends among reef sites along the south shore, we found predominately negative trends among the coral reefs along the north DR shore (i.e., six negative and zero positive). The East Sector, which is the Punta Cana region, had, in 2015, the lowest coral cover, highest algal abundance, relatively few herbivores, and few recruiting corals. Most of the variables reflect reefs that continue in poor condition (i.e., three negative and zero positive trends).

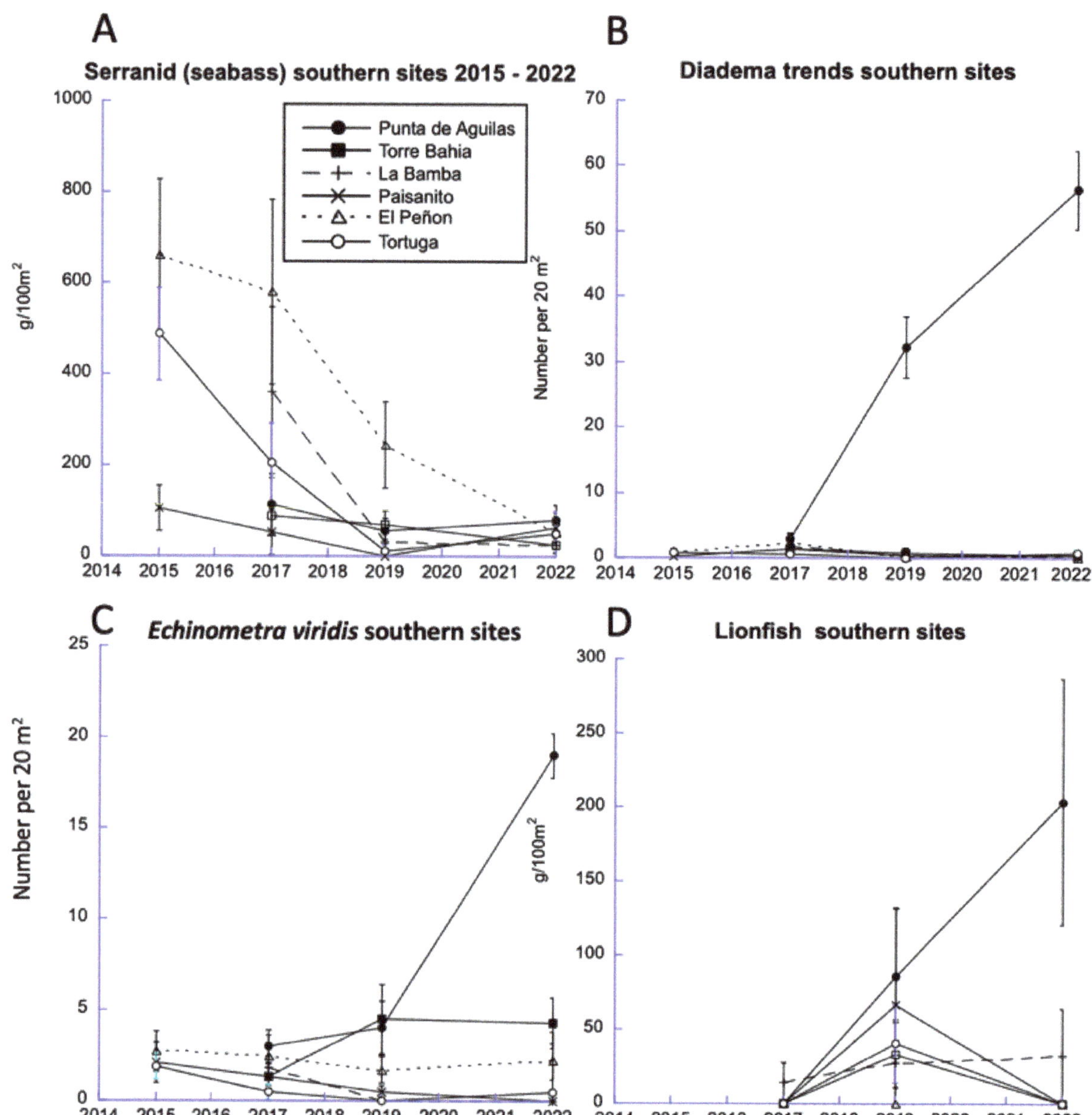

Figure 11. Trends in seabass (Serranidae such as groupers, coney, and graysby) abundance and their potential prey across all years: (**A**) changes in seabass abundance for all southern sites (note that Punta Aguilas had low seabass abundance all years of monitoring), and (**B,C**) *Diadema* and *Echinometra viridis* and (**D**) lionfish.

The curved arrows from coral recruits to coral cover in Figure 12 denote the role of coral recruitment in maintaining coral abundances. The management for improved coral recruitment, also called recovery resilience [3], is designed to facilitate the recovery of coral reef ecosystems following a disturbance. This concept for management was previously advanced [5,8]; however, there has been little attention paid to the consequences of which species to recruit to the reef. In other words, while coral recruitment will augment coral cover, it does not mean that the new corals will be functionally the same as those that make up the coral reef framework.

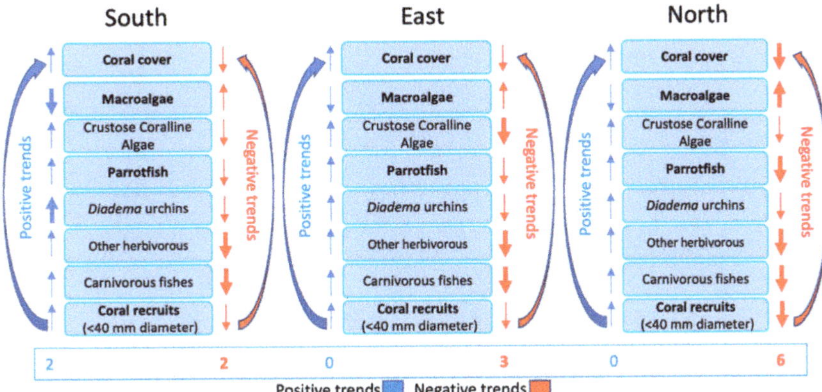

Figure 12. Trends in major coral reef drivers (good condition or positive on left in blue; poor condition or negative on right in red) were evaluated using the monitoring protocol from [25]. Most important reef characteristics are in boldface. North Sector DR reefs were clearly trending most negatively, possibly because of multiple stresses (coral bleaching, hurricanes, and overfishing).

The composition of the DR's coral reefs is changing. The juvenile corals are dominated by the relatively flat lettuce and mustard corals (*Agaricia agaricites* and *Porites astreoides*; Figure 13 left), whereas the single-most abundant adult coral is the star coral *Orbicella faveolata* (a coral reef framework species) (Figure 13, right). All coral species from the juvenile coral quadrats and the adult coral transects show a declining shift in the relative abundance of framework species and the increasing recruitment of weedy (juvenile) coral species (Figure 14). Corals that recruit readily reach reproductive maturity relatively quickly but lack vertical structure and are operationally called weedy corals (sensu [26]). As a result, the structure of the DR's reefs may be becoming increasingly flat and, therefore, lacking habitat for reef fish and recruiting corals [20].

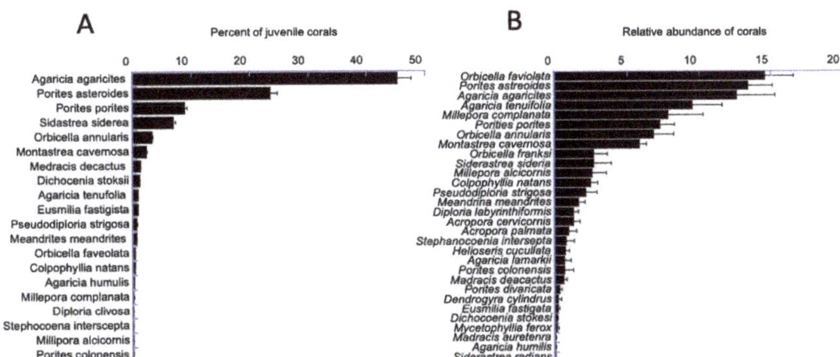

Figure 13. Relative abundances for juvenile (**A**) and adult (**B**) coral species. Juvenile (N, 20 spp.) and adult (N, 30 spp.) coral averages are for all sites and all years.

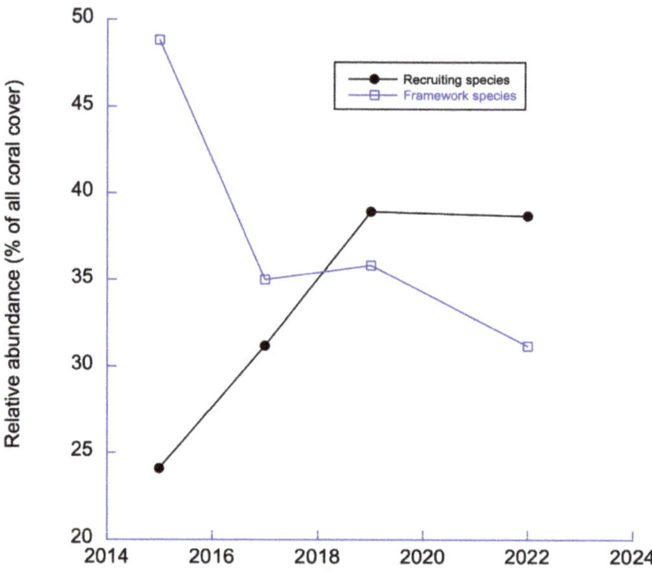

Figure 14. Trends in recruiting ("weedy") and framework species.

Lower figure (Figure 14) shows that there has been a steady decline in reef-building corals (framework species) and an increase of weedy low-profile (recruiting) coral species. The weedy species are low in stature but most readily colonize coral reefs [26]. This results in the flattening of coral reefs [27,28].

4. Conclusions

Since monitoring began in 2015, the coral reefs along the DR's north shore have declined the most steeply. The greatest changes have occurred in the Montecristi Region. Those reefs, in 2015, were among the best in the Caribbean in terms of high coral cover, low algal biomass, abundant parrotfish, and abundant juvenile corals (compared with Eastern Caribbean and Belize [6,7], respectively). In 2015, there were large stands of acroporid corals (elkhorn and staghorn corals) along with branching leafy coral (*Agaricia tenuifolia*) (see Figure S2). However, in late summer of 2016 and September of 2017, a coral bleaching event and two hurricanes, respectively, killed corals in the Montecristi region (Figure S2). Those events, along with sustained fishing pressure (Figure 8B), seem to have contributed to the declines in corals and herbivorous fishes and increases in macroalgae among north-shore reefs. In addition, the fish fauna country-wide is below that of most other fished reefs in the Caribbean (Figure 2C).

Coral disease was evident in most reefs in the Dominican Republic. Of particular concern is the recent appearance of the Stony Coral Tissue Loss (SCTL) disease that has devastated corals in other areas of the Caribbean, such as the Florida Keys. The reported relatively high levels of SCTL disease in the Las Galeras region [18] did not correspond to a decline in coral cover in that region (i.e., Figure 2C). We observed disease among several coral species in the South Sector's Bayahibe region (at the Tortuga and El Peñon sites); however, that region had the lowest reported coral disease mortality [18]. Nevertheless, we recorded a significant decline at the Tortuga site (Figure 2B).

In summary, the coral reefs of the Dominican Republic have been degraded from a variety of sources. Coral reef recovery from disturbances, such as hurricanes, disease, or coral bleaching, requires the settlement and growth of juvenile corals. Macroalgae, which remain relatively abundant on the DR's coral reefs, need to be reduced because they prevent coral colonization and growth. The decline in reef-building framework corals and the increased relative abundance of low-profile weedy corals now dominate the DR's coral reefs. The resulting flatter reefs have fewer spaces where reef fish and lobsters can hide and grow upward more slowly, which could compromise their capacity to keep up with increasing rates of sea-level rise.

Supplementary Materials: The following supporting information can be downloaded at: https://www.mdpi.com/article/10.3390/d15030389/s1, Figure S1: Trends for each site over all years. Table S1: Photographs from Montecristi region showing the effects of coral bleaching (2016) and hurricanes (September 2017).

Author Contributions: R.S.S. and R.T. selected research areas and monitoring protocol. R.S.S. and R.T. collected all data. R.S.S. performed most of the data analyses and wrote the first draft. All authors have read and agreed to the published version of the manuscript.

Funding: All funding was from PropaGas Foundation (Dominican Republic).

Institutional Review Board Statement: No organism was touched and thus does not require ethical approval of any kind.

Data Availability Statement: All data are available upon request to Steneck@maine.edu.

Acknowledgments: We received considerable help in the field from Ramón de León Barrios (Reef Support BV) Doug Rasher (Bigelow Laboratories for Ocean Sciences), students in Dominican Republic of Universidad Autónoma de Santo Domingo, Pontificia Universidad Católica Madre y Maestra, and el Instituto Tecnológico de Santo Domingo, and students in the University of Maine's School of Marine Sciences. Specifically (2017: Elise Hartill, Makaila Kowalsky, Katie Liberman, and Tiega Martin; 2019: Gretchen Grebe, Hannah Kerrigan, Mackenzie Menard, and Grace McDermott).

Conflicts of Interest: The authors declare no conflict of interest.

References

1. Hughes, T.P.; Graham, N.A.; Jackson, J.B.; Mumby, P.; Steneck, R.S. Rising to the challenge of sustaining coral reef resilience. *Trends Ecol. Evol.* **2010**, *25*, 633–642. [CrossRef]
2. Connell, J.H. Disturbance and recovery of coral assemblages. *Coral Reefs* **1997**, *16*, S101–S113. [CrossRef]
3. Steneck, R.S.; Arnold, S.N.; Boenish, R.; De León, R.; Mumby, P.J.; Rasher, D.B.; Wilson, M.W. Managing Recovery Resilience in Coral Reefs Against Climate-Induced Bleaching and Hurricanes: A 15 Year Case Study from Bonaire, Dutch Caribbean. *Front. Mar. Sci.* **2019**, *6*, 265. [CrossRef]
4. Jackson, J.B.C.; Donovan, M.K.; Cramer, K.L.; Lam, V.V. *Status and Trends of Caribbean Coral Reefs. Global Coral Reef Monitoring Network*; IUCN: Gland, Switzerland, 2014; pp. 1970–2012.
5. Graham, N.A.; Bellwood, D.R.; Cinner, J.E.; Hughes, T.P.; Norström, A.V.; Nyström, M. Managing resilience to reverse phase shifts in coral reefs. *Front. Ecol. Environ.* **2013**, *11*, 541–548. [CrossRef]
6. Steneck, R.S.; Mumby, P.J.; MacDonald, C.; Rasher, D.B.; Stoyle, G. Attenuating effects of ecosystem management on coral reefs. *Sci. Adv.* **2018**, *4*, eaao5493. [CrossRef]
7. Mumby, P.J.; Steneck, R.S.; Roff, G.; Paul, V.J. Marine reserves, fisheries ban, and 20 years of positive change in a coral reef ecosystem. *Conserv. Biol.* **2021**, *35*, 1473–1483. [CrossRef]
8. Mumby, P.J.; Steneck, R.S. Coral reef management and conservation in light of rapidly evolving ecological paradigms. *Trends Ecol. Evol.* **2008**, *23*, 555–563. [CrossRef]
9. Arnold, S.N.; Steneck, R.S.; Mumby, P. Running the gauntlet: Inhibitory effects of algal turfs on the processes of coral recruitment. *Mar. Ecol. Prog. Ser.* **2010**, *414*, 91–105. [CrossRef]
10. Lang, J.C.; Marks, K.W.; Kramer, P.A.; Kramer, P.R.; Ginsburg, R.N. 2010 AGRRA Protocols Version 5.4; Atlantic Gulf Rapid Reef Assessment. Available online: https://www.researchgate.net/profile/Philip-Kramer-3/publication/265148106_Agrra_protocols_version_54/links/5553ba6a08ae980ca6085ac3/Agrra-protocols-version-54.pdf (accessed on 6 March 2023).
11. Steneck, R.S.; Dethier, M.N. A Functional Group Approach to the Structure of Algal-Dominated Communities. *Oikos* **1994**, *69*, 476. [CrossRef]
12. Box, S.; Mumby, P. Effect of macroalgal competition on growth and survival of juvenile Caribbean corals. *Mar. Ecol. Prog. Ser.* **2007**, *342*, 139–149. [CrossRef]
13. Rasher, D.B.; Hay, M.E. Chemically rich seaweeds poison corals when not controlled by herbivores. *Proc. Natl. Acad. Sci. USA* **2010**, *107*, 9683–9688. [CrossRef]
14. Sale, P.F.; Douglas, W.A. Precision and accuracy of visual census technique for fish assemblages on coral patch reefs. *Environ. Biol. Fishes* **1981**, *6*, 333–339. [CrossRef]

15. Bohnsack, J.A.; Harper, D.E. *Length-Weight Relationships of Selected Marine Reef Fishes from the Southeastern United States and the Caribbean*; U.S. Department of Commerce, National Oceanic and Atmospheric Administration, National Marine Fisheries Service, Southeast Fisheries Center: Miami, FL, USA, 1988.

16. Baums, I.B.; Paris, C.B.; Chérubin, L.M. A bio-oceanographic filter to larval dispersal in a reef-building coral. *Limnol. Oceanogr.* **2006**, *51*, 1969–1981. [CrossRef]

17. Alvarez-Filip, L.; Estrada, N.; Pérez-Cervantes, E.; Molina-Hernández, A.; González-Barrios, F.J. A rapid spread of the stony coral tissue loss disease outbreak in the Mexican Caribbean. *PeerJ* **2019**, *7*, e8069. [CrossRef]

18. Croquer, A.; Zambrano, S.; King, S.; Reyes, A.; Sellares-Blanco, R.; Trinidad, A.V.; Villalpando, M.; Rodriguez-Jerez, Y.; Vargas, E.; Cortes-Useche, C.; et al. Stony Coral Tissue Loss Disease and Other Diseases Affect Adults and Recruits of Major Reef Builders at Different Spatial Scales in the Dominican Republic. *Gulf Caribb. Res.* **2022**, *33*, GCFI1–GCFI13. [CrossRef]

19. Steneck, R.; Arnold, S.; Mumby, P. Experiment mimics fishing on parrotfish: Insights on coral reef recovery and alternative attractors. *Mar. Ecol. Prog. Ser.* **2014**, *506*, 115–127. [CrossRef]

20. Alvarez-Filip, L.; Dulvy, N.; Gill, J.A.; Côté, I.M.; Watkinson, A.R. Flattening of Caribbean coral reefs: Region-wide declines in architectural complexity. *Proc. R. Soc. B Boil. Sci.* **2009**, *276*, 3019–3025. [CrossRef]

21. Levitan, D.R. Community Structure in Time Past: Influence of Human Fishing Pressure on Algal-Urchin Interactions. *Ecology* **1992**, *73*, 1597–1605. [CrossRef]

22. Steneck, R.S. Regular sea urchins as drivers of shallow benthic marine community structure. *Develop. Aquacult. Fish. Sci.* **2020**, *43*, 255–279. [CrossRef]

23. Mumby, P.J.; Harborne, A.R.; Brumbaugh, D.R. Grouper as a Natural Biocontrol of Invasive Lionfish. *PLoS ONE* **2011**, *6*, e21510. [CrossRef]

24. Carpenter, R.C.; Edmunds, P.J. Local and regional scale recovery of Diadema promotes recruitment of scleractinian corals. *Ecol. Lett.* **2006**, *9*, 271–280. [CrossRef] [PubMed]

25. Steneck, R.S.; McClanahan, T.; Arnold, S.N.; Brown, J.B. *A Report on the Status of the Coral Reefs of Bonaire in 2005 with Advice on a Monitoring Program*; Unpublished report to the Bonaire Marine National Park (STINAPA); University of Maine Darling Marine Center: Walpole, ME, USA, 2005.

26. Darling, E.S.; Alvarez-Filip, L.; Oliver, T.A.; McClanahan, T.R.; Côté, I.M. Evaluating life-history strategies of reef corals from species traits. *Ecol. Lett.* **2012**, *15*, 1378–1386. [CrossRef] [PubMed]

27. Alvarez-Filip, L.; Dulvy, N.K.; Côté, I.M.; Watkinson, A.R.; Gill, J.A. Coral identity underpins architectural complexity on Caribbean reefs. *Ecol. Appl.* **2011**, *21*, 2223–2231. [CrossRef] [PubMed]

28. Mumby, P.J.; Steneck, R.S. The Resilience of Coral Reefs and Its Implications for Reef Management. In *Coral Reefs: An Ecosystem. In Transition*; Springer: Berlin, Germany, 2010; pp. 509–519. [CrossRef]

Article

Fantastic Flatworms and Where to Find Them: Insights into Intertidal Polyclad Flatworm Distribution in Southeastern Australian Boulder Beaches

Louise Tosetto [1], Justin M. McNab [1,2], Pat A. Hutchings [1,3,*], Jorge Rodríguez [1,3] and Jane E. Williamson [1]

1   Marine Ecology Group, School of Natural Sciences, Macquarie University, Sydney, NSW 2109, Australia
2   Department of Applied Biosciences, Macquarie University, Sydney, NSW 2109, Australia
3   Australian Museum Research Institute, Australian Museum, Sydney, NSW 2010, Australia
*   Correspondence: pat.hutchings@australian.museum

Abstract: There is a rapid and extensive decline of our marine biodiversity due to human impacts. However, our ability to understand the extent of these effects is hindered by our lack of knowledge of the occurrence and ecology of some species groups. One such group of understudied organisms are marine flatworms of the order Polycladida, a conspicuous component of southeastern Australia's marine ecosystems that has received little attention over the years. Intertidal boulder beaches support a diverse range of polyclad flatworms in other countries, but the role of these environments in maintaining biodiversity is not well understood. In this study, we identified hotspots of flatworm occurrence by assessing the diversity and overall abundance of flatworms at boulder beaches along the southeast Australian coast. Bottle and Glass, Sydney Harbour, was found to be the most diverse site for flatworms. We also identified a higher occurrence of flatworms under large boulders and less exposed beaches and noted an increased presence of flatworms at higher latitudes. Probable influences on these patterns such as the requirement for shelter and protection are discussed. This study contributes to our knowledge of Australia's coastal biodiversity and can be used to assist in the management and conservation of our marine environments.

Keywords: Polycladida; ecology; biodiversity; marine ecosystems

Citation: Tosetto, L.; McNab, J.M.; Hutchings, P.A.; Rodríguez, J.; Williamson, J.E. Fantastic Flatworms and Where to Find Them: Insights into Intertidal Polyclad Flatworm Distribution in Southeastern Australian Boulder Beaches. *Diversity* **2023**, *15*, 393. https://doi.org/10.3390/d15030393

Academic Editor: Thomas J. Trott

Received: 20 January 2023
Revised: 6 March 2023
Accepted: 6 March 2023
Published: 9 March 2023

1. Introduction

Marine systems are facing substantial and rapid declines in biodiversity due to human impacts such as urbanisation, climate change, overharvesting and pollution [1–4]. One issue with understanding the extent and rate of decline is a basic lack of inventory of the diversity and complexity of marine communities and systems [5]. This is particularly pertinent for Australian marine systems where large gaps exist in our baseline knowledge of some species groups and their distributions. Such gaps hinder our ability to understand ecosystem functioning, and the magnitude of biodiversity loss in response to perturbations [6].

For rocky intertidal regions, the type of substratum can substantially influence the presence and success of the diversity of inhabiting organisms, particularly if the substratum is dynamic. Intertidal and shallow subtidal rocky areas of unstable rock substrata comprising pebbles, cobbles and boulders (referred to as 'boulder beaches' here) are dynamic in that they create disturbance events when the substrata move in response to high energy events [7,8]. These boulder beaches typically support a broad range of biota, including rare species [9–12]. While the ecological processes of animals inhabiting intertidal boulder beaches in southeastern Australia have been assessed for some species [13–15], the role of boulder beaches in maintaining biodiversity is not well understood in this area due to the lack of inventory of organisms, especially less abundant species. This is particularly concerning for this area of southeastern Australia, which is a known biodiversity hotspot for numerous other groups of marine intertidal organisms [16–19]. Many of these organisms are cryptic in habit, such as those occurring in crevices or under rocks, and/or with external features and colouration patterns that mimic the substrate they inhabit, making species identification difficult [20]. Such paucity in our understanding of community structure and functioning of these habitats in southeast Australia hinders development of ecologically driven resource management plans in such habitats vulnerable to anthropogenic perturbations [21].

One such group of understudied organisms are marine flatworms of the Order Polycladida. Polyclads have the potential to substantially impact their communities by predating on a range of invertebrates such as crustaceans, corals and molluscs, including some commercially valuable species, such as bivalves [22–25]. Recent studies have shown the rich biodiversity of polyclad flatworms in boulder beaches across southeastern Australian coasts [26,27]; however, their abundance and geographical distribution are only eclectic and poorly understood.

Records from overseas show that polyclad flatworms usually reside on the underside of rocks on boulder beaches [28–32]. In southeastern Australia, boulder beaches occur along the coast and in estuaries, presenting different levels of wave exposure [33]. Due to the abundance of these habitats and the lack of data on polyclads in temperate southeastern Australia, an assessment of the suitability of boulder beaches as habitat for polyclad flatworms was done. This is the first assessment of polyclad flatworm diversity and abundance in temperate Australian waters.

Here, we document the diversity and abundance of polyclad flatworms (hereafter called 'flatworms') in intertidal beaches along over 1367 km of the southeastern coastline of Australia and covering a range of beaches with different wave exposures. Specifically, the research asked the following questions: (1) Do flatworms inhabit the underside of boulders on these intertidal beaches? (2) Are there areas with higher species diversity (i.e., biodiversity hotspots) along the coastline than others? (3) Does the size of the boulder influence the abundance or diversity of flatworms? (4) Does beach exposure influence the relationship between rock size and the diversity and abundance of flatworms?

2. Materials and Methods

The demography of intertidal polyclad flatworms in boulder beaches along 532.45 km of the southeastern coastline of Australia, from 29°49′01.6″ S, 153°17′34.4″ E to 38°30′24.6″ S, 145°07′33.8″ E, was assessed (Figure 1). Boulder beaches, defined as those with a mean rock diameter of >256 mm [34], are distinct sedimentary coastal features with unique morphological characteristics [33,35], typically occurring in higher wave-energy environments [36]. Appropriate boulder beaches were selected by assessing maps and from discussions with other researchers and the local communities. Sites suitable for flatworms were then chosen based on the requirement of the presence of gravel to boulders (between 8.1 mm and 100 cm, respectively) and accessibility by foot at low tide (Figure 1).

Rocks on beaches suitable for flatworms were classified using a Wentworth scale according to an expanded version of Oak (1984) [33], as outlined in Table 1. Beach exposure was calculated using Baardseth's wave exposure index which involved counting the number of 9° sectors that contained a fetch of greater than 7.5 km [37,38]. In this index, beaches are classified from 0–9, where 0 is least exposed and 9 is most exposed. According to this classification system, Bottle and Glass, Chowder Bay and Shelly Beach were classified as index 0, Phillip Island and San Remo were classified as index 3, Port Macquarie was classified as index 4, Diggers Camp and Foster as index 8 and Boat Harbour was classified as index 9.

Figure 1. Locations surveyed during this study (**left**) and examples of boulder beaches: Boat Harbour, Gerroa, NSW (**top right**), Inverloch, VIC (**bottom right**). Photos: ESRI ArcGIS (**left**), Jorge Rodríguez (**right**).

Table 1. Overview of sampled rock size categories from boulder beaches in southeastern Australia.

Boulder Size Range (cm)	Wentworth Category
0.10–0.40	Coarse sand
0.41–0.80	Fine gravel
0.81–1.60	Coarse gravel
1.61–3.20	Medium gravel
3.21–6.40	Cobble
6.41–12.80	Coarse cobble
12.81–25.60	Small boulder
25.61–51.20	Medium boulder
51.21–102.40	Large boulder

Sampling each beach involved two hours of continuous searching in which rocks of varying sizes were lifted and the underside inspected for flatworms. The longest length of the rock was then measured as a proxy for boulder size (Table 1). Following inspection and measurement, the rocks were returned to their original upright position. Beaches were methodically sampled to ensure that rocks of all sizes were sampled, and only sampled once. Sampling started 1.5–2 h prior to a daytime low tide, depending on the weather conditions and tidal heights (Table 2). To characterise the range and abundance of boulder sizes on each beach (Figure 2), three radial 2 m plots were also assessed at a subset of the beaches (Boat Harbour, Bottle and Glass, Chowder Bay, Port Macquarie, San Remo and Shelly Beach, Eden), where the longest length of all bounders within the circumference of each plot was measured. Radial plots were chosen for their representativeness of the boulder beach habitat that had been searched. Finally, a range of other substrata at each beach were also searched for flatworms to ensure the flatworms collected were representative of the species occurring at that location.

Table 2. List of sites and sampling times.

State	Locality	Latitude	Longitude	Date	Sampling Time	Baardseth's Wave Exposure Index
New South Wales	Diggers Camp Beach, Diggers Camp	29°49′01.6″ S	153°17′34.4″ E	8 December 2019	12:00 p.m.–14:00 p.m.	8
	Shelly Beach, Port Macquarie	31°27′27.7″ S	152°56′04.4″ E	7 January 2020	12:30 p.m.–14:30 p.m.	4
	Pebbly Beach, Forster	32°10′46.0″ S	152°31′10.6″ E	6 December 2019	09:00 a.m.–11:00 a.m.	8
	Chowder Bay, Sydney Harbour	33°50′19.8″ S	151°15′16.2″ E	20 February 2020	13:30 p.m.–15:30 p.m.	0
	Bottle and Glass, Sydney Harbour	33°50′54.0″ S	151°16′13.1″ E	25 October 2019; 21 February 2020	12:00 p.m.–14:00 p.m.; 14:30 p.m.–16:30 p.m.	0
	Boat Harbour, Gerroa	34°45′02.0″ S	150°49′56.5″ E	7 July 2018	07:30 a.m.–8:30 a.m. *	9
	Shelly Beach, Eden	37°04′22.0″ S	149°54′45.6″ E	10 July 2018	11:20 a.m.–12:20 p.m. *	0
Victoria	San Remo	38°31′11.9″ S	145°22′02.2″ E	13 July 2018	07:30 a.m.–8:30 a.m. *	3
	Cats Bay, Phillip Island	38°30′24.6″ S	145°07′33.8″ E	12 July 2018	16:00 p.m.–17:00 p.m. *	3

* Sites sampled simultaneously by two teams during a one hour interval.

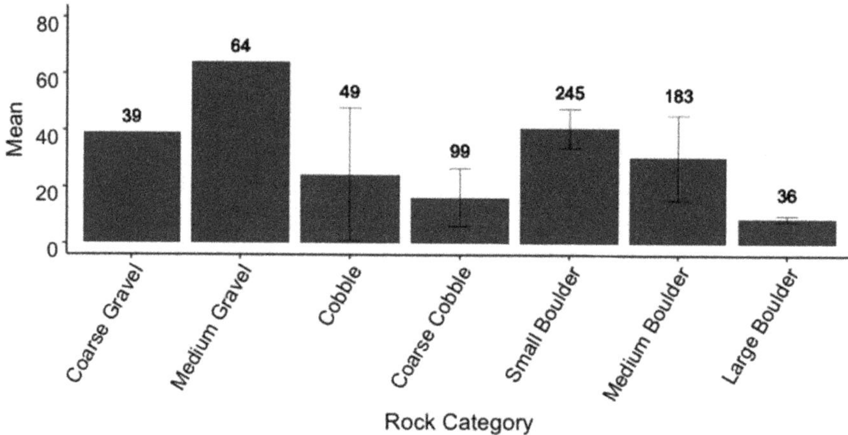

Figure 2. Mean (±SE) number of rocks measured in the representative radial plots in each of the boulder categories across the subset of sampled sites. Numbers above each bar refer to the frequency of rocks counted in each category. Refer to Table 1 for definitions of rock sizes. *Note: There is no SE for coarse and medium gravel as these rock sizes were only counted in the radial plots at Port Macquarie.*

Once a flatworm was seen, it was removed from the substratum with a fine paintbrush and kept in a separate container filled with seawater in a portable cooler for live transport to a fieldwork laboratory. This was necessary to ensure correct identification to species [26]. In the laboratory, animals were fixed with either 10% buffered formalin or Bouin's liquid after a small tissue sample was taken for sequencing, then stored in 70% ethanol. Once back at the University laboratory, specimens were dehydrated in an ethanol series from the original 70% to 90% and finally 100%, cleared in benzyl benzoate, embedded in paraffin wax using a Leica EG1150 H Paraffin Embedding Station, and sagittally sectioned in serial using an American Optical Spencer Rotary Microtome 820 at a thickness between 7 and 10 μm, depending on the size of the individual. Sections were stained with AZAN (trichrome staining method), mounted on glass slides in DPX (Dibutylphthalate Polystyrene Xylene) and observed and photographed under an Olympus BX53 compound microscope for species identification. For details regarding sequencing, see [26].

Statistical Analyses

All statistical analyses were completed in R [39]. Bottle and Glass was sampled at two timepoints (25 October 2019; 21 February 2020). To ascertain if data could be pooled for this site, we assessed differences in abundance between the two sampling timepoints using a generalised linear model (glm) [40]. Because flatworm abundance is count data, the model was fitted with a Poisson distribution. Flatworm abundance was included as the response variable and the sampling timepoint and boulder size included as the explanatory variables with an interaction term between them. There was no significant interaction ($p = 0.961$), nor was there any significance between the two timepoints ($p = 0.722$) or boulder size ($p = 0.401$), thus flatworm counts for the two sampling points at Bottle and Glass were pooled for any further abundance analysis.

To detect any beaches with increased biodiversity, three measures of alpha diversity were obtained: Shannon's H′, species richness and species evenness (Pielou's Eveness/J). Diversity measures for Shannon's H and richness were obtained using *diversity* and *specnumber* functions in the vegan package [41]. Pielou's evenness was obtained by dividing Shannon's H by its log. To explore relationships between alpha diversity (Shannon's H′) and environmental variables we ran simple linear models that included Shannon's H′ as the response variable and mean rock size and beach exposure as explanatory variables. We included an interaction term between rock size and exposure. Where the interaction term was not significant, the main effects in the model were analysed.

To identify any hotspots of occurrence in overall abundance of flatworms, a glm with Poisson distribution was constructed using the lme4 package [40]. Flatworm abundance was included as the response variable with site as the explanatory variable. Pairwise comparisons between sites were run using the *emmeans* function in the emmeans package [42] and the fitted model checked for overdispersion and excess zeros. To establish whether there was a relationship between flatworm abundance and boulder size, a glm with a Poisson distribution was fitted with the number of flatworms as the response variable and boulder size as the explanatory variable. To establish if there was a particular category of boulder that flatworms were more likely to be found on, a second glm was fitted with flatworm number as the response variable and rock size category (according to Table 1) as the explanatory variable. Pairwise comparisons between rock size categories were run using the emmeans function in the emmeans package. The models were checked for overdispersion and excess zeros.

To test whether beach exposure level influenced the distribution of flatworms on rock sizes, a glm with Poisson distribution was constructed with flatworm abundance as the response variable and the main effects of rock size and exposure, with an interaction between rock size and exposure. The model was not over-dispersed, nor did the fitted model have excess zeros compared to the number of zeros in the data. Significance of interactions were obtained from an analysis of deviance table obtained using a Chi-squared test. To interpret the coefficients of the rock size and exposure, we obtained slopes of rock size by exposure and pairwise differences between slopes using the *emtrends* function in the emmeans package. We used the *emmip* function to create the interaction plots of estimated marginal means based on the fitted model described above [42].

3. Results

3.1. Species of Flatworms Identified

We found that flatworms found on temperate boulder beaches inhabit the underside of boulders in all cases. No flatworms occurred on the top or sides of any of the rock size categories that we assessed in this study. A range of species from several families were identified as inhabiting intertidal boulder beaches in southeastern Australia (Table 3).

Species occurrence varied among species, both within and between families. For example, *Echinoplana celerrima* Haswell, 1907, occurred in all sites sampled except San Remo, whereas the closely related *Ceratoplana falconerae* Rodriguez et al., 2021, was extremely rare in our sampling and occurred at only one site. *Notoplana australis* (Schmarda, 1859) rarely occurred at sites but was common at the sites where it did occur (Table 3).

Table 3. Species and number of individuals found in intertidal boulder beaches in southeastern Australia.

Family	Species	Diggers Camp	Port Macquarie	Forster	Chowder Bay	Bottle and Glass	Boat Harbour	Shelly Eden	San Remo	Phillip Island
Acotylea										
Gnesiocerotidae	*Echinoplana celerrima* Haswell, 1907	6	10	7	4	9	4	4		4
	Ceratoplana falconerae Rodriguez et al., 2021									1
Notocomplanidae	*Notocomplana distincta* (Prudhoe, 1982)									1
Notoplanidae	*Notoplana australis* (Schmarda, 1859)							17		3
	Notoplana felis (Rodriguez et al., 2021)									1
	Notoplana longiducta Hyman, 1959				1	7				
Pseudostylochidae	*Tripylocelis typica* Haswell, 1907		2	2				2		
Planoceridae	*Planocera edmondsi* Prudhoe, 1982								1	
	Planocera sp.					1				
Stylochidae	*Leptostylochus victoriensis* Beveridge, 2017						2			
	Stylochus sp.				2	4				
Cotylea										
Cestoplanidae	*Cestoplana rubrocincta* (Grube, 1840)					2			2	
Euryleptididae	*Eurylepta* sp.					1				
Prosthiostomidae	*Enchiridium* sp.	1								
Pseudocerotidae	*Pseudoceros* sp.					1				

3.2. Hotspots of Abundance and Biodiversity

Polyclad flatworms occurred in all boulder beaches assessed. Bottle and Glass hosted the highest abundance of flatworms, with significantly higher numbers compared with Boat Harbour ($z = 3.803$, $p = 0.004$), Chowder Bay ($z = 3.076$, $p = 0.054$), Phillip Island ($z = 3.977$, $p = 0.002$), Port Macquarie ($z = 3.103$, $p = 0.049$) and San Remo ($z = 3.954$, $p = 0.002$). Shelly Beach, Eden, also had a significantly higher abundance of flatworms compared with San Remo ($z = 0.342$, $p = 0.018$) (Figure 3).

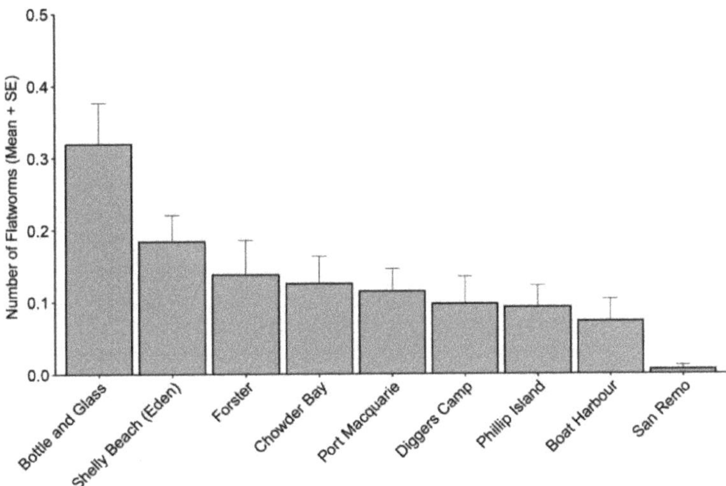

Figure 3. Differences in the average (+SE) abundance of flatworms found at each site sampled along the southeastern coast of Australia.

Patterns of flatworm diversity were similar to those of abundance. Bottle and Glass, Chowder Bay and Phillip Island all scored high across all diversity measures, while Boat Harbour, Diggers Camp, Forster, Macquarie Port, San Remo and Shelly Beach (Eden) presented lower diversity values (Figure 4).

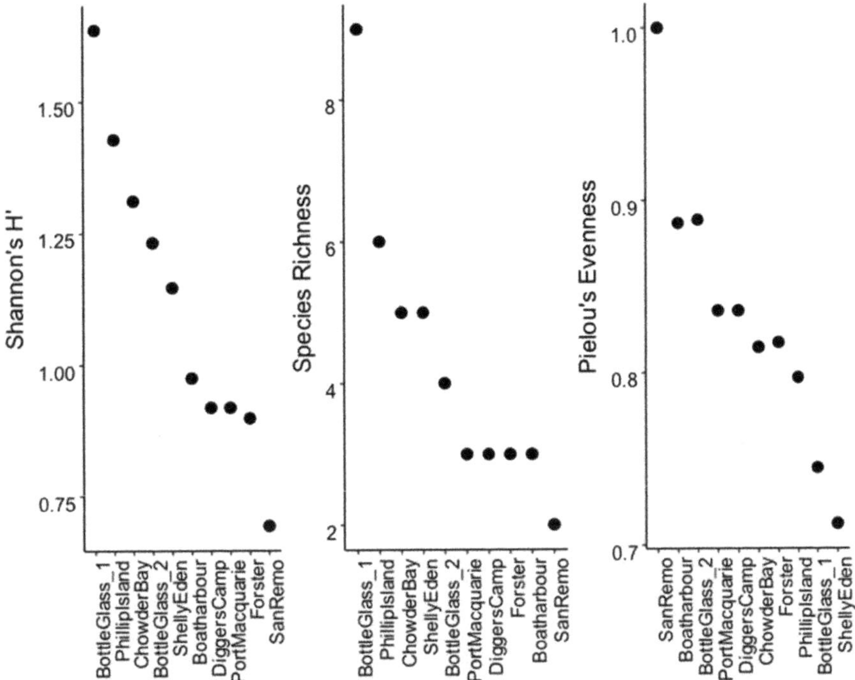

Figure 4. Different diversity indices for flatworms at each site. Bottle and Glass was sampled on two occasions and as such each sampling event has been separated in the figure.

3.3. Influence of Boulder Size and Beach Exposure on Flatworm Diversity

There was no significant interaction between mean rock size and exposure (F = 3.664, p = 0.214) on the diversity of flatworms. Examination of the main effects also showed no effect of mean rock size (F = 0.048, p = 0.838) or exposure (F = 0.551, p = 0.711) on flatworm diversity.

3.4. Relationship between Rock Size and Flatworm Abundance

We found that increasing boulder size had a significant effect on flatworm abundance (z = 5.543, p < 0.001). Across all sites, the large boulder category had significantly higher abundance of flatworms when compared with the medium boulder (z = 3.435, p = 0.011), small boulder (z = 5.869, p < 0.001) and coarse cobble (z = 4.141, p < 0.001) categories. There was no significant difference in the abundance of flatworms found between the coarse cobble, small boulders or medium boulders (Figure 5).

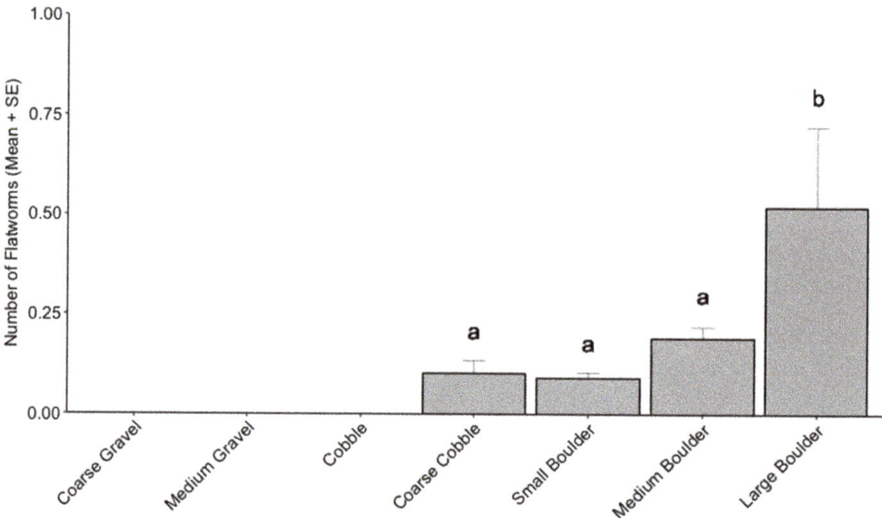

Figure 5. Average (+SE) number of flatworms under each rock size pooled across all boulder beaches. Differences in lower case letters indicate significant difference in flatworm abundance.

Within a beach, flatworm distribution did not mirror the relative distribution of rock sizes. For example, flatworms from Port Macquarie occurred under rock of medium boulder to coarse cobble size classes but not under the broad range of smaller rock sizes. No flatworms occurred under cobble, despite the size class occurring in four of the beaches and being the predominant rock category at Phillip Island and Shelly Beach, Eden.

3.5. Influence of Beach Exposure on Preferred Boulder Size for Flatworms

There was a significant interaction between rock size and exposure (F = 6.470, p = 0.0015) on the abundance of flatworms. The mean numbers of flatworms increased with boulder size on exposure level 3 when compared with exposure levels 8 and 9. There was no other significant interaction between exposure levels and boulder sizes (Figure 6).

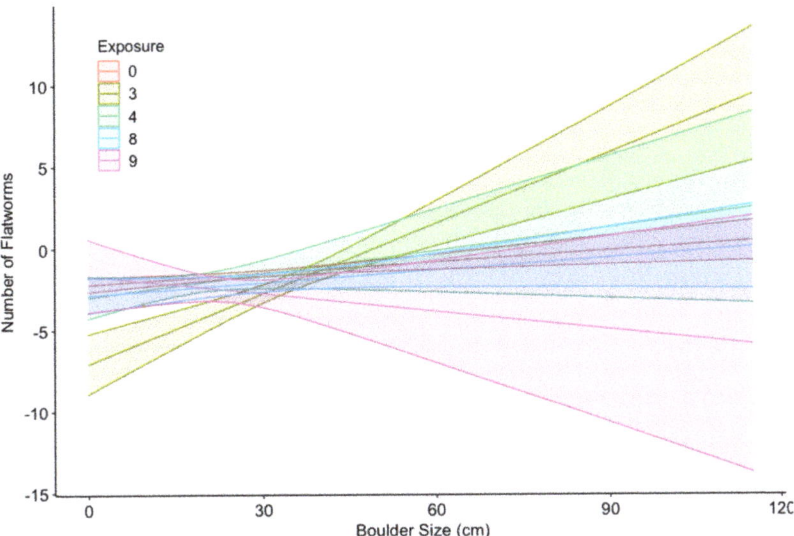

Figure 6. Interaction plot comparing the estimated marginal means of flatworms across boulder sizes and different levels of exposure (0 being a more protected beach and 9 being more exposed). Non-parallel lines indicate an interaction between factors. Ribbons represent 95% confidence interval limits.

4. Discussion

This is the first study to document the occurrence of polyclad flatworm species on intertidal boulder beaches in southeastern Australia, and the first to provide an ecological context for the distribution of polyclad flatworms in such habitats in the southern hemisphere. Fifteen species of flatworms from 10 families were observed on the boulder beaches, which spanned 9 degrees of latitude and 1367 km of coastline. Some species regularly occurred in the sampling, whereas the occurrence of others was rare. A higher number of flatworms occurred under rock sizes of medium boulders to coarse cobble, even though a diversity of other rock size classes occurred on the beaches sampled. We also observed that more flatworms occurred under larger rock size classes at more protected beaches.

Bottle and Glass from Sydney Harbour contained the highest abundance and the most flatworm diversity of all beaches sampled, and thus appears to be a hotspot for polyclad flatworms. Chowder Bay, less than 2 km away from Bottle and Glass in Sydney Harbour, also scored a high diversity measure; however, so did Phillip Island, which is over 1000 km away via the coastline. Sydney Harbour is a known global hotspot for marine and estuarine diversity, with a relative greater number of species and habitats represented than most of the harbours and estuaries in Australia and worldwide [43]. Sydney Harbour is a hotspot for other benthic invertebrate species including molluscs, crustaceans, polychaetes and echinoderms [16]; it is thus not surprising that a higher diversity and abundance of polyclad flatworms is found in these locations. While geographically distant from Sydney Harbour, Phillip Island resides on the edge of Bass Strait, a body of water also known for its unique biodiversity [2]. All three sites are more likely to be affected by anthropogenic perturbations than other sites sampled due to their proximity to major cities, making this diversity an interesting phenomenon leading to speculation as to whether polyclad flatworms in more disturbed areas have an advantage over other species in under-rock assemblages. Unfortunately, we were unable to directly assess this concept in the present study.

The concept of natural disturbance impacts on polyclad flatworms was examined by screening our research for impacts associated with wave exposure and rock size. Flatworms occurred under rock diameters of 6.4 cm or more at all beaches. Larger boulder sizes supported more individuals than smaller rocks, despite the greater abundance of smaller rock sizes compared to larger boulders. Since polyclad species found in the intertidal are usually small (5 to 30 mm body length) and fragile, their more common occurrence in larger boulders and less exposed beaches could be attributed to a need for shelter from high wave action. Larger boulders could also provide better protection against potential predators such as some fish and crustacean species [44,45]. Furthermore, many flatworms feed on sessile organisms [44,45], and it is possible that larger boulders may have greater abundance of their preferred prey. Future studies should investigate if distribution of flatworms is also related to the availability of sessile fauna on boulders. Most polyclad species may also be photophobic and actively hide from the light during the day, which would explain the higher occurrence of flatworms under larger boulders where it is presumably darker. Similar nocturnal behaviours have been reported from other intertidal and shallow subtidal marine invertebrates such as chitons [46], gastropods, sea urchins and sea cucumbers [47,48] that appear cryptic during the day and emerge at night-time to feed.

Our observations do not support predictions of the Intermediate Disturbance Hypothesis (IDH), which states that species diversity should be highest at intermediate levels of disturbance (i.e., intermediate boulder sizes) [49]. While the IDH traditionally refers to sessile organisms, polyclad flatworms are unlikely to move beyond the underside of their rocks in areas of high wave action (Rodriguez, personal observation) and can thus be considered semi-sessile in these situations. According to the IDH, highly disturbed areas, analogous to smaller rock sizes that would be more tossed about on exposed beaches, should support less diversity because organisms do not have the opportunity to successfully colonise in the harsh environment. Similarly, less disturbed areas, analogous to the underside of rocks that do not move, should harbour less diversity due to competitive exclusion by dominant species [50]. However, neither was the case in this study. Polyclad flatworms form only one part of the under-rock species assemblage at these beaches, and it is likely that more complex interactions at play may impact or mask effects of the IDH. More research on the movement of flatworms in relation to wave energy, and the composition of the under-rock communities are needed to tease apart such patterns.

Another obvious hypothesis to explain the absence of pattern associated with the IDH is the low sampling effort in this study. With the exception of Bottle and Glass, which was sampled twice, all sampling consisted of a single survey per site. It is highly possible that our snapshot of diversity was not at an appropriate temporal scale to measure such ecological patterns. We are confident that our biodiversity snapshot is rigorous. Patterns of biodiversity on other intertidal invertebrates have been successfully done along the Australian coastline using similar single standardised time searches as ours [51], and with experienced researchers as with our team. We are not confident that the diversity of flatworms is static over time; however, and that our diversity estimates are comprehensive. It is far more likely that our flatworm diversity estimates grossly underestimate the diversity of flatworms at each beach. It is well known that intertidal species vary in species occurrence and abundance over seasons and years [52]. We therefore propose a future study that assesses the diversity and abundance of flatworms on these boulder beaches at least seasonally over several years to glean an understanding of processes that may affect polyclad flatworm demography.

A trend of increasing abundance of flatworms with increasing latitude was observed in these data (see supplementary Figure S1). However, given that lower latitudes were sampled in summer time and higher latitudes sampled in winter, it is not possible to substantiate this trend until a more comprehensive study over multiple seasons is undertaken. The trends observed at different latitudes may be driven by the different seasons that sampling was undertaken. The eastern coast of Australia is susceptible to East Coast Lows, a dangerous weather system which can bring gales and heavy rain. While these low-pressure systems can occur at any time of the year, they are much more common in Autumn and Winter [53]. Given that storm intensity can influence intertidal communities [54,55], it is possible that flatworms are also affected by winter storms. While higher average population richness at higher latitudes has been observed in other studies [56], our results are biased by the increased presence of 17 *N. australis* in Shelly Beach (Eden). Further research needs to be done to ascertain if there is a trend in flatworm abundance and those of other under-boulder communities at higher latitudes or if there is a seasonal influence.

Echinoplana celerrima was the most common polyclad species found in intertidal boulder beaches on the southeastern Australia coast. This species occurred at all boulder beaches except for San Remo, while other polyclad species occurred in only one to three of each of the nine studied boulder beaches. *Echinoplana celerrima* presents the common acotylean body plan, with a small size (10 to 25 mm long, 5 to 10 mm wide), light brown colouration, eyes arranged in two elongate groups, ruffled pharynx located in the middle of the body and genital systems found in the posterior body third [26]. None of these characteristics indicate at first sight why this particular species is so successful in southeastern Australia boulder beaches compared to other taxa such as *Notocomplana longiducta* Hyman, 1959 or *Notocomplana distincta* (Prudhoe, 1982) which present similar anatomical traits and habits. Polyclad flatworms are generally highly selective in prey choice; however, some species exhibit different dietary preferences related to the abundance of suitable prey in a particular locality [57] and others have been reported to feed on a wide variety of invertebrates [58]. On shores where the preferred mussel prey is abundant, the Mediterranean flatworm *Stylochus mediterraneus* Galleni, 1976 feeds almost exclusively on these, while in locations where this primary prey species is rare and the oysters are widespread, flatworms feed on the latter ([57] and references within). It is thus possible that *E. celerrima* is an opportunistic predator and able to feed on a range of prey present, or switch between preferred prey species, compared to the other studied species. Without knowledge of prey preference and feeding habits for *E. celerrima* in relation to the other under-boulder species on the rocks they inhabit, it is difficult to understand any mechanisms underlying the relative importance of ecological processes such as feeding and competition on this group of species.

The only other location where *E. celerrima* has been documented to occur in addition to southeastern Australia is the Mediterranean and Black Sea. It is unclear whether the species occurs in areas between these two regions due to the lack of research targeted at flatworms. As this distance is so great, the most parsimonious explanation of this widespread occurrence is human-induced transportation between the two regions, possibly through ballast water, attached to the hull of a ship or carried with oysters or other animals. If *E. celerrima* is indeed an opportunistic predator with no strict preferred prey, it would explain its ability to settle away from its original habitat. Similar remarks were made for both *E. celerrima* and *Euplana gracilis* (Girard, 1850) [59]. Prior to being discovered in Port Phillip Bay (Victoria, Australia) by Prudhoe [59], *Euplana gracilis* was only described for the Atlantic coast of North America. Bennet and Pope [60] regarded the Victorian coast as a cold-temperate region, similar to that of the places where it was first found.

Many aspects of the biology of Australian polyclads remain unknown, hindering our ability to discern the processes driving their distribution patterns on boulder beaches. Characteristics such as the presence of a larval stage during the developing process, dietary habits, dispersion and seasonality could have major impacts on the distribution, richness and abundance of these species. Although the most common mode of development in polyclads is direct development (where the embryo develops directly into a form resembling the young adult), there are many species that develop indirectly through a planktonic phase with transient larval features [61]. Our knowledge on these matters is severely lacking; however, with the mode of development having been described for less than 8% of known polyclad species [61]. All of these characteristics are likely to have an impact on flatworm abundance and distribution, yet the lack of such knowledge hinders our ability to completely analyse these patterns.

While polyclad research has seen a resurgence in interest over the last decade, most studies are taxonomic and systematic in nature, or focus on natural products and other aspects of the flatworm's biology [62,63] and do not include data on the ecology of the investigated species. Future studies in southeastern Australian intertidal boulder beaches should focus on (1) continuing sampling of the boulder beaches to obtain an understanding of the temporal variability of flatworms, (2) assessing sessile fauna alongside flatworms to determine if there are similar distributions, (3) developing culturing techniques for flatworm larvae to close the life history loop for key species, (4) assessing the diets of key polyclad species, (5) gathering genetic data of all sampled species to study population connectivity at the intraspecific level, and (6) understanding the impacts of anthropogenic disturbances on under-boulder community diversity and abundance, including polyclad flatworms. Such information will create a strong baseline of information on polyclad flatworms and their communities, which can help inform conservation and management efforts of our coastal marine environments and contribute to our knowledge of Australia's biodiversity.

5. Conclusions

In this study, the abundance and diversity of polyclad flatworms across southeastern Australia is investigated. It is the first study to record the occurrence as well as the ecological context of flatworm species on intertidal beaches in this region. We identified 15 species of flatworms from 10 families on intertidal boulder beaches with hotspots of abundance and diversity at those sites most likely to be influenced by anthropogenic disturbance. There was higher abundance of flatworms on larger boulders at more protected beaches, which is possibly attributed to a need for shelter from high wave action, predation and daytime light. This study lacks a high degree of sampling effort over multiple time scales and future studies that assess abundance and diversity of flatworms on these beaches will obtain insights into processes driving their occurrence. Future directions for studies in southeastern Australian boulder beaches are provided so that a baseline of information on flatworms and their communities can be documented. This study is an important contribution to the knowledge of Australia's coastal marine systems.

Supplementary Materials: The following supporting information can be downloaded at https://www.mdpi.com/article/10.3390/d15030393/s1, Figure S1: Mean flatworms (estimated marginal means) across differing latitudes and boulder sizes.

Author Contributions: J.E.W., J.R. and P.A.H. conceived the study. J.E.W., P.A.H. and J.R. acquired the funding for the research. J.E.W., L.T., J.M.M. and J.R. conducted the fieldwork. L.T. and J.M.M. analysed the data. J.E.W. and J.R. wrote the first draft of the manuscript. L.T., J.E.W., P.A.H. and J.M.M. produced the final version of the manuscript. All authors have read and agreed to the published version of the manuscript.

Funding: This research was funded by the School of Natural Sciences at Macquarie University and the Australian Museum. Our thanks go to this institution for their continued support for biological research.

Institutional Review Board Statement: Not applicable.

Data Availability Statement: The datasets generated and analysed during the current study are our own. Datasets are available from the corresponding author on reasonable request.

Acknowledgments: Thanks to Audrey Falconer, Leon Altoff and the members of the Field Naturalists' Club of Victoria for their assistance collecting samples and their financial support through the FNCV Environment Fund. We thank Sue Lindsay for access to the facilities of the microscopy laboratory of Macquarie University and logistical support. We also extend our gratitude to Patrick Burke, Ryan Nevatte, Audrey Watson, Stephanie Bagala and Tristan Guillemin from Macquarie University and Anthony Abi-Saab from Port Macquarie for their help during fieldwork. The authors thank the School of Natural Sciences at Macquarie University for their institutional and financial support, and the Australian Museum Research Institute and the members of the Marine Invertebrates and Malacology Departments for providing access to their facilities and laboratories and assisting in fieldwork.

Conflicts of Interest: The authors declare no conflict of interest.

References

1. Sala, E.; Knowlton, N. Global marine biodiversity trends. *Annu. Rev. Environ. Resour.* **2006**, *31*, 93–122. [CrossRef]
2. Butler, A.J.; Rees, T.; Beesley, P.; Bax, N.J. Marine biodiversity in the Australian region. *PLoS ONE* **2010**, *5*, e11831. [CrossRef]
3. Dirzo, R.; Young, H.S.; Galetti, M.; Ceballos, G.; Isaac, N.J.B.; Collen, B. Defaunation in the Anthropocene. *Science* **2014**, *345*, 401–406. [CrossRef]
4. McCauley, D.J.; Pinsky, M.L.; Palumbi, S.R.; Estes, J.A.; Joyce, F.H.; Warner, R.R. Marine defaunation: Animal loss in the global ocean. *Science* **2015**, *347*, 6219. [CrossRef]
5. McClenachan, L.; Ferretti, F.; Baum, J.K. From archives to conservation: Why historical data are needed to set baselines for marine animals and ecosystems. *Conserv. Lett.* **2012**, *5*, 349–359. [CrossRef]
6. Hutchings, P.A. Potential loss of biodiversity and the critical importance of taxonomy—An Australian perspective. *Adv. Mar. Biol* **2021**, *88*, 3–16.
7. Sousa, W.P. Disturbance in marine intertidal boulder fields: The nonequilibrium maintenance of species diversity. *Ecology* **1979**, *60*, 1225–1239. [CrossRef]
8. McGuinness, K.A. Disturbance and organisms on boulders. II. Causes of patterns in diversity and abundance. *Oecologia* **1987**, *71*, 420–430. [CrossRef]
9. Chapman, M.G. Molluscs and echinoderms under boulders: Tests of generality of patterns of occurrence. *J. Exp. Mar. Biol. Ecol.* **2005**, *325*, 65–83. [CrossRef]
10. Chapman, M.G. Intertidal boulder-fields: A much neglected, but ecologically important, intertidal habitat. *Oceanogr. Mar. Biol. Ann. Rev.* **2017**, *55*, 35–53.
11. Chapman, M.G.; Underwood, A.J.; Browne, M.A. An assessment of the current usage of ecological engineering and reconciliation ecology in managing alterations to habitats in urban estuaries. *Ecol. Eng.* **2018**, *120*, 560–573. [CrossRef]
12. Liversage, K.; Kotta, J. Unveiling commonalities in understudied habitats of boulder-reefs: Life-history traits of the widespread invertebrate and algal inhabitants. *Mar. Biol. Res.* **2018**, *14*, 655–671. [CrossRef]
13. Smith, K.A.; Otway, N.M. Spatial and temporal patterns of abundance and the effects of disturbance on under-boulder chitons. *Molluscan Res.* **2013**, *18*, 43–57. [CrossRef]
14. Liversage, K.; Cole, V.J.; McQuaid, C.D.; Coleman, R.A. Intercontinental tests of the effects of habitat patch type on the distribution of chitons within and among patches in intertidal boulder field landscapes. *Mar. Biol.* **2012**, *159*, 2777–2786. [CrossRef]
15. Liversage, K.; Cole, V.; Coleman, R.; McQuaid, C. Availability of microhabitats explains a widespread pattern and informs theory on ecological engineering of boulder reefs. *J. Exp. Mar. Biol. Ecol.* **2017**, *489*, 36–42. [CrossRef]
16. Hutchings, P.A.; Ahyong, S.T.; Ashcroft, M.B.; McGrouther, M.A.; Reid, A.L. Sydney Harbour: Its diverse biodiversity. *Aust. Zool.* **2013**, *36*, 257–320. [CrossRef]
17. Liversage, K.; Benkendorff, K. A preliminary investigation of diversity, abundance, and distributional patterns of chitons in intertidal boulder fields of differing rock type in South Australia. *Molluscan Res.* **2013**, *33*, 24–33. [CrossRef]
18. Nimbs, M.; Willan, R.; Smith, S. Is Port Stephens, eastern Australia, a global hotspot for biodiversity of Aplysiidae (Gastropoda: Heterobranchia)? *Molluscan Res.* **2016**, *37*, 45–65. [CrossRef]
19. Ahyong, S.T. Biogeography of Australian Marine Invertebrates. In *Handbook of Australasian Biogeography*; CRC Press and Taylor & Francis: Boca Raton, FL, USA, 2017; pp. 81–99.
20. Nimbs, M.J.; Davis, T.R.; Holmes, S.P.; Hill, L.; Wehmeyer, S.; Prior, A.; Williamson, J.E. The Taming of Smeagol? A New Population and an Assessment of the Known Population of the Critically Endangered Pulmonate Gastropod *Smeagol hilaris* (Heterobranchia, Otinidae). *Diversity* **2023**, *15*, 86. [CrossRef]
21. Aldana, M.; Maturana, D.; Pulgar, J.; García-Huidobro, M.R. Predation and anthropogenic impact on community structure of boulder beaches. *Sci. Mar.* **2016**, *80*, 543–551. [CrossRef]
22. Merory, M.; Newman, L. A new stylochid flatworm (Platyhelminthes, Polycladida) from Victoria, Australia and observations on its biology. *J. Nat. Hist.* **2005**, *39*, 2581–2589. [CrossRef]
23. Lee, K.-M.; Beal, M.A.; Johnston, E.L. A new predatory flatworm (Platyhelminthes, Polycladida) from Botany Bay, New South Wales, Australia. *J. Nat. Hist.* **2006**, *39*, 3987–3995. [CrossRef]
24. Gammoudi, M.; Ahmed, R.B.; Bouriga, N.; Ben-Attia, M.; Harrath, A.H. Predation by the polyclad flatworm *Imogine mediterranea* on the cultivated mussel *Mytilus galloprovincialis* in Bizerta Lagoon (northern Tunisia). *Aquac. Res.* **2016**, *1*, 10. [CrossRef]
25. Bolton, D.K.; Clark, G.F.; Johnston, E.L. Novel in situ predator exclusion method reveals the relative effects of macro and mesopredators on sessile invertebrates in the field. *J. Exp. Mar. Biol. Ecol.* **2019**, *513*, 13–20. [CrossRef]
26. Rodríguez, J.; Hutchings, P.A.; Williamson, J.E. Biodiversity of intertidal marine flatworms (Polycladida, Platyhelminthes) in southeastern Australia. *Zootaxa* **2021**, *5024*, 1–63. [CrossRef]
27. Newman, L.J.; Cannon, L.R. *Marine Flatworms: The World of Polyclads*; CSIRO Publishing: Collingwood, Australia, 2003; 97p.
28. Holleman, J. Some New Zealand Polyclads (Platyhelminthes, Polycladida). *Zootaxa* **2007**, *1560*, 1–17. [CrossRef]
29. Bahia, J.; Padula, V.; Delgado, M. Five new records and morphological data of polyclad species (Platyhelminthes: Turbellaria) from Rio Grande do Norte, Northeastern Brazil. *Zootaxa* **2012**, *3170*, 31–44. [CrossRef]
30. Dixit, S.; Sivaperuman, C.; Raghunathan, C. Three new records of polyclad flatworms from India. *Mar. Biodivers. Rec.* **2015**, *8*, E29. [CrossRef]

31. Noreña, C.; Rodríguez, J.; Pérez, J.; Almon, B. New Acotylea (Polycladida, Platyhelminthes) from the east coast of the North Atlantic Ocean with special mention of the Iberian littoral. *Zootaxa* **2015**, *4039*, 157–172. [CrossRef]

32. Oya, Y.; Kajihara, H.A. New Species of *Phaenoplana* (Platyhelminthes: Polycladida) from the Ogasawara Islands. *Species Divers* **2019**, *24*, 1–6. [CrossRef]

33. Oak, H.L. The Boulder Beach: A Fundamentally Distinct Sedimentary Assemblage. *Ann. Am. Assoc. Geogr.* **1984**, *74*, 71–82. [CrossRef]

34. Wentworth, C.K. A scale of grade and class terms for clastic sediments. *J Geol.* **1922**, *30*, 377. [CrossRef]

35. Carter, R.W.G.; Orford, J.D. The morphodynamics of coarse clastic beaches and barriers: A short-and long-term perspective. *J. Coast. Res.* **1993**, *15*, 158–179.

36. McKenna, J. Boulder Beaches. In *Encyclopedia of Coastal Science*; Encyclopedia of Earth Science Series; Schwartz, M.L., Ed.; Springer: Dordrecht, The Netherlands, 2005. [CrossRef]

37. Baardseth, E.M. Square-scanning, two-stage sampling method of estimating seaweed quantities. *Rep. Norw. Inst Seaweed Res.* **1970**, *33*, 1–41.

38. Wernberg, T.; Connell, S.D. Physical disturbance and subtidal habitat structure on open rocky coasts: Effects of wave exposure, extent and intensity. *J. Sea Res.* **2008**, *59*, 237–248. [CrossRef]

39. RStudio Team. *RStudio: Integrated Development for R. RStudio*; PBC: Boston, MA, USA, 2020; Available online: http://www.rstudio.com/ (accessed on 10 January 2023).

40. Bates, D.; Maechler, M.; Bolker, B.; Walker, S. Fitting Linear Mixed-Effects Models Using lme4. *J. Stat. Softw.* **2015**, *67*, 1–48. [CrossRef]

41. Oksanen, J.; Guillaume Blanchet, F.; Friendly, M.; Kindt, R.; Legendre, P.; McGlinn, D.; Minchin, P.R.; O'Hara, R.B.; Simpson, G.L.; Solymos, P.; et al. Vegan: Community Ecology Package. R Package Version 2.5-7. 2020. Available online: https://CRAN.R-project.org/package=vegan (accessed on 10 January 2023).

42. Russell, V.; Lenth, R. emmeans: Estimated Marginal Means, aka Least-Squares Means. R Package Version 1.7.2. 2022. Available online: https://CRAN.R-project.org/package=emmeans (accessed on 10 January 2023).

43. Johnston, E.L.; Mayer-Pinto, M.; Hutchings, P.A.; Marzinelli, E.M.; Ahyong, S.T.; Birch, G.; Booth, D.J.; Creese, R.G.; Doblin, M.A.; Figueira, W.; et al. Sydney Harbour: What we do and do not know about a highly diverse estuary. *Mar. Freshw. Res.* **2015**, *66*, 1073–1087. [CrossRef]

44. Barton, J.; Humphrey, C.; Bourne, D.G.; Hutson, K.S. Biological controls to manage *Acropora* eating flatworms in coral aquaculture. *Aquac. Environ. Interact.* **2020**, *12*, 61–66. [CrossRef]

45. Itoi, S.; Sato, T.; Takei, M.; Yamada, R.; Ogata, R.; Oyama, H.; Teranishi, S.; Kishiki, A.; Wada, T.; Noguchi, K.; et al. The planocerid flatworm is a main supplier of toxin to tetrodotoxin-bearing fish juveniles. *Chemosphere* **2020**, *249*, 126217. [CrossRef]

46. Herrera, A.; Bustamante, R.H.; Shepherd, S. The fishery for endemic chitons in the Galapagos Islands. *Not. Galápagos* **2003**, *62*, 24–28.

47. Rogers, C.N.; Williamson, J.E.; Carson, D.G.; Steinberg, P.D. Diel vertical movement by mesograzers on seaweeds. *Mar. Ecol. Prog. Ser.* **1998**, *166*, 301–306. [CrossRef]

48. Shepherd, S.; Toral-Granda, V.; Edgar, G.J. Estimating the abundance of clustered and cryptic marine macro-invertebrates in the Galápagos with particular reference to sea cucumbers. *Not. Galápagos* **2003**, *62*, 36–39.

49. Connell, J.H. Diversity in Tropical Rain Forests and Coral Reefs. *Science* **1978**, *199*, 1302–1310. [CrossRef]

50. Dial, R.; Roughgarden, J. Theory of marine communities: The intermediate disturbance hypothesis. *Ecology* **1988**, *79*, 1412–1424. [CrossRef]

51. Benkendorff, K.; Davis, A.R. Identifying hotspots of molluscan species richness on rocky intertidal reefs. *Biodivers. Conserv.* **2002**, *11*, 1959–1973. [CrossRef]

52. Underwood, A.J.; Chapman, M.G. Spatial analyses of intertidal assemblages on sheltered rocky shores. *Aust. J. Ecol.* **1998**, *23*, 138–157. [CrossRef]

53. Pepler, A.; Coutts-Smith, A.; Timbal, B. The role of East Coast Lows on rainfall patterns and inter-annual variability across the East Coast of Australia. *Int. J. Climatol.* **2014**, *34*, 1011–1021. [CrossRef]

54. Corte, G.N.; Schlacher, T.A.; Checon, H.H.; Barboza, C.A.; Siegle, E.; Coleman, R.A.; Amaral, A.C.Z. Storm effects on intertidal invertebrates: Increased beta diversity of few individuals and species. *PeerJ* **2017**, *5*, e3360. [CrossRef]

55. Mieszkowska, N.; Burrows, M.T.; Hawkins, S.J.; Sugden, H. Impacts of pervasive climate change and extreme events on rocky intertidal communities: Evidence from long-term data. *Front. Mar. Sci.* **2021**, *8*, 642764. [CrossRef]

56. Lawrence, E.R.; Fraser, D.J. Latitudinal biodiversity gradients at three levels: Linking species richness, population richness and genetic diversity. *Glob. Ecol. Biogeogr.* **2020**, *29*, 770–788. [CrossRef]

57. Galleni, L.; Tongiorgi, P.; Ferrero, E.; Salghetti, U. *Stylochus mediterraneus* (Turbellaria: Polycladida), predator on the mussel Mytilus galloprovincialis. *Mar. Biol.* **1980**, *55*, 317–326.

58. Rawlinson, K.A.; Gillis, J.A.; Billings, R.E.; Borneman, E.H. Taxonomy and life history of the *Acropora*-eating polyclad flatworm: *Amakusaplana acroporae* nov. sp. (Polycladida, Prosthiostomidae). *Coral Reefs* **2011**, *30*, 693–705. [CrossRef]

59. Prudhoe, S. Polyclad turbellarians from the southern coasts of Australia. *Rec. Aust. Mus.* **1982**, *18*, 361–384.

60. Bennett, I.; Pope, E.C. Intertidal zonation of the exposed rocky shores of Victoria, together with a rearrangement of the biogeographical provinces of temperate Australian shores. *Aust. J. Mar. Freshw. Res.* **1960**, *4*, 105–159. [CrossRef]

61. Rawlinson, K.A. The diversity, development and evolution of polyclad flatworm larvae. *EvoDevo* **2014**, *5*, 9. [CrossRef]
62. McNab, J.M.; Rodríguez, J.; Karuso, P.; Williamson, J.E. Natural products in polyclad flatworms. *Mar. Drugs* **2021**, *19*, 47. [CrossRef]
63. McNab, J.M.; Briggs, M.T.; Williamson, J.E.; Hoffmann, P.; Rodriguez, J.; Karuso, P. Structural Characterization and Spatial Mapping of Tetrodotoxins in Australian Polyclads. *Mar. Drugs* **2022**, *20*, 788. [CrossRef]

Article

Large-Scale Variation in Diversity of Biomass-Dominating Key Bryozoan Species in the Seas of the Eurasian Sector of the Arctic

Nina V. Denisenko * and Stanislav G. Denisenko

Zoological Institute of the Russian Academy of Sciences (ZIN RAS), Saint Petersburg 199034, Russia
* Correspondence: ndenisenko@zin.ru

Abstract: An analysis of archival and literary materials, as well as recently collected data in coastal areas at 14 locations in the Eurasian seas showed that the diversity of biomass-dominating key bryozoan species is low, totaling 26 species, less than 1/15 of the total bryozoan fauna richness. Their number decreases eastward from 17 species with an average total biomass of >16 g/m^2 in the Barents Sea to three species with an average biomass of about 3 g/m^2 in the East Siberian Sea. In the Chukchi Sea, their number and average biomass increase to 10 species and ~12 g/m^2, respectively. Average biomass strongly correlates with the number of species in each sea. Furthermore, variation in biomass is significantly correlated with the composition of bottom sediments and, in some locations, with depth. The marked decrease in the number of key species along the vector from Barents→Kara→Laptev→East Siberian Sea is due to a decline in the number of boreal and boreal–Arctic bryozoans of Atlantic origin. In contrast, the appearance of boreal and boreal–Arctic Pacific species is responsible for the increase in key species in the Chukchi Sea.

Keywords: bryozoa; biomass-dominating key species; biogeographic affiliation; distribution; depth; sediments grain size; coastal area; Arctic seas

1. Introduction

Bryozoans, such as polychaetes, crustaceans and molluscs, are one of the most diverse taxonomic groups of the Arctic [1], but their role in the formation of the total zoobenthos bioresources in this region of the world ocean has been a matter of contention. Some workers believe that this entire group has only patchy distributed aggregations in the Arctic seas [2–8]. The results of the studies conducted over a broad timeframe, however, indicate that stable mass aggregations of bryozoans have existed in the Arctic seas over a long period of time [9–13]. Furthermore, some workers argue that bryozoans are one of the background groups that contribute significantly to the total biomass of zoobenthos communities on the continental shelf [12–18]. These diametrically opposed views on the importance of bryozoans in zoobenthos communities have provided a stimulus for quantitatively assessing the distribution and abundance of bryozoans on the continental shelf of the Barents Sea [19]. Mapping of then-available data showed that an increase in the total biomass of this group tended to coincide with sea regions having bottom sediments dominated by stones, gravel, and other coarse-grained fractions, a relationship confirmed by statistical analysis [19].

It was therefore established by the early 1990s that the distribution and abundance of bryozoans in the open part of the Barents Sea is highly variable and controlled by environmental parameters [11,14]. However, insufficient information about bryozoans in the coastal regions of seas where environmental parameters undergo the most abrupt changes hindered the identification of such patterns in shallow waters.

Several expeditions conducted by the Zoological Institute of the Russian Academy of Sciences in the 1970–2000s collected data on bryozoan biomass in the coastal regions of the Eurasian seas. Some of this information has been published, but the results presented

Citation: Denisenko, N.V.; Denisenko, S.G. Large-Scale Variation in Diversity of Biomass-Dominating Key Bryozoan Species in the Seas of the Eurasian Sector of the Arctic. *Diversity* **2023**, *15*, 604. https://doi.org/10.3390/d15050604

Academic Editors: Thomas J. Trott and Bert W. Hoeksema

Received: 19 March 2023
Revised: 24 April 2023
Accepted: 26 April 2023
Published: 28 April 2023

in specialized studies on bryozoans were mostly concerned with their biogeographical and species composition [20–24]. Some observations on the quantitative representation of this group can be found in papers dealing with descriptions of benthic communities in general [25–29], but this information is insufficient to identify possible causes that can explain changes in the distribution and biomass of bryozoans.

The purpose of this study was to characterize the distribution and biomass of bryozoans in the Eurasian Arctic seas based on an integrated analysis of the available archival and literature data together with recently collected biomass data from surveys conducted in the upper littoral of the coastal regions. We hypothesize that both the diversity and biomass of key bryozoan species, species represented by a biomass >1 g/m^2, must decrease eastward in parallel with similar changes in species richness of the overall bryozoan fauna and that the key bryozoan species that have originated in the Atlantic are replaced by Arctic or Pacific species.

2. Materials and Methods

2.1. Study Area

The study area covered 14 coastal localities in the five seas of the Arctic, which vary in their environmental conditions [30]: Barents, Kara, Laptev, East Siberian, and Chukchi (Figure 1). Three of these locations were situated in the Barents Sea. One of these, Yarnyshnaya Bay, lies in the south-western part of the sea and is greatly influenced by Atlantic waters carried from the west with the Murmansk near-shore current [31]. Sea water temperature in this bay is normally above zero throughout the year, with surface water warming to 7–9 °C in summer while not exceeding 4–5 °C deeper in the water column. Salinity in the bay is 33.4–34.5 psu, which is close to that in the ocean [32]. Bottom sediment composition varies greatly from the mouth to the head of the bay and from the littoral to deeper areas, ranging from stony bottoms to soft silty sands (Table 1) [33].

The second Barents Sea location is in the waters of the Franz Josef Land Archipelago (FJL), in particular off Hayes Island and the straits around it (Figure 1). Arctic waters influence this area with negative sea water temperatures throughout the year at depths below 5 m and a steady summer temperature of +2 °C in the surface water layers [30,32]. Bottom sediment composition at sampling sites varies from boulder-covered bottoms to silty sands or mixed gravel–sandy–clayey sediments in the upper littoral [25,33] (Table 1).

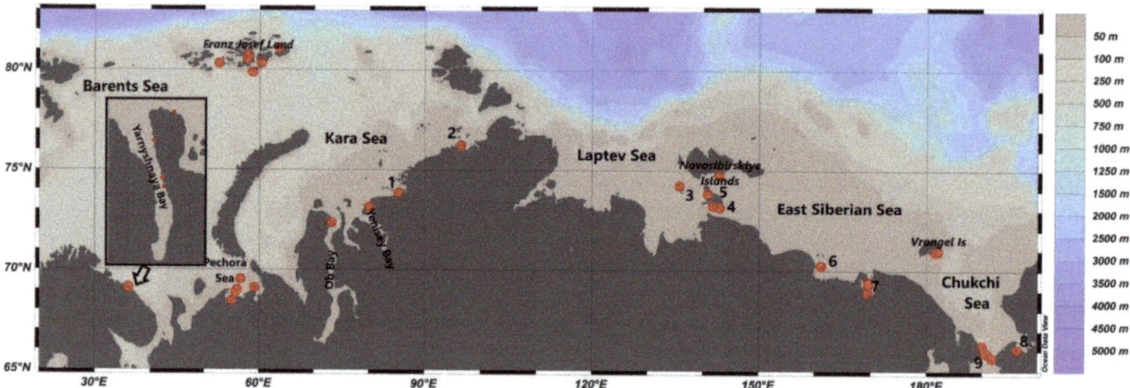

Figure 1. Map of the study area. Sample locations indicated in red. Numbers on the map correspond to the names of locations. 1—Pyasina Bay; 2—Nordenskjold Archipelago; 3—Stolbovoy Island; 4—Large and Small Liakhovskiye Islands; 5—Bunge Land Island; 6—East Siberian Sea coastal area; 7—Chaun Bay; 8—Kotzebue Sound; 9—Bering Strait.

Table 1. Sampling stations and environmental parameters in the study area. Sediment type: C, clay; M, mud; Ms, muddy sand; Fs, fine sand; S, sand; Sis, silt with shells; s, shells; SiS, silty sand; G, gravel; P, pebble; B, boulders; R, rock.

Expedition/ Research Vessel	Date	Station	Long.	Lat.	Gear	Sample Area (m^2)	Samples No.	Depth (m)	Sediments	°C	Bottom Salinity (psu)
Barents Sea											
Franz Josef Land area											
R/V Dalniye Zelentsy	20 August 1992	4	63.5383	81.0899	Grab	0.25	3	75	Ms	−0.2	34.9
R/V Dalniye Zelentsy	20 August 1992	6	58.6763	79.8749	—»—	0.25	3	43	PSS	−0.5	34.9
R/V Dalniye Zelentsy	20 August 1992	8	57.8913	80.7363	—»—	0.25	3	29	FsP	−0.2	34.9
R/V Dalniye Zelentsy	25 August 1992	15	60.3333	80.3333	—»—	0.25	3	37	Sis	−0.2	34.9
R/V Dalniye Zelentsy	24 August 1992	16	58.8263	79.8753	—»—	0.25	3	43	PsM	−0.2	34.9
R/V Dalniye Zelentsy	25 August 1992	17	58.0883	80.6186	—»—	0.25	3	27	PsM	−0.2	34.9
R/V Dalniye Zelentsy	26 August 1992	18	52.6333	80.3166	—»—	0.25	3	29	P	−0.2	34.9
Hayes Island	25 July 1984	Section 1	57.6688	80.5258	Quadrat	0.1	9	0–15	P	0.4	33.8
Pechora Sea coastal area											
R/V J. Smirnitskiy	10 August 1995	7	55.431	68.362	Box corer	0.1	3	7	MsC	10.1	18
R/V Geophizik	23 June 1995	1	56.0075	69.0044	Grab	0.1	3	7	Ms	4.1	21.4
R/V Geophizik	23 June 1995	2	55.7098	68.9035	—»—	0.1	3	9.7	Fs	4.3	27.1
R/V Dalniye Zelentsy	27 September 1999	4	56.6063	69.5351	—»—	0.1	3	27	Ms	4.5	29.1
R/V Prof. Kuznetsov	6 September 2016	10	59.1688	69.1186	—»—	0.1	3	11	GS	3.2	32.7
SW Barents Sea											
Yarnyshnaya Bay	9 July 1987	Section 2	36.0537	69.1123	Quadrat	0.1	12	0–25	BP	5.0	33.5
Yarnyshnaya Bay	10 August 1987	Section 3	36.0541	69.1121	—»—	0.1	15	0–65	BP, S	5.0	33.5
Yarnyshnaya Bay	20 July 1987	Section 4	36.0461	69.1258	—»—	0.1	12	0–25	RB	5.0	33.5
Yarnyshnaya Bay	1 August 1987	Section 5	36.0633	69.1350	—»—	0.1	10	0–20	BP	5.0	33.5
Kara Sea											
R/V J. Smirnitskiy	11 September 1995	80	96.417	76.263	Box corer	0.1	3	32	GSS	0.2	30.2
R/V J. Smirnitskiy	12 September 1995	82	85.280	73.591	—»—	0.1	3	17	CS	5.5	17
R/V J. Smirnitskiy	14 September 1995	83	80	73.17	—»—	0.1	3	32	CS	−0.4	30
R/V J. Smirnitskiy	16 September 1995	92	73.198	72.367	—»—	0.1	3	19	CS	1.5	25.5
Laptev Sea											
Golikov et al., 1990 [26]	August 1973	Lyakhovskiye Islands	141.7132	73.2446	Quadrat	0.1	3	5	SMP	−1.2	
Golikov et al., 1990 [26]	August 1973	—«—	142.9409	73.1668	—»—	0.1	3	11	SMP	−1.2	
Golikov et al., 1990 [26]	August 1973	—«—	140.7986	73.8596	—»—	0.1	3	10	SMP	−1.2	
Golikov et al., 1990 [26]	August 1973	Zemlia Bunge	142.8732	74.7953	—»—	0.1	3	15	SMP	−1.2	
Golikov et al., 1990 [26]	August 1973	Stolbovoy Isl	135.7764	74.2733	—»—	0.1	3	22–25	SMP	−1.2	
Golikov et al., 1990 [26]	August 1973	—«—	135.7764	74.2733	—»—	0.1	3	12–15	PG	−1.2	

Table 1. *Cont.*

Expedition/ Research Vessel	Date	Station	Long.	Lat.	Gear	Sample Area (m^2)	Samples No.	Depth (m)	Sediments	°C	Bottom Salinity (psu)
East Siberian Sea											
R/V J. Smirnitskiy	24 August 1995	34	169.513	69.476	Box corer	0.1	3	15	SiS	−0.2	34
R/V Ivan Kireev	5 September 2004	66	161.122	70.27	Grab	0.1	3	9	SiS	1.9	22.9
R/V Ivan Kireev	7 September 2004	81	169.3877	68.5943	—»—	0.1	3	13	SiS	3.0	20.8
R/V Ivan Kireev	8 September 2004	88	169.5527	69.3318	—»—	0.1	3	8	SiS	2.5	19.6
Chukchi Sea											
Feder, Jevet 1978 [34])	11 September 1976	Kotzebue Snd	197.3215	66.1046	Grab	0.1		25	SiS		
R/V Alpha Helix	1 September 1995	33	190.971	66.3002	Grab	0.1	3	45	SiS		
Vrangel Isl	15 August 1976	Section 1	182	71	Quadrat	0.1; 1.0	9	5–20	GM	0.5	29.1
Vrangel Isl	20 August 1976	Section 2	181.5333	70.9666	—»—	0.1; 1.0	8	4–25	PM	−1.2	28.0
Vrangel Isl	25 August 1976	Section 3	181.5333	70.9666	—»—	0.1	4	2–5	PS	0.4	30.1
R/V Sever	22 August 2005	6	191.7042	65.6777	Grab	0.1	3	49	PS	2.6	32.7
R/V Sever	22 August 2005	8	190.9167	65.8781	—»—	0.1	3	54	PS	2.7	32.1
R/V Sever	22 August 2005	10	190.3767	66.0005	—»—	0.1	3	53.7	PS	4.6	31.9

The third locality is in the south-eastern Barents Sea (Pechora Sea) near the mainland, in shallow waters with depths less than 27 m. It is influenced by transformed Atlantic waters desalinated by the runoff from the Pechora River and is characterized by sandy–fine-grained sandy–clayey sediments. The water is strongly mixed by winds and tidal currents. Negative temperatures persist throughout the entire water column in winter and warm to 5–8 °C during the summer [35].

Unlike the positions of the Barents Sea locations, all the Kara Sea ones are in its south. The first two are situated across the Ob' and Yenisei gulfs with depths of up to 35 m. They are strongly affected by freshwater runoff from the Ob' and Yenisei rivers. Consequently, the salinity in the near-bottom water layer does not exceed 21 psu [36]. The gradient of near-bottom water temperature ranges from −2 to −1 °C at depths as shallow as less than 5 m [36]. Because of the significant amount of organic and inorganic matter transported by rivers, the bottom sediments are relatively soft and consist primarily of fine-grained sand with river silt deposits [27].

The third and fourth locations are in the south-eastern part of the Kara Sea: Pyasina Bay and south of the Nordenskjold Archipelago. Coarse-grained rocks with a predominance of gravel comprise bottom sediments at both locations [33]. Water temperature gradients are similar to those in the Ob' and Yenisei bays, but salinity is much higher and exceeds 32 psu (Table 1) [36].

The Laptev Sea includes three study areas within the Severnaya Zemlya Archipelago that have non-uniform habitats characterized by soft sandy–aleuritic deposits mixed with clay alternating with gravel–rocky ridges [26]. As in the Kara Sea, the near-bottom salinity varies greatly from 17 to 32.5 psu depending on depth and season of the year, reaching maximum values in winter [30]. Near-bottom water temperature in the southern part of the sea varies just within several degrees, ranging from −1.8 to 1 °C [30].

The habitats in the localities chosen for the East Siberian Sea are very similar to those in the Laptev Sea. They are characterized by the predominance of soft-bottom deposits alternating with coarse-grained rocks that are patchily distributed [33]. Near-bottom water temperatures usually remain subzero throughout the year, but the surface water layer in the shallow Chaun Bay sometimes warms up in summer to 10 °C. Salinity does not rise above 32 psu [37].

The combined impact of Arctic and Pacific water masses causes a greater variation in habitats across the Chukchi Sea [38] and determines the oceanographic parameters of the water column. Near-bottom temperatures in the southern localities of the Bering Strait are above zero throughout the year (3–5 °C), and salinity exceeds 32 psu. Bottom sediments are coarse grained [33]. The northern locality lies near Wrangel Island in waters that are typical of the Arctic: summer surface water temperature does not exceed 2 °C, and at depths below 5 m in summer and throughout the water column in winter, water temperature remains below zero. Salinity stays about 32 psu throughout the year. In Kotzebue Sound (Alaska Peninsula waters of the SE Chukchi Sea) and the coastal waters of the Chukchi Peninsula, the temperature regime of the coastal waters is close to that of Bering Strait, but the composition of bottom sediments varies, consisting primarily of silty sands with an admixture of stones [38].

2.2. Material

The dataset for this study was assembled from the authors' own research, published literature [27–29], and from collections and data of catalogs of the Zoological Institute of the Russian Academy of Sciences (Saint Petersburg, Russia) archived from 1983 to 2014 and collected prior to the onset of steady warming of the Arctic, which began in 2010 [39]. The dataset included information on bryozoans from the 14 coastal localities in the Barents, Kara, Laptev, East Siberian, and Chukchi seas. In total, the material examined comprised 177 samples collected from 33 stations and 5 transects. Sampling depths ranged from the littoral to 55 m.

Sampling sites differed both in temperature and depth, as well as in the bottom sediment composition (Table 1). On transects, three replicates were collected at depths ranging from 5 to approximately 25 m with 0.1 m^2 or 1 m^2 quadrats using SCUBA with divers deployed from a rubber boat (Table 1). At the stations located far from shore (Figure 1), the material was collected from research vessels by van Veen (sampled surface: 0.1 m^2) or a Russian modification of the Petersen grab called "Ocean" (sampled surface: 0.25 m^2) grab samplers, or box core (sampled surface: 0.1 m^2). Three replicate grab samples were taken at each station, washed through a 1 mm mesh sieve to remove the substrate, and then fixed with 4% formaldehyde in sodium tetraborate. In the laboratory, bryozoans were separated from the remaining substrate, identified, counted, weighed, and preserved in 75% ethanol. In some cases, colonies were not counted because flexible, bush-like colonies are often interlaced and clumped together in a dense mass, making their counting difficult.

Biomass was used to measure the abundance of all species. To determine biomass, bryozoans were removed from their substrate and weighed. Encrusting species firmly attached to substrates were scraped off and weighed. Bush-like bryozoans were detached at their point of attachment and weighed. All weights include colony exoskeletons. Among samples, species with a biomass >1 g/m^2 were designated as key species in the bryozoan taxocenes. The number of colonies of each of these species was counted and used as a second estimate of abundance. In a few cases, colonies of matted bush-like species were interlaced, so the entire clump was treated as a single colony. The study used valid species names according to the World Register of Marine Species (WoRMS) (https://www.marinespecies.org, 20 March 2023 accessed). Information on the biogeographic affiliation of bryozoan species was taken from the literature [19,23,24,40,41].

2.3. Analytical Methods

Average total biomass, standard error (SE), and standard deviation (SD) were used to estimate variation among different seas and variation among individual masses of colonies of a species. The total average biomass of key bryozoan species in each sea was calculated as the sum of their weights among samples within a sea divided by the number of those samples. The average biomass of each key bryozoan species within each sea was calculated individually as the sum of its biomass among all locations divided by the total number of samples within a sea.

Relationships between the biomass and selected environmental variables were evaluated using the ridge multiple regression model that minimizes collinearity between the predictor variables. The analysis was performed using Statistica 6.0 (StatSoft Russia, Moscow, Russia). The Ocean data view (AWI) software package was used to draw a map of the study region.

The Chekanovsky–Sørensen similarity index (Cz) [42–44] was used to assess the similarity among samples using data of species presence/absence. For each resulting dendrogram, locations that were less than 50% similar were regarded as distinct. Faunistic differences were also evaluated using the pair-group method with arithmetic means [45–47]. Statistically significant differences ($p < 0.05$) within and among seas in the resulting groups of key species composition were evaluated using the one-way ANOSIM test (Primer 6) [48].

3. Results

3.1. Composition of Biomass-Ranked Key Bryozoan Species

In total, 26 bryozoan species were found in the coastal regions of the Eurasian seas, with their biomass in assemblages exceeding 1 g/m^2 (Table 2). All species belong to the class Gymnolaemata, one-fifth of which are representatives of the order Ctenostomatida and lack the calcareous skeleton, while the remaining species belong to the order Cheilostomatida, whose colonies are calcified to a varying degree. The taxonomic diversity of biomass-ranked key bryozoan species ranged from three in the East Siberian Sea to 17 species in the Barents Sea (Figure 2). The diversity of key species in the regarded areas is influenced to a certain extent by the water temperature in the near-bottom layer because the correlation is very close to significant (Pearson correlation where R = 0.63; *p* = 0.05) (Figure 3).

The cluster analysis based on comparisons of Cz similarity indices on data of species presence/absence collected at different localities of the Barents Sea indicated the key bryozoan species composition in the south-western (Yarnyshnaya Bay), south-eastern (coastal areas of the Pechora Sea), and north-eastern (FJL) parts of this sea are distinct because the differences in similarity among them exceeded 60% (Figure 4).

Table 2. Individual average biomass (g/m^2) with standard deviation (B ± SD) and abundance (col/m^2) with standard deviation (D ± SD), and biogeographic affiliation of biomass-dominating key bryozoan species within each Eurasian sea of the Arctic. A—Arctic species occurring in Arctic waters; B, At and BA, At—boreal and boreal–Arctic species of Atlantic origin occurring in the Atlantic Ocean and in the Atlantic waters in the Arctic; B, P and BA, P—boreal and boreal–Arctic species of Pacific origin occurring in Pacific Ocean and in Pacific waters in the Arctic; BA, ws—widespread boreal–Arctic species occurring in both Atlantic and Pacific Oceans.

Taxon	Biogeo-Graphic Affiliation	Barents Sea		Kara Sea		Laptev Sea		East Siberian Sea		Chukchi Sea	
Order Ctenostomatida		Biomass	D	Biomass	D	Biomass	D	Biomass	D	Biomass	D
Alcyonidium disciforme	A	25.6 ± 7.5	45 ± 10	17.2 ± 13.7	18 ± 9	4.7 ± 7.7	40 ± 16	2.0 ± 0.5	56 ± 43	20.7 ± 1.5	20 ± 16
Alcyonidium gelatinosum	BA, ws	24.2 ± 16.2	8 ± 6							6.67 ± 1.2	4 ± 1
Alcyonidium hirsutum	B, At	33.7 ± 17.7	90 ± 38								
Alcyonidium vermiculare	B, At									43.6 ± 3.8	2 ± 1
Flustrellidra gigantea	B, P									22.5 ± 3.6	5 ± 2
Flustrellidra hispida	B, At	14.0 ± 7.6	175 ± 135								
Order Cheilostomatida											
Bugulopsis peachii	BA, ws	3.1 ± 1.1	>100			4.7 ± 3.2	>100				
Celleporina ventricosa	BA, At	8.1 ± 4.2	16 ± 8	6.5 ± 2.9	10 ± 4					5.6 ± 4.2	25 ± 10
Celleporina surcularis	BA, At	2.4 ± 0.9	4 ± 2								
Cystisella saccata	BA, ws			4.5 ± 2.9	36 ± 18	0.2 ± 0.0	8 ± 2			4.9 ± 1.8	29 ± 12
Dendrobeania flustroides	BA, P									5.0 ± 1.8	>100
Escharella ventricosa	BA, At	1.4 ± 0.5	17 ± 9								
Eucratea loricata	BA, ws	23.9 ± 16.2	>100	0.9 ± 0.5	48 ± 20	9.4 ± 6.6	>100	1.8 ± 0.1	25 ± 10	7.6 ± 1.4	>100
Escharella dijmphnae	A	2.1 ± 2.9	10 ± 6								
Hippoporella fastigatoavicularis	BA, P									3.0 ± 0.5	18 ± 5
Leieschara subgracilis	BA, ws	3.3 ± 1.3	4 ± 2								
Microporella ciliata	B, ws	5.3 ± 1.5	145 ± 87								
Myriozoella costata	BA, At	2.3 ± 1.2	17 ± 15								
Parasmittina jeffreysi	BA, ws			1.2 ± 0.7	10 ± 4						
Porella tumida	BA, P									2.71 ± 0.15	42 ± 7
Posterula sarsii	BA, At	5.1 ± 1.6	9 ± 7								
Pseudoflustra solida	A	1.2 ± 0.2	10 ± 2	1.7 ± 1.3	10 ± 5	1.3 ± 0.5	6 ± 2				
Ragionula rosacea	BA, ws	1.9 ± 0.3	3 ± 1								
Serratiflustra serrulata	BA, ws			4.7 ± 2.5	24 ± 6			5.6 ± 2.1	13 ± 8		
Terminoflustra membranaceotruncata	BA, ws			12.5 ± 5.6	8 ± 2	5.5 ± 4.1	4 ± 1				
Tricellaria arctica	BA, ws	1.7 ± 0.9	20 ± 9								

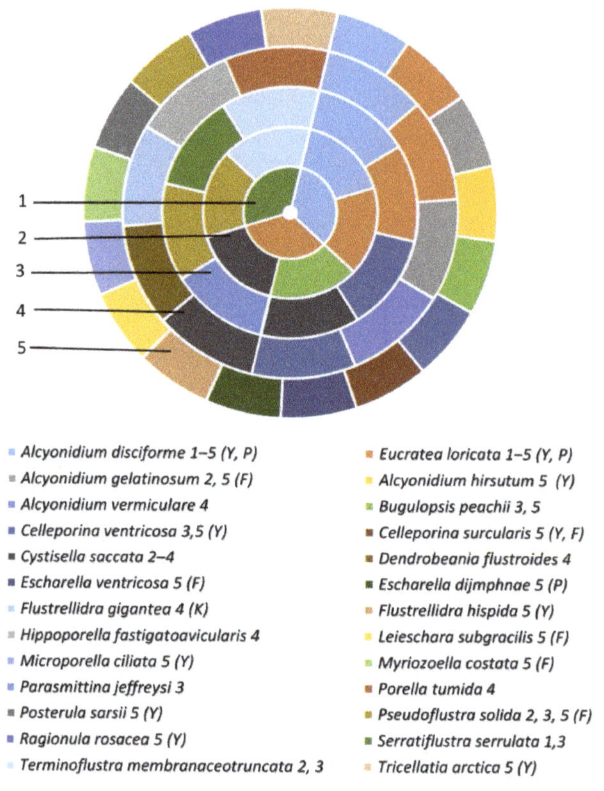

Figure 2. Taxonomic composition of biomass-dominating key species in the coastal areas of the Eurasian Arctic seas. 1—the East Siberian Sea; 2—the Laptev Sea; 3—the Kara Sea; 4—the Chukchi Sea; 5—the Barents Sea. Letters near the species name in legend: Y—Yarnyshnaya Bay; F—Franz Josef Land area; P—Pechora Sea coastal area; K—Kotzebue Sound.

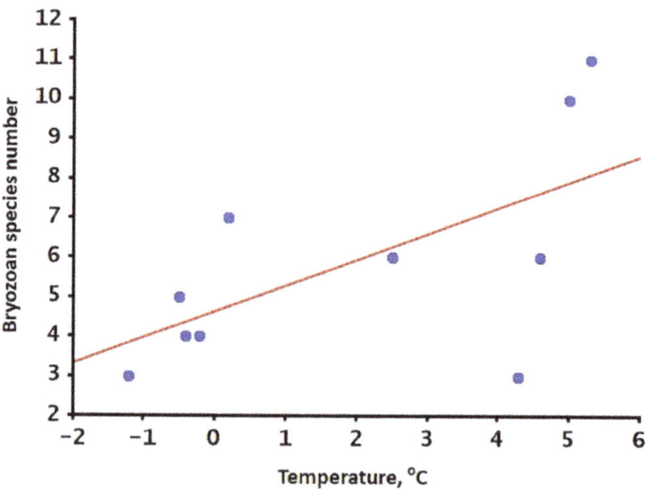

Figure 3. Relationship between the near-bottom water temperature and the number of biomass-dominating key bryozoan species in the Eurasian Arctic seas.

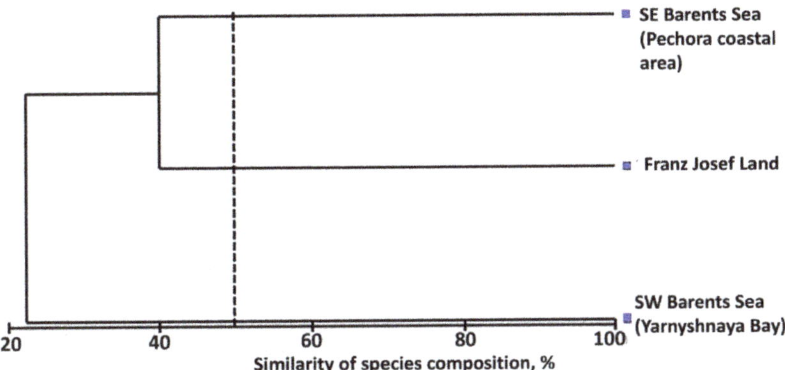

Figure 4. Dendrogram of the similarity of the composition of biomass-dominating key bryozoan species between coastal localities within the Barents Sea. The vertical dashed line indicates 50% similarity.

Similar calculations based on such data collected in the Siberian seas and in each area of the Barents Sea (Figure 5) indicated high similarity (>50%) in key species composition in the Kara and Laptev seas and the SE Barents Sea of the study region. Therefore, these areas were united into a single group. Low similarity among the SW Barents, Franz Josef Land areas, and Chukchi and East Siberian Seas indicates distinct species composition. A statistically significant difference in key species composition between the FJL, SW Barents area, East Siberian and Chukchi seas, on the one side, and the complex of the SE Barents, Kara, and Laptev seas, on the other, was confirmed by the ANOSIM test (Global R = 0.84; $p = 0.03$).

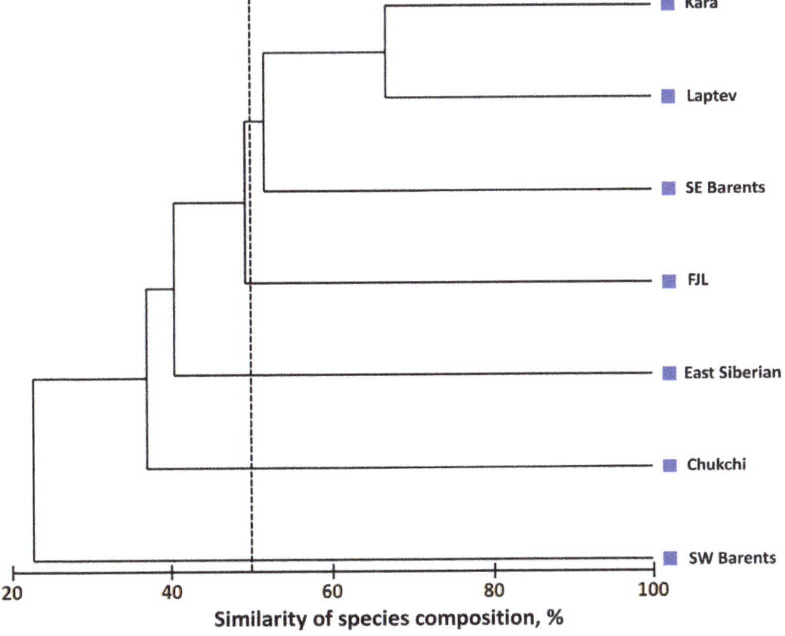

Figure 5. Dendrogram of the similarity of the composition of biomass-dominating key bryozoan species in the coastal areas of the studied seas. The vertical dashed line indicates 50% similarity.

3.2. Biogeographic Composition of Biomass-Dominating Key Bryozoan Species

The key bryozoan species identified in this study belong to six biogeographic categories (Table 2). Some of these are autochthonous Arctic species and widely distributed boreal–Arctic species living not only in the Arctic, but also in temperate zones of the world ocean. There are also boreal–Arctic and boreal species of Atlantic origin distributed primarily in temperate latitudes of the Atlantic Ocean and in the regions influenced by the Atlantic water masses of the Arctic Basin, as well as boreal and boreal–Arctic species of Pacific origin living in the Arctic in the regions influenced by the Pacific waters.

The analysis of proportions of species with different biogeographic affinities suggests that in the Barents Sea, there is a large proportion of species of Atlantic origin, with a progressive eastward decrease in this proportion toward the Kara and then the Laptev seas. Among the key species of the Chukchi Sea, a fairly large proportion is represented by the Pacific bryozoans, while in the East Siberian Sea, these two biogeographic categories, both Atlantic and Pacific, are absent. The proportion of Arctic species increases gradually eastward from the Barents to the East Siberian Sea and then drops significantly in the Chukchi Sea (Table 2, Figure 6).

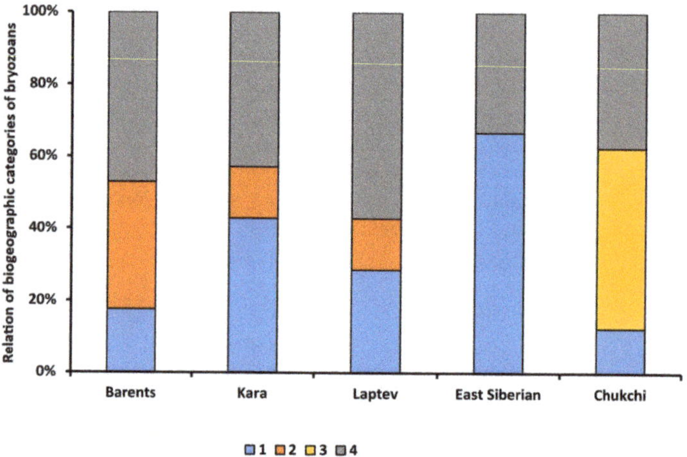

Figure 6. Variation in the proportion of key in biomass bryozoan species of different biogeographic affiliation found in the coastal area within the study area. 1—Arctic species; 2—boreal and boreal–Arctic species of Atlantic origin; 3—boreal and boreal–Arctic species of Pacific origin; 4—widespread boreal–Arctic species.

3.3. Biomass Characteristics

The key biomass species showed fairly large variation in weights among samples within seas and in average values among seas. In the Barents Sea, the variation of this characteristic between different species was as large as two orders of magnitude ranging from 1.42 to 133.72 g/m^2 (Table 2). In the Kara and Laptev seas, it differed by more than ten-fold varying from 1.09 to 17.18 g/m^2, and in the East Siberian Sea, all dominant species had comparable average biomasses (1.77–5.60 g/m^2). In the Chukchi Sea, variation in average individual biomasses rose again from 2.71 to 43.60 g/m^2 (Table 2).

The total average biomass of all key bryozoan species calculated for each of the seas was also markedly variable ranging from 3.13 ± 2.14 to 16.67 ± 30.71 g/m^2 (mean ± SE) (Figure 7). Its minimum value was observed in the East Siberian Sea, where it was comparable to that in the Laptev Sea (3.83 ± 3.24 g/m^2), but in the Kara Sea the average biomass was already twice as high (6.09 ± 5.65 g/m^2). The average biomass of the key species calculated for the Chukchi Sea was comparable to that in the Barents Sea, but was slightly lower (12.05 ± 2.72 g/m^2).

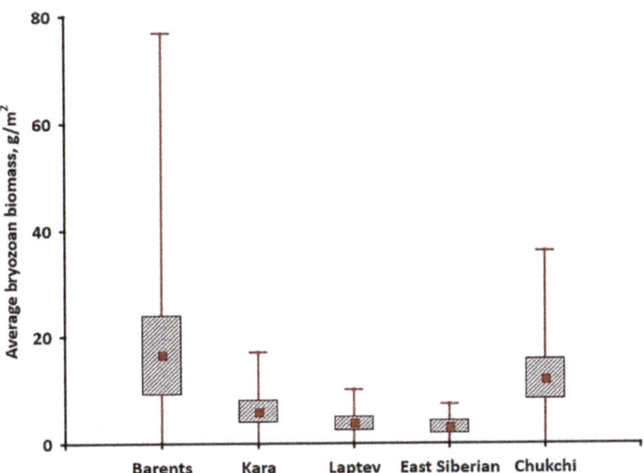

Figure 7. Variations of average biomass of key bryozoan species in the coastal areas of the studied seas. Symbology: solid square, mean; box, standard error of the mean; capped lines, standard deviation.

The average biomass correlated fairly well with the number of key species that composed this biomass in each sea (Pearson correlation where R = 0.92; $p < 0.05$).

Of the environmental parameters examined in this study: depth, temperature, and grain-size distribution of bottom sediments, only the latter was a strong influence on variation in biomass for the dominant bryozoan species (Pearson correlation where R = 0.82; $p = 0.03$) (Figure 8). Depth and temperature did not have a significant effect on the characteristics of bryozoans in the coastal locations studied (R = 0.03; $p = 0.07$ and R = 0.17; $p = 0.9$, respectively) (Figure 9). The exception was the Pechora Sea, where the biomass showed a significant association with depth (Pearson correlation where R = 0.98; $p = 0.02$), a relationship that becomes only a trend for the waters around the FJL (Pearson correlation R = 0.71; $p = 0.08$) (Figure 9B,C).

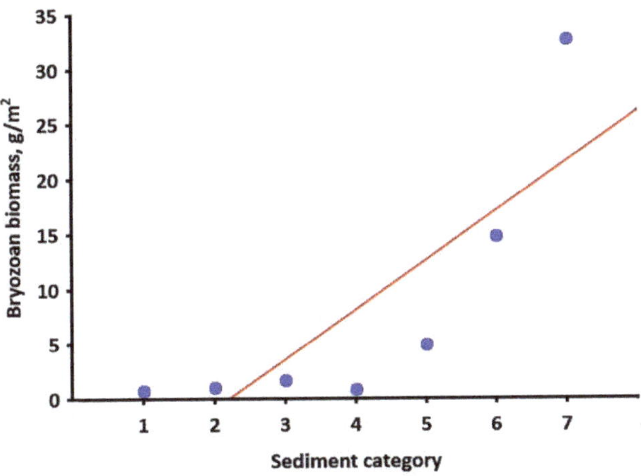

Figure 8. Relationship between the bottom sediment composition and biomass of key bryozoan species in the coastal area of the Eurasian Arctic seas. 1—clay with sand and muddy sand; 2—silty sand and fine sand; 3—silty sand with shells; 4—sand, silt with gravel; 5—pebble, shells with mud; 6—pebble with shells and gravel with sand; 7—boulders, pebble, rocks.

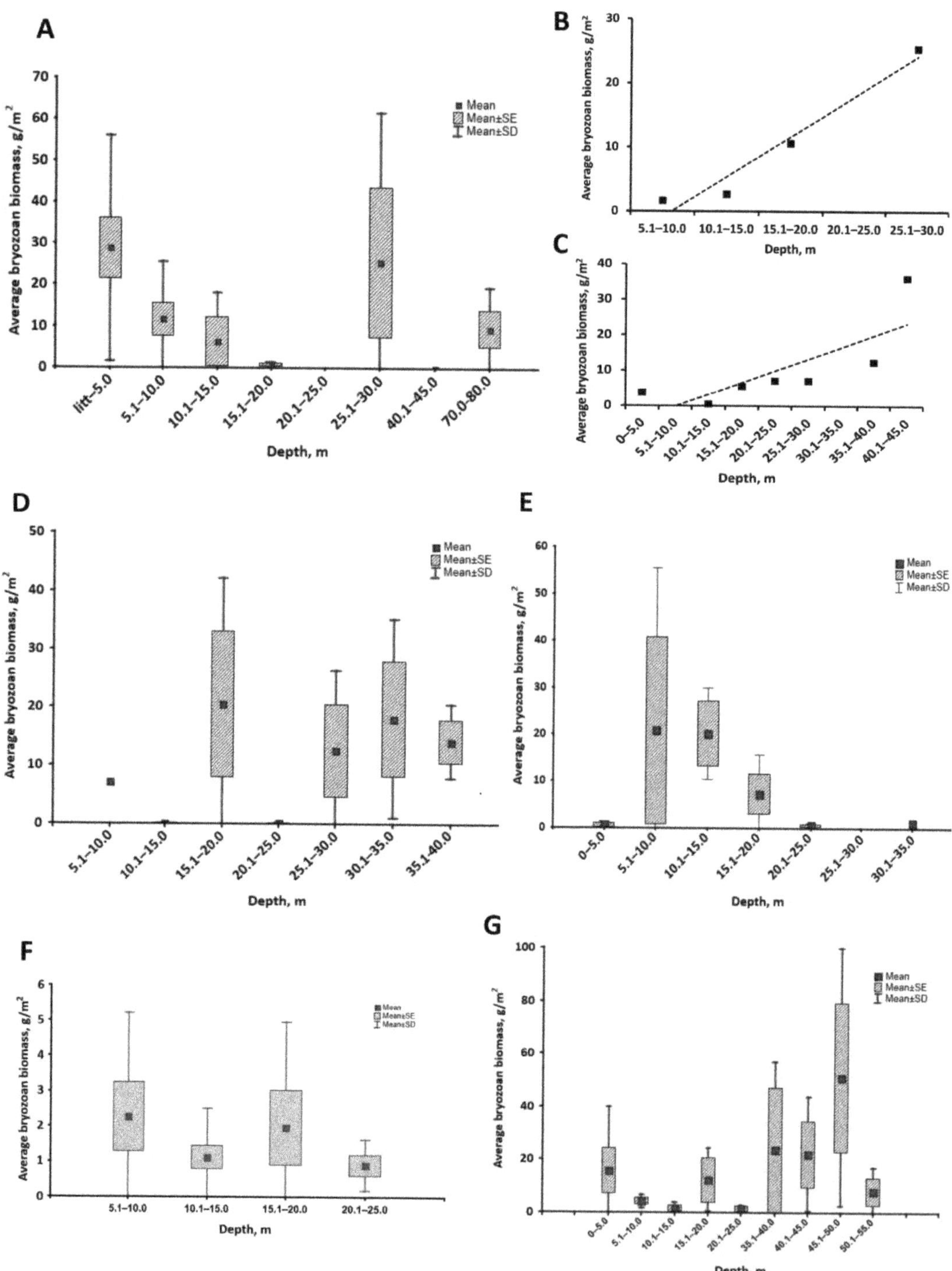

Figure 9. Variation of total average biomass of key bryozoan species with depth in the studied areas. Barents Sea: (**A**)—SW (Yarnyshnaya Bay), (**B**)—SE (Pechora sea coastal area), (**C**)—NE (Franz Josef Land Archipelago area); (**D**)—the Kara Sea; (**E**)—the Laptev Sea; (**F**)—the East Siberian Sea; (**G**)—the Chukchi Sea.

4. Discussion

4.1. Diversity of Biomass-Dominating Key Bryozoan Species and Their Biogeographic Affiliation

Despite a long history of zoobenthos research in the Arctic, the faunistic composition and distribution of bryozoans in this region remain insufficiently studied, even though about 90% of the species inhabiting the seas in this region are currently known [49]. Previous studies in the Barents Sea have shown that only about 40–50 bryozoan species formed dense assemblages [19], which accounted for less than 10% of overall bryozoan species richness in the Arctic [50]. The diversity of species that contribute appreciably to the total zoobenthos biomass in the coastal waters was even much lower. Moreover, while the biomass-dominating species in the entire area of the sea comprised representatives of all orders of marine bryozoans (Cyclostomatida, Ctenostomatida, and Cheilostomatida) [19], those inhabiting the coastal areas lacked significant cyclostome assemblages.

Among the 26 biomass-dominating key species of bryozoans (with biomass exceeding 1 g/m^2 in the Eurasian seas), the maximum number of species occurs in the shallow waters of the Barents Sea. In the Siberian seas, key species richness gradually decreases in parallel to that of the overall bryozoan fauna [50,51], reaching its minimum in the East Siberian Sea (three species). The Chukchi Sea, however, shows an increased number of biomass-dominating key species, again in line with a similar trend for the total species richness of the bryozoan fauna [50].

It is currently known that in the Eurasian sector of the Arctic, bryozoans are represented by two distinctly different faunistic complexes [51], and the exact composition of biomass-dominating key species is different in most seas. The exceptions are such typical Arctic seas as the Laptev and East Siberian seas, in which the composition of biomass-dominating species differs by no more than 47%. The similarity in composition of biomass-dominating key species in these seas is readily explained by similarity in habitats, in particular by the predominance of soft-bottom sediments in the coastal zone [33] together with persistent negative temperatures near the seafloor and decreased salinity [30].

Only two of the species designated as biomass-dominating key species, the Arctic *Alcyonidium disciforme* and the more widely distributed boreal–Arctic *Eucratea loricata*, form settlements with a biomass greater than 1 g/m^2 in all seas of the study area. Some species, notably *Alcyonidium gelatinosum*, *Cystisella saccata*, *Pseudoflustra solida*, and *Serratiflustra serrulata*, form sufficiently dense assemblages in the shallow waters of two or three seas. However, most biomass-dominating key bryozoan species were reported from the coastal regions of only one sea.

It should be noted that dense assemblages in the coastal zone are formed both by eurybathic (for instance, the cheilostome *Leieschara subgracilis*, *Posterula sarsi*, *Myriozoella costata*, and *Microporella ciliata*, or the ctenostome *Alcyonidium gelatinosum*) and stenobathic forms. The latter include *Alcyonidium hirsutum* and *Flustrellidra hispida*, the obligate inhabitants of the littoral zone. It is noteworthy that differences in species composition between the biomass-dominating bryozoans were observed not only across seas, but also within a single sea. For instance, in the Barents Sea, the overlap in species composition of biomass-dominating bryozoans between different localities is less than 40%. The lack of settlements of species such as *Microporella ciliata*, *Alcyonidium hirsutum*, and *Flustrellidra hispida* in the northern and south-eastern parts of the seas and abundant occurrence of *Myriozoella costata* and *Escharella ventricosa* only in the shallow waters around the FJL and *E. dijmphnae* in the Pechora Sea are easily explainable by their different tolerance to temperature and salinity factors. The first three species are boreal in their biogeographic range and are unable to reproduce at negative water temperatures. The next two species are Atlantic boreal–Arctic in origin and reproduce successfully under a wider range of temperature regimes. However, they are very sensitive to desalination of the water column, which is typical for the Arctic water mass. *E. dijmphnae* belongs to the Arctic forms, which prefer lower salinities, are intolerant to temperatures higher than 4–5 °C, and reproduce at near-zero temperatures [19]. Biological requirements of each representative of the bryozoan fauna, including the presence of a short-lived lecithotrophic larva, are factors that cause a high degree of settlement aggregation of many bryozoan species. This conclusion is confirmed by observations of bryozoan settlements made in the littoral zone of the same bay at different years over a time period of 30–35 years. For instance, widely distributed *E. loricata* had dense settlements in Yarnyshnaya Bay both in 2014 [52] and 1982–1988 [53]. In contrast, the settlements of *Alcyonidium hirsutum* and *Flustrellidra hispida*, whose mean biomass in aggregations in the south-western part of the Barents Sea near the Kola peninsula had reached several hundred grams in the 1980s [53], were not reported in 2014 [52]. Given that the last two species belong to the boreal fauna [19], any influence of the steady warming of the Arctic that began in the 2010s [39] can be ruled out. These observations lead to the conclusion that the absence of these species in the material collected in 2014 is the result of differences in sampling, or otherwise, they have disappeared due to certain autoecological processes. It should be mentioned once again that some of the aforementioned species have aggregations that remain stable throughout a long time period. These species include *Microporella ciliata*, which was recorded in the south-western part of the Barents Sea, and *Alcyonidium disciforme* from the Pechora Sea, whose mass settlements had been found in these locations as early as the first half of the 20th century [10,11,54]. The latter species was previously categorized as dominant in the shallow-water zoobenthos communities of the Laptev Sea [55] and around the FJL [25]. In the 1970s, high biomass values were reported for *A. gelatinosum* in the vicinity of the FJL [25] and for *Eucratea loricata* near Wrangel Island (Chukchi Sea) [56]. The latter species was recently noted as biomass-dominating in the waters off West Spitsbergen [57].

The established biogeographic composition of the bryozoan fauna of the Arctic seas was formed in the geological past after the Quaternary glaciation as a result of the impact of Atlantic and Pacific waters characterized by different intensity inputs [19,51]. The observed differences in the biogeographic composition of key species in the coastal areas of regarded seas are a consequence of the preference for species with different biogeographic affiliations to water masses of a certain origin. In particular, boreal Atlantic species are found mainly in the areas of the modern influence of the Atlantic water masses and observed in the south-western part of the Barents Sea [31,32], and as this influence weakens and water masses transform by cooling, the proportion of boreal–Arctic Atlantic species with high biomass values also drops, gradually decreasing eastward to the Laptev Sea. This trend is accompanied by an increase in the proportion of Arctic elements among the biomass-dominant bryozoans; it is especially high in the shallow waters of the East Siberian Sea. In areas under the predominant influence of Arctic water masses [19,24,58], boreal species are absent. However, boreal–Arctic Atlantic species of bryozoans are still found in the coastal regions of the Kara Sea and the Laptev Seas. as well as in the Chaun Bay of the East Siberian Sea [22], where they do not form dense aggregations. In the Chukchi Sea, which is under the influence of the Pacific waters flowing into this sea across Bering Strait [37], a significant proportion of biomass-dominating species of bryozoans have a Pacific origin. Only the widely distributed boreo–Arctic species that make up about 40% of all key species do not show any propensity for waters of a certain origin. The variation in the biogeographic composition of biomass-dominating key species described in this study is predictably consistent with the trends observed for variation in the proportion of different biogeographic groups in the overall bryozoan fauna [19,23,24,57,58]. Boreal–Arctic Atlantic and Pacific species penetrated into the Siberian seas as a result of their dispersion in the geological past [50].

4.2. Biomass of Key Species

The average biomass of ctenostome bryozoan species that lack the calcareous skeleton is often one order of magnitude greater than the calcified cheilostome, a result that is mostly explained by the individual weight of the colony. For instance, in the coastal waters of the Kola Peninsula, the high biomass of the pancake-shaped colonies of *A. hirsutum* and *F. hispida* is the consequence of the high density of their colonies [53]. In contrast, the branching shrubby colonies of *Alcyonidium gelatinosum* and *Flustrellidra gigantea* that reach

15–20 cm in height and the ribbon-shaped colonies of *A. vermiculare* that grow up to 80 cm in height have a high biomass primarily because of their size despite the low number of colonies.

Previous studies focused on the biomass estimation of bryozoans that were conducted in the Barents Sea showed a significant variation in bryozoan biomass across this water body. The variation was statistically coupled with the structure of bottom sediments and the patchiness in the distribution of hard bottoms and mixed sediments [11,14,19]. It is the patchiness of bottom sediments and the extent of settlement aggregation of different bryozoan species that result in differences in quantitative estimations of bryozoans in different seas of the study region [2–6,9,10,14–18]. Another reason for the different reported biomass values of bryozoans in bottom communities was the use of collection tools with different surface areas. Locally aggregated assemblages of bryozoans can be efficiently sampled by grab samplers with a small area, while the use of trawls offered better chances in collecting colonies over a larger area of the seafloor covered with scattered stones and gravel.

The patchiness of bottom sediments prevented us from identifying any patterns of variation of bryozoan biomass with depth. However, in those regions that had uniform bottom sediments, such as the region located in the Pechora Sea, the relationship between the bryozoan biomass and depth was statistically significant. On the other hand, despite the apparent lack of any pattern in variation in biomass with depth, a feature common to the Siberian seas (Kara, Laptev, and East Siberian) and the south-eastern part of the Barents Sea, was a very slow growth of bryozoan colonies at depths of 0 to 5 m, where the seafloor is dominated by silty–clay sediments and water that contains a high concentration of inorganic particles due to the intensive erosion of the coastal permafrost [33,59]. The first factor limits the availability of substrates suitable for attachment, while the second has an adverse effect on the filter-feeding tentacular apparatus of bryozoans and inflicts injuries to the tentacles. These observations support the conclusion that depth is likely a secondary influence that indirectly affects bryozoan biomass through the structure of bottom sediments and some other factors such as water turbidity.

The average bryozoan biomass for the coastal regions was more than five times as great as the average biomass of this group previously reported for the open part of the Barents Sea [60]. This supports the idea that the proportion of bryozoans in the total zoobenthos biomass for the coastal regions can be much higher than the same proportion for the entire sea, which was previously reported to be approximately 4% [60].

It is currently impossible to determine the position that the bryozoans occupy among other biomass-forming groups of the zoobenthos in the coastal zone of the Siberian seas because of the lack of quantitative data in the literature [5–8]. The contribution of bryozoans to the total zoobenthos biomass in this region is probably even lower than is currently estimated, because the average bryozoan biomass is one order of magnitude lower there than in the Barents Sea.

A lower average bryozoan biomass in the coastal regions of the Siberian seas compared to the Barents and Chukchi Seas is the consequence of their harsher environmental conditions [30,36,37,59]. Subzero temperatures in the near-bottom water layer and a lower salinity, together with the predominance of fine-grained fractions in the bottom sediments of the Siberian seas, result in a decrease in the diversity of species dominating in biomass. This decline is a consequence of the elimination of less tolerant bryozoan species in this region of the Arctic and the lower rates of colony growth in widely tolerant bryozoans [19].

5. Conclusions

An integrated quantitative study of large-scale variation in bryozoan biomass in the Eurasian seas allowed the assessment of the distribution of dominant species, i.e., key biomass species, in the coastal zone of this Arctic region. It was shown that as the influence of Atlantic and Pacific waters weakens, boreal species are excluded from the fauna, and the density of colonies of Atlantic and Pacific boreal–Arctic bryozoans become lower. This is likely due to the limitation of their growth rates leading also to the decline in the diversity of key species. This brings about a decrease in the individual average biomass of each species and more generally in the average biomass of the whole group. Lower average biomass values in the coastal zone of the Siberian seas are directly related to the increase in the proportion of soft-bottom sediments in this region (Figure 10). The correspondence of bryozoan species with Atlantic and Pacific origin to the expansion of Atlantic and Pacific water masses in the geological past, respectively, indicates their relatively recent occupation of the studied seas.

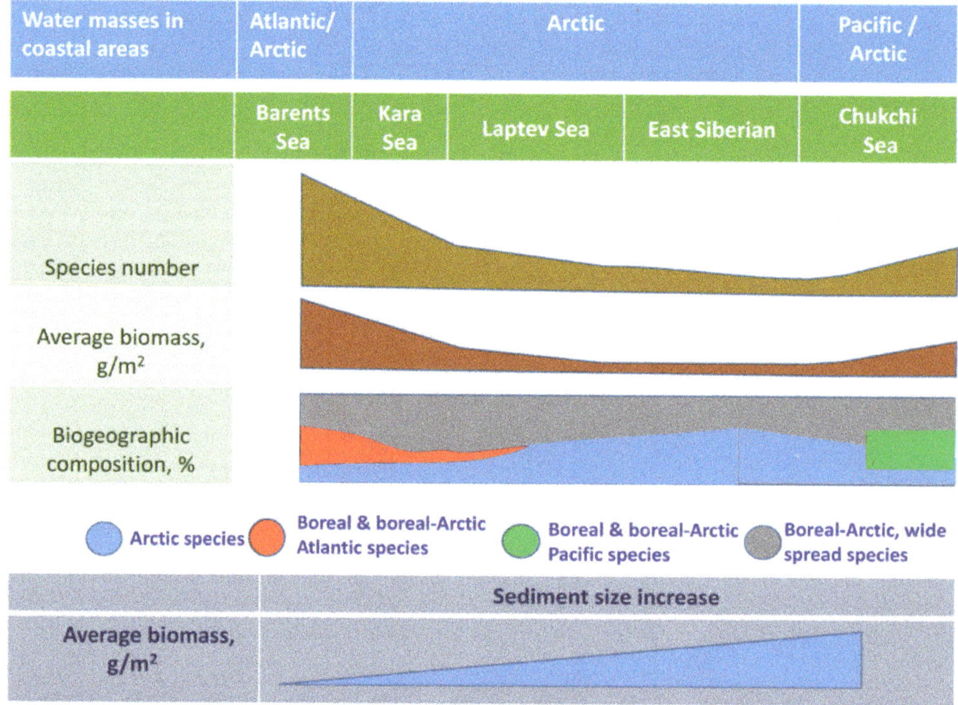

Figure 10. Integrated scheme of large-scale variations of diversity among biomass-dominating key bryozoan species in the Eurasian Arctic Seas.

Author Contributions: Conceptualization, N.V.D.; data curation, N.V.D. and S.G.D.; formal analysis, N.V.D. and S.G.D.; methodology, N.V.D. and S.G.D.; project administration, N.V.D.; software, N.V.D. and S.G.D.; validation, N.V.D. and S.G.D.; visualization, N.V.D. and S.G.D.; writing—original draft, N.V.D.; writing—review and editing, N.V.D. and S.G.D. All authors have read and agreed to the published version of the manuscript.

Funding: This study was funded by the Ministry of Science and Higher Education of the Russian Federation (project 'Taxonomy, biodiversity and ecology of invertebrates in Russian and adjacent waters of the World Ocean, continental water bodies and humid areas' № 122031100275-4).

Institutional Review Board Statement: Not applicable.

Informed Consent Statement: Not applicable.

Data Availability Statement: The data are available on request from the corresponding author.

Acknowledgments: We are grateful to the employees of the Zoological Institute of RAS for their help in collecting some of the samples. We thank T. Trott and the reviewers for the thoughtful comments that improved the manuscript.

Conflicts of Interest: The authors declare no conflict of interest.

References

1. Jørgensen, L.L.; Archambault, P.; Blicher, M.; Denisenko, N.; Guðmundsson, G.; Iken, K.; Roy, V.; Sørensen, J.; Anisimova, N.; Behe, C.; et al. Benthos. In *State of the Arctic Marine Biodiversity*; Barry, T., Price, C., Olsen, M., Christensen, T., Frederiksen, M., Eds.; Conservation of Arctic Flora and Fauna International Secretariat: Akureyri, Iceland, 2017; pp. 85–107.
2. Brotskaya, V.F.; Zenkevich, L.A. Quantitative distribution of the bottom fauna of the Barents Sea. *Proc. VNIRO* **1939**, *4*, 3–150. (In Russian)
3. Filatova, Z.F.; Zenkevich, L.A. Quantitative distribution of the bottom fauna of the Kara Sea. *Trans. All Union Hydrobiol. Soc.* **1957**, *8*, 3–62. (In Russian)
4. Sirenko, B.I.; Bluhm, B.; Iken, K.; Crane, K.; Gladish, V. Some results of investigations of composition and quantitative distribution of epifauna in the Chukchi Sea. In *Ecosystems and Biological Resources of the Chukchi Sea and Adjacent Areas*; Sirenko, B.I., Ed.; Explorations of the Fauna of the Seas; ZIN RAS Press: St. Petersburg, Russia, 2009; Volume 64, pp. 200–212.
5. Denisenko, S.G.; Sirenko, B.I.; Gagaev, S.Y.; Petriashhov, V.V. Bottom communities structure and spatial distribution in the East Siberian Sea at depth more than 10 m. In *Fauna of the East Siberian Sea, Distribution Patterns and Structure of Bottom Communities*; Sirenko, B.I., Denisenko, S.G., Eds.; Explorations of the Fauna of the Seas; ZIN RAS Press: St Petersburg, Russia, 2010; Volume 66, pp. 130–143. (In Russian)
6. Galkin, S.V.; Vedenin, A.A. Macrobenthos of the Yenisey Bay and adjacent Kara Sea shelf. *Oceanology* **2015**, *55*, 668–676.
7. Vedenin, A.A.; Galkin, S.V.; Kozlovskiy, V.V. Macrobenthos of the Ob Bay and adjacent Kara Sea shelf. *Polar Biol.* **2015**, *38*, 829–844. [CrossRef]
8. Denisenko, S.G.; Grebmeier, J.M.; Cooper, L.W. Assessing bioresources and standing stock of zoobenthos (key species; high taxa; trophic groups) in the Chukchi Sea. *Oceanography* **2015**, *28*, 146–157. [CrossRef]
9. Kuznetsov, V.V. Dynamic of the biocenosis of *Microporella ciliata* (Pallas) in the Barents Sea. *Proc. Zool. Inst. USSR Ac. Sci.* **1941**, *7*, 114–139. (In Russian)
10. Zenkevich, L.A. Quantitative estimation of the bottom fauna of the Pechora region of the Barents and White Seas. *Proc. Float. Mar. Sci. Inst.* **1927**, *2*, 3–64. (In Russian)
11. Denisenko, N.V. Structure and distribution of the bryozoan fauna of the south–eastern part of the Barents Sea. In *Biogeocenoses of Glacial Shelfs of Western Arctic*; Matishov, G.G., Tarasov, G.A., Denisenko, S.G., Galaktionov, K.V., Eds.; KSC RAS Press: Apatity, Russia, 1996; pp. 224–237. (In Russian)
12. Denisenko, N.V. Fauna of bryozoans of Cheshskaya Bay: Its structure and distribution. In *Current Benthos of the Barents and Kara Seas*; Matishov, G.G., Denisov, V.V., Chinarina, A.D., Mitina, E.G., Anisimova, N.A., Eds.; KSC RAS Press: Apatity, Russia, 2000; pp. 203–219. (In Russian)
13. Denisenko, N.V. Bryozoans (Bryozoa) of Yarnyshnaya Bay (Eastern Murman): Fauna, distribution, seasonal changes. In *Current Benthos of the Barents and Kara Seas*; Matishov, G.G., Denisov, V.V., Chinarina, A.D., Mitina, E.G., Anisimova, N.A., Eds.; KSC RAS Press: Apatity, Russia, 2000; pp. 219–227. (In Russian)
14. Denisenko, N.V.; Denisenko, S.G.; Lehtonen, K.K.; Andersin, A.-B.; Sandler, H.R. Zoobenthos of the Cheshskaya Bay (southeastern Barents Sea): Spatial distribution and community structure in relation to environmental factors. *Pol. Biol.* **2007**, *30*, 735–746. [CrossRef]
15. Filatova, Z.A. Quantitative estimation of the bottom fauna of the south–western part of the Barents Sea. *Proc. PINRO* **1938**, *2*, 3–58. (In Russian)
16. Idelson, M. Materials on quantitative estimations of the bottom fauna of the Spitzbergen Bank (Barents Sea). *Proc. Mar. Sci. Inst.* **1938**, *4*, 27–46. (In Russian)
17. Zatsepin, V.I. Communities of the bottom fauna of the Murman Coast of the Barents Sea. Part 2. *Proc. All Union Hydrobiol. Soc. USSR* **1962**, *12*, 245–344. (In Russian)
18. Pergament, T.S. Benthos distribution in coastal zone of the Eastern Murman. *Proc. Murmansk Mar. Station ZIN Ac. Sci. USSR* **1957**, *3*, 75–89. (In Russian)
19. Denisenko, N.V. *Distribution and Ecology of Bryozoans of the Barents Sea*; KSC RAS Press: Apatity, Russia, 1990. (In Russian)
20. Androsova, E.I. Bryozoans (Bryozoa) in biocenoses of the Arctic Ocean (in the Franz Josef Land Archipelago area). In *Biocenoses of Shelf of Franz Josef Land and Adjacent Waters*; Golikov, A.N., Ed.; Explorations of the Fauna of the Seas; Nauka: Leningrad, Russia, 1977; Volume 14, pp. 194–204. (In Russian)
21. Gontar, V.I. Bryozoans (Bryozoa) of the Laptev Sea and New Siberian Shoals. In *Ecosystems of the New Siberian Shoals and Fauna of the Laptev Sea and Adjacent Waters of the Arctic*; Golikov, A.N., Ed.; Explorations of the Fauna of the Seas; Nauka: Leningrad, Russia, 1990; Volume 37, pp. 130–138. (In Russian)
22. Gontar, V.I. Distribution and ecology of bryozoans (Bryozoa) of Chaun Bay of the East Siberian Sea. In *Ecosystems, the Flora and Fauna of Chaun Bay of the East Siberian Sea. Part 1*; Golikov, A.N., Ed.; Explorations of the Fauna of the Seas; ZIN RAS Press: St Petersburg, Russia, 1994; Volume 47, pp. 144–147. (In Russian)
23. Denisenko, N.V. Bryozoans of the Chukchi Sea and Bering Strait. In *Fauna and Zoogeography of Benthos of the Chukchi Sea*; Sirenko, B.I., Vasilenko, S.V., Eds.; Explorations of the Fauna of the Seas; ZIN RAS Press: St Petersburg, Russia, 2008; Volume 61, pp. 163–198. (In Russian)

24. Denisenko, N.V. Bryozoans (Bryozoa) of the East Siberian Sea. In *Fauna of the East Siberian Sea, distribution Patterns and Structure of Bottom Communities*; Sirenko, B.I., Denisenko, S.G., Eds.; Explorations of the Fauna of the Seas; ZIN RAS Press: St Petersburg, Russia, 2010; Volume 66, pp. 89–129. (In Russian)

25. Golikov, A.N.; Averintsev, V.G. Biocenoses of the upper regions of the shelf of the Franz Josef Land Archipelago and some regularities of their distribution. In *Biocenoses of the Shelf of Franz Josef Land and the Fauna of Adjacent Waters*; Golikov, A.N., Ed.; Explorations of the Fauna of the Seas; Nauka: Leningrad, Russia, 1985; Volume 14, pp. 5–54. (In Russian)

26. Golikov, A.N.; Scarlato, O.A.; Averitsev, V.G.; Menshutkina, T.V.; Novikov, O.K.; Sheremetevsky, A.M. Ecosystems of the New Siberian Shoals, their distribution and function. In *Ecosystems of the New Siberian Shoals and fauna of the Laptev Sea and Adjacent Waters of the Arctic*; Golikov, A.N., Ed.; Explorations of the Fauna of the Seas; Nauka: Leningrad, Russia, 1990; Volume 37, pp. 4–79. (In Russian)

27. Golikov, A.N.; Gagaev, S.Y.; Galtsova, V.V.; Golikov, A.A.; Dunton, K.; Menshutkina, T.V. Ecosystems and the flora and fauna of the Chaun Bay of the East–Siberian Sea. In *Ecosystems, the Flora and Fauna of Chaun Bay of the East Siberian Sea. Part 1*; Golikov, A.N., Ed.; Exploration of the Fauna of the Seas; ZIN RAS Press: St Petersburg, Russia, 1994; Volume 47, pp. 4–110. (In Russian)

28. Golikov, A.N.; Anisimova, N.A.; Golikov, A.A.; Denisenko, N.V.; Kaptilina, T.V.; Menshutkin, V.V.; Menshutkina, T.V.; Novikov, O.K.; Panteleyeva, N.N.; Frolova, E.A. *Bottom Communities and Biocenoses of the Yarnishnaya Inlet of the Barents Sea and Their Seasonal Dynamics*; KSC RAS Press: Apatity, Russia, 1993. (In Russian)

29. Luppova, E.N.; Anisimova, N.A.; Denisenko, N.V.; Frolova, E.A. The bottom biocenoses of Franz Josef Land. In *The Environment and Ecosystems of Franz Josef Land (Achipelago and Shelf)*; Matishov, G.G., Ed.; KSC RAS Press: Apatity, Russia, 1993; pp. 118–142. (Translated into English)

30. Dobrovolsky, A.D.; Zalogin, B.S. *Seas of the USSR*; Moscow State University Press: Moscow, Russia, 1982. (In Russian)

31. Loeng, H. Features of the physical oceanographic conditions of the Barents Sea. *Pol. Res.* **1991**, *10*, 5–18. [CrossRef]

32. Pisarev, S.V. Review of oceanographic condition of the Barents Sea. In *System of the Barents Sea*; Lisitsyn, A.P., Ed.; GEOS: Moscow, Russia, 2021; pp. 153–166. (In Russian)

33. Kosheleva, V.A.; Yashin, D.S. *Bottom Sediments of the Arctic Seas of Russia*; VNII Oceangeologia Press: St Petersburg, Russia, 1999. (In Russian)

34. Feder, H.M.; Jewett, S.C. *Survey of the Epifaunal Invertebrates of Northern Sound, Southeastern Chukchi Sea, and Kotzebue Sound*; Report of Institute of Marine Science No R78–1; University of Alaska Press: Fairbanks, AK, USA, 1978.

35. Adrov, N.M.; Denisenko, G.S. Oceanographic characteristics of the Pechora Sea. In *Biogeocenoses of Glacial Shelf of the Western Arctic Seas*; Matishov, G.G., Tarasov, G.A., Denisenko, S.G., Denisov, V.V., Galaktionov, K.V., Eds.; KSC RAS Press: Apatity, Russia, 1996; pp. 166–179. (In Russian)

36. Burenkov, V.I.; Vasilkov, A.P. The influence of runoff from land on the distribution of hydrologic characteristics of the Kara Sea. *Oceanology* **1994**, *34*, 591–599.

37. Dudarev, O.V.; Semiletov, I.P.; Charkin, A.N.; Botsul, A.I. Deposition settings on the continental shelf of the East Siberian Sea. *Trans. RAS* **2006**, *409*, 822–827. (Translated into English) [CrossRef]

38. Pisareva, M.N.; Pickart, R.S.; Iken, K.; Ershova, E.A.; Grebmeier, J.M.; Cooper, L.W.; Bluhm, B.A.; Nobre, C.; Hopcroft, R.R.; Hu, H.; et al. The relationship between patterns of benthic fauna and zooplankton in the Chukchi Sea and physical forcing. *Oceanography* **2015**, *28*, 68–83. [CrossRef]

39. Karsakov, A.L.; Borovkov, V.A.; Sentyabov, E.V.; Ivshin, V.A.; Balyakin, G.G.; Abolmasova, Z.V. Oceanographic conditions in the sea of the North European basin and North Atlantic in 2014–2015 and their influence on distribution of commercially important fishes. In *Environmental Conditions of Aquatic Biological Resources: Characteristics of Oceanographic Conditions in 2014–2015 in the Major Harvesting Areas of the Russian Fleet Operation*; Trans. Russ. Federal Inst. Fish. Oceanogr.; VNIRO Press: Moscow, Russia, 2016; Volume 164, pp. 5–21. (In Russian)

40. Gontar, V.I.; Denisenko, N.V. Arctic Ocean bryozoans. In *The Arctic Seas. Climatology, Oceanography, Geology, and Biology*; Herman, Y., Ed.; Van Nostrand Reinhold: New York, NY, USA, 1989; pp. 341–372.

41. Denisenko, N.V.; Blicher, M.E. Bryozoan diversity, biogeographic patterns, and distribution in Greenland waters. *Mar. Biodivers.* **2021**, *51*, 73. [CrossRef]

42. Czekanowski, J. Zur differential Diagnose der Neandertalgruppe. *Korresp. Dtsch. Ges. Anthropol.* **1909**, *40*, 44–47.

43. Sørensen, T.A. A new method of establishing groups of equal amplitude in plant sociology based on similarity of species content and its application to analysis of vegetation on Danish commons. *Kogelige Dan. Selsk. Biol. Skr.* **1948**, *622*, 1–34.

44. Magurran, A. *Measuring Biological Diversity*; Blackwell Publishing: Oxford, UK, 2004.

45. Pesenko, Y.A. *Rules and Methods of Quantitative Analyses in Faunistic Investigations*; Nauka: Moscow, Russia, 1982. (In Russian)

46. Pielou, E.C. *The Interpretation of Ecological Data*; John Wiley and Sons Inc.: New York, NY, USA, 1984.

47. Gray, J.S.; Elliott, M. *Ecology of Marine Sediments: Science to Management*, 2nd ed.; Oxford University Press: London, UK, 2009.

48. Clarke, K.R.; Warwick, R.M. *Changes in Marine Communities: An Approach to Statistical Analysis and Interpretation*, 2nd ed.; PRIMER-E: Plymouth, UK, 2001.

49. Denisenko, N.V. Species richness and the level of knowledge of the bryozoan fauna of the Arctic region. *Proc. Zool. Inst. Russ. Ac. Sci.* **2020**, *324*, 353–363. [CrossRef]

50. Denisenko, N.V. Species richness and distribution patterns of bryozoans of the Arctic region: List of bryozoan fauna. In *Annals of Bryozoology 7: Aspects of History of Research of Bryozoans*; Wyze Jackson, P.N., Spenser Jones, M.E., Eds.; International Bryozoology Association: Dublin, Ireland, 2022; pp. 121–143.

51. Denisenko, N.V. Taxonomic structure and endemism of the bryozoan fauna in the Arctic region. *Paleont. J.* **2022**, *56*, 774–792. [CrossRef]

52. Evseeva, O.Y.; Ishkulova, T.G.; Dvoretsky, A.G. Environmental drivers of an intertidal bryozoan community in the Barents Sea: A case study. *Animals* **2022**, *12*, 552. [CrossRef] [PubMed]

53. Denisenko, N.V. Species composition and distribution of bryozoans in intertidal zone of some inlets of the Eastern Murman. In *Benthos of the Barents Sea. Distribution, Ecology and Population Structure*; Semenov, V.N., Averintsev, V.G., Galaktionov, K.V., Eds.; Kola Branch AS USSR Press: Apatity, Russia, 1984; pp. 71–79. (In Russian)

54. Denisenko, S.G.; Denisenko, N.V.; Lehtonen, K.K.; Andersin, A.-B.; Laine, A.O. Macrozoobenthos of the Pechora Sea (SE Barents Sea): Community structure and spatial distribution in relation to environmental conditions. *Mar. Ecol. Prog. Ser.* **2003**, *258*, 109–123. [CrossRef]

55. Petryashov, V.V.; Golikov, A.A.; Schmidt, M.; Rachor, E. Macrozoobenthoc of the Laptev Sea shelf. In *Fauna and Ecosystems of the Laptev Sea and Adjacent Deep–Water Areas of the Arctic Basin. Part I*; Sirenko, B.I., Ed.; Explorations of the Fauna of the Seas; ZIN RAS Press: St Petersburg, Russia, 2004; Volume 54, pp. 9–27. (In Russian)

56. Golikov, F.N.; Sirenko, B.I.; Petriashev, V.V.; Gagaev, S.Y. Distribution of bottom communities in the Chukchi Sea from results of diving investigations. In *Ecosystems and Bioresources of the Chukchi Sea and Adjacent Areas*; Sirenko, B.I., Ed.; Explorations of the Fauna of the Seas; ZIN RAS Press: St Petersburg, Russia, 2009; Volume 64, pp. 56–62. (In Russian)

57. Evseeva, O.Y.; Dvoretsky, A.G. Shallow–water bryozoan communities in a glacier fjord of West Svalbard; Norway: Species composition and effects of environmental factors. *Biology* **2023**, *12*, 185. [CrossRef]

58. Denisenko, N.V. Bryozoans of the Kara Sea: Species diversity, changes, and peculiarities of biogeographic composition. *Proc. Zool. Inst. Russ. Acad. Sci.* **2021**, *325*, 217–234. (In Russian) [CrossRef]

59. Prischepa, B.F. (Ed.) *Ecosystem of the Kara Sea*; PINRO Press: Murmansk, Russia, 2008. (In Russian)

60. Denisenko, S.G. *Biodiversity and Bioresources of Macrozoobenthos in the Barents Sea: Structure and Long–Term Changes*; Nauka: St Petersburg, Russia, 2013. (In Russian)

Article

Artificial Seaweed Substrates Complement ARMS in DNA Metabarcoding-Based Monitoring of Temperate Coastal Macrozoobenthos

Barbara R. Leite [1,2,3], Sofia Duarte [1,2], Jesús S. Troncoso [3,4] and Filipe O. Costa [1,2,*]

1   Centre of Molecular and Environmental Biology (CBMA) and ARNET-Aquatic Research Network, Department of Biology, University of Minho, Campus de Gualtar, 4710-057 Braga, Portugal
2   Institute of Science and Innovation for Bio-Sustainability (IB-S), University of Minho, Campus de Gualtar, 4710-057 Braga, Portugal
3   UVIGO Marine Research Centre (CIM-UVIGO), ECIMAT Marine Station, Toralla Island, S/N, 36331 Vigo, Spain
4   Department of Ecology and Animal Biology, Marine Sciences Faculty, University of Vigo, Campus Lagoas-Marcosende, 36310 Vigo, Spain
*   Correspondence: fcosta@bio.uminho.pt

Abstract: We used DNA metabarcoding to compare macrozoobenthic species colonization between autonomous reef monitoring structures (ARMS) and artificial seaweed monitoring systems (ASMS). We deployed both substrates in two different locations (Ría de Vigo and Ría de Ferrol, NW Iberian coast) and collected them after 6, 9, and 12 months to assess species composition of the colonizing communities through high-throughput sequencing of amplicons within the barcode region of the mitochondrial cytochrome c oxidase I (COI-5P) and the V4 domain of the 18S rRNA genes. We observed a consistently low similarity in species composition between substrate types, independently of sampling times and sites. A large fraction of exclusive species was recorded for a given substrate (up to 72%), whereas only up to 32% of species were recorded in both substrates. The shape and structural complexity of the substrate strongly affected the colonization preferences, with ASMS detecting more exclusive crustacean and gastropod species and a broader diversity of taxonomic groups (e.g., Entoprocta and Pycnogonida were detected exclusively in ASMS). We demonstrate that despite the customary use of ARMS for macrozoobenthos monitoring, by using ASMS we complemented the recovery of species and enlarged the scope of the taxonomic diversity recorded.

Keywords: substrate type; artificial reef monitoring structures (ARMS); artificial seaweed monitoring system (ASMS); environmental DNA; COI; 18S

Citation: Leite, B.R.; Duarte, S.; Troncoso, J.S.; Costa, F.O. Artificial Seaweed Substrates Complement ARMS in DNA Metabarcoding-Based Monitoring of Temperate Coastal Macrozoobenthos. *Diversity* **2023**, *15*, 657. https://doi.org/10.3390/d15050657

Academic Editors: Thomas J. Trott and Michael Wink

Received: 29 March 2023
Revised: 28 April 2023
Accepted: 4 May 2023
Published: 12 May 2023

1. Introduction

Species interactions within marine communities are responsible for the maintenance of a biological network (i.e., producers, predators, and decomposers) highly important in ecosystem processes (e.g., energy flow, primary and secondary production, nutrient recycling [1,2]). However, the functioning and ability of marine ecosystems to provide services can be severely compromised due to the effects of global impacts (e.g., multiple stressors, and other human pressures [3,4]).

The Lusitanian biogeographic province [5] constitutes a particularly interesting spot for marine research since it harbors a high diversity of macrofauna from various adjacent regions, and many species have their northern or southern range limits in this area [6]. Monitoring these communities is particularly relevant to assess the impact of global change on marine biodiversity and ecosystems (e.g., shifts in species range expansions or alteration of dispersal patterns) and changes in species interactions [1,7].

However, large-scale biodiversity assessment and hard-bottom community sampling are challenging, mostly due to difficulties in the assessment and retrieval of organisms for identification [8,9]. Implementing innovative and standardized methods is, thus, essential [10,11] to make species data more accessible and facilitate spatial-temporal comparisons [12]. Technical advances in monitoring through the implementation of innovative molecular approaches [13], namely DNA metabarcoding, provide an opportunity to rapidly improve the accuracy and throughput for marine biodiversity assessment and monitoring [14,15]. The combined employment of artificial substrates together with DNA metabarcoding, may be a valuable replicable and standardized methodology to monitor marine macrozoobenthos, using a more cost-effective and less challenging approach (i.e., faster and with greater throughput and accuracy) [16]. Although artificial substrates have already been used to promote colonization and monitor marine communities (e.g., [17–19]), the implementation of such a strategy in large-scale comparisons is difficult due to the low level of standardization of the monitoring methodologies [20].

Autonomous reef monitoring structures (ARMS), originally developed to mimic coral reef diversity, have a structure with cavities influenced by high and low light spaces and various flow regimes [21]. These characteristics provide shelter for small invertebrates (e.g., protecting against predation) and surfaces for sessile organisms' settlement [22,23]. Deployed over the long term, ARMS allow to assess and interpret the diversity, distribution, and structure of hard-bottom marine communities [24], and has been frequently applied in the assessment of diversity in a variety of geographic regions (e.g., Caribbean and Indo-Pacific [25], Singapore [26], Europe [27], Iberian Coast [28]). An artificial seaweed monitoring system (ASMS) is an alternative artificial substrate that mimics macroalgae, and that has been previously employed to study macrozoobenthos colonization in Ría de Ferrol (NW Iberian Peninsula) [29,30]. Using morphological approaches to compare the taxa composition between dendritic substrates (ASMS) and ARMS deployed side-by-side, the authors concluded that the substrates supported different and complementary macrofauna assemblages [30]. The observed differences reflect the differential attractiveness of the more complex tridimensional structure of ASMS to shelter a distinct set of species, particularly highly mobile fauna [30].

Thus, the combined monitoring using ARMS and ASMS could potentially provide more comprehensive and comparable assessments of a broader spectrum of the biodiversity of hard-bottom communities. However, the research effort required for such a monitoring scheme using morphology-based approaches would be probably too demanding and logistically unfeasible. Alternatively, the employment of DNA metabarcoding for species identifications would allow high-throughput monitoring and greater accuracy, including the capability to discriminate cryptic species, or specimens that can be damaged during sample processing, as well as taxa more recalcitrant to identification through morphology [11,18,31–33]. Despite the increasing implementation of ARMS in different geographical locations for hard-bottom marine monitoring using molecular approaches [27,34], the comparison between different types of artificial substrates and their influence on the assessment of macroinvertebrate species has not been performed yet. In this study, we compared the macrozoobenthic species colonizing ARMS and ASMS, in order to investigate the impact of their shape complexity on the recovery of species and the thoroughness of coastal monitoring. To this end, we deployed ARMS and ASMS side-by-side in two locations on the NW Iberian coast, and using DNA metabarcoding we monitored changes in species composition at three time points over a period of 12 months.

2. Materials and Methods

2.1. Study Area

This study was carried out at two locations on the NW Iberian Atlantic coast: Bajo Tofiño (42°13′42.3″ N 8°46′43.2″ W, Ría de Vigo, Spain) and San Cristovo (43°27′53.8″ N 8°18′00.7″ W, Ría de Ferrol, Spain). Ría de Vigo and Ría de Ferrol are fully marine environments structurally composed of semi-enclosed bays, which include both hard and soft substrata and have high primary productivity [35–37]. Both regions are busy harbors and ports directly impacted by human activities (e.g., sewage runoff or harvesting [38,39]).

2.2. Sampling Design

We selected two types of artificial substrates: ARMS (Figure 1a), which are small, tiered platforms, composed of 9 piled-up plates (23 × 23 cm) of grey type I PVC separated by spacers and affixed to the seafloor, and ASMS (Figure 1b), which are plastic commercial artificial plants (IKEA, Delft, The Netherlands), with 28 cm height and composed by green polyethylene with a complex structure formed by different orientation of the plant branches (Figure 1). Three replicates of the two substrates were deployed in June 2018 anchored to a cement plate (60 × 60 cm) and fixed to the bottom (approximately 11 m of depth), in the two study sites. After 6, 9, and 12 months of deployment, one replicate of each substrate was collected in both study sites. Previous studies have shown that one replicate processed by metabarcoding, and using two molecular targets, is sufficient to capture the same, or higher, macrozoobenthos species diversity when compared with the cumulative number of species detected using triplicates and morphology-based assessments e.g., [33,40]. Therefore, one replicate of each substrate was used at each site/sampling time combination.

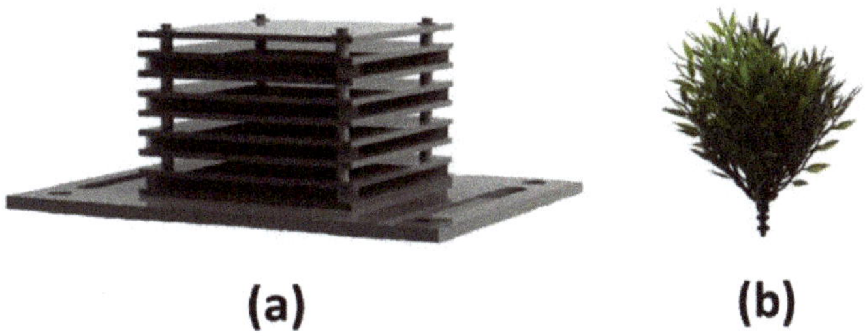

Figure 1. Artificial substrates used for marine macrozoobenthic colonization: (**a**) ARMS and (**b**) ASMS.

2.3. ARMS and ASMS Collection and Processing

In order to limit the loss of motile organisms, ARMS were enclosed in a labeled plastic box by scuba divers that lifted it onto the boat. For ASMS, each sample was carefully enclosed in a 500 µm mesh bag and then introduced in a hermetic plastic bag before being released from the substratum with a scraper.

At the laboratory, samples were photographed and then processed. We disassembled ARMS plate by plate following the procedure of Leray and Knowlton [22], and each branch of ASMS was also detached and processed individually. Then, each sample (plate or branch) was carefully washed using filtered sea water and shaken vigorously, and the representative mobile and sessile fauna were separated. The mobile fauna was brushed and sieved (500 µm). After collecting the mobile fauna, the sessile fauna was scraped with a spatula into a tray. The water in the container of each substrate was also sieved (500 µm), and the retained organisms were preserved with mobile fauna. All samples were then preserved in absolute ethanol and stored at −20 °C until further analysis.

2.4. DNA Metabarcoding

Samples collected from both locations after 6 (T1), 9 (T2), and 12 (T3) months were used to assess the species composition of the colonizing communities in both substrates, through high-throughput sequencing (HTS) of amplicons from the mitochondrial cytochrome c oxidase I (COI) and the 18S rRNA (18S) genes.

The mobile and sessile fauna were processed, amplified, and sequenced individually. DNA extraction procedures were adapted from Ivanova et al. [41] silica-based method, as described by Steinke et al. [42]. Ethanol-preserved samples were first filtered to retain the biomass and the ethanol was discarded. Then, based on the wet weight of each sample [42], an appropriate volume of a lysis buffer solution was added (100 mM NaCL, 50 mM Tris-

HCL pH 8.0, 10 mM EDTA, 0.5% SDS) and each sample was incubated overnight at 56 °C, while gently mixed on a shaker (60 rpm). Negative controls were included through all DNA extraction procedures. To maximize diversity recovery, two aliquots of each lysate were used, totaling two DNA extractions per sample. After extraction, the aliquots of genomic DNA for the same sample were pooled together.

The production of amplicon libraries and the HTS were carried out at Genoinseq (Cantanhede, Portugal), as described below. For COI, the primer pair mICOIintF (5′-GGWACWGGWTGAACWGTWTAYCCYCC-3′) [43] and LoboR1 (5′-TAAACYTCWGGRTGWCCRAARAAYCA-3′) [44] was selected to amplify an internal segment with 313 bp. The primer pair TAReuk454FWD1 (5′-CCAGCASCYGCGGTAATTCC-3′) and TAReukREV3 (5′-ACTTTCGTTCTTGATYRA-3′) [45,46] was used to amplify 400 bp of the V4 region of the 18S rRNA gene. PCR reactions were performed using KAPA HIFI HotStart PCR Kit according to manufacturer instructions, 0.3 µM of each PCR primer and 50 ng of template DNA in a total volume of 25 µL. For the mICOIintF/LoboR1 primer pair, the PCR conditions involved a 3 min denaturation at 95 °C, followed by 35 cycles of 98 °C for 20 s, 60 °C for 30 s, and 72 °C for 30 s and a final extension at 72 °C for 5 min. For the TAReuk454FWD1/TAReukREV3 primer pair, the PCR conditions involved a 3 min denaturation at 95 °C, followed by 10 cycles of 98 °C for 20 s, 57 °C for 30 s, and 72 °C for 30 s and 25 cycles of 98 °C for 20 s, 47 °C for 30 s, and 72 °C for 30 s, and a final extension at 72 °C for 5 min.

Second PCR reactions added indexes and sequencing adapters to both ends of the amplified target region according to manufacturer's recommendations [47]. PCR products were then one-step purified and normalized using SequalPrep 96-well plate kit (ThermoFisher Scientific, Waltham, MA, USA) [48], pooled and pair-end sequenced in an Illumina MiSeq® sequencer with the V3 chemistry, according to manufacturer's instructions (Illumina, San Diego, CA, USA).

PCR amplification failed with the COI primers for the following samples and fractions: T1, for ARMS in Ría de Ferrol (both mobile and the sessile fauna) and in Ría de Vigo (sessile fauna), and ASMS in Ría de Vigo (mobile fauna); T2, for ARMS in Ría de Vigo (sessile fauna) (Table S1). With the exception of T1 for ARMS in Ría de Ferrol, where only 18S data were produced, for all other samples at least one macrozoobenthic fraction (i.e., either sessile or mobile) was successfully recovered with COI. Therefore, for all samples, except T1/ARMS/Ferrol, we opted to use merged lists of species obtained together with 18S and COI, in subsequent analyses.

2.5. Data Processing

Raw reads, extracted from Illumina MiSeq® System in fastq format, were size- (<100 bp for COI region and <150 bp for 18S) and quality filtered to remove sequencing adapters (PRINSEQ v.0.20.4 [49]). Bases with an average quality lower than Q25 in a window of 5 bases were trimmed. The filtered forward-R1 and reverse-R2 reads were merged (make.contigs function) by overlapping pair-end reads using mothur 1.39.5 [50,51] and primers sequences were removed (trim.seqs function).

For COI, the usable reads were then processed and submitted to mBrave-Multiplex Barcode Research and Visualization Environment (www.mbrave.net, accessed on 28 August 2020, [52]) for filtering (maximum 313 bp, minimum 150 bp, QV > 10) and subsequent queries using the sample batch function which is linked with the Barcode of Life Data System (BOLD) [53]. Curated regional reference libraries e.g., [54,55] were given priority for taxonomic assignments in mBRAVE. Reads were taxonomically assigned at species level using 97% similarity threshold and were only retained for further analysis when an OTU was composed of at least eight sequences.

The 18S usable reads were processed and quality controlled (maximum 400 bp, >150 aligned nucleotides, <2% ambiguities or homopolymers, 50 alignment identity, 40 alignment score), aligned using the SILVA Incremental Aligner (SINA v1.2.10 for ARB SVN (revision 21008) [56] against the SILVA SSU rRNA SEED, and analyzed in SILVAngs database (https://ngs.arb-silva.de/silvangs/, accessed on 27 February 2023 [57]), to generate the OTU tables and taxonomic assignments. After these initial steps, identical reads were identified (dereplication), the unique reads were clustered (OTUs) on a per-sample basis, and the reference read of each OTU was then taxonomically classified. VSEARCH (version 2.15.1; https://github.com/torognes/vsearch, accessed on 27 February 2023) [58] was used for dereplication and clustering, applying identity criteria of 1.00 and 0.99, respectively. The taxonomic classification was performed using BLASTn (2.2.30+; http://blast.ncbi.nlm.nih.gov/Blast.cgi, accessed on 27 February 2023) [59] with standard settings and the non-redundant version of the SILVA SSU Ref dataset (release 138.1; http://www.arb-silva.de, accessed on 27 February 2023). The taxonomic classification of each OTU reference read was mapped onto all reads that were assigned to the respective OTU using 99% similarity threshold. Reads without any or weak classifications, where the function "(% sequence identity + % alignment coverage)/2" did not exceed the value of 70, remained unclassified, and were assigned to "No Taxonomic Match".

For both markers, the obtained reads were analyzed separately for mobile and sessile fauna samples, and then combined for data analysis. The status of the species names was verified in the World Register of Marine Species (WoRMS) database (http://www.marinespecies.org/, accessed on 7 march 2023).

2.6. Statistical Analyses

Only OTUs with matches at species level (>97% for COI and >99% for 18S) and composed by at least eight sequences were retained for further data analyses. Any cases of ambiguous assignments (e.g., more than one species name >97% or >99% threshold) were inspected and resolved individually. Only assignments to morphospecies were used to allow accurate and standardized comparisons, given that OTUs may not always correspond with morphospecies.

For each marker, the number of reads from OTUs with a match to the same species were summed up together, and presence(1)/absence(0) species tables were constructed in Microsoft Excel (for Windows) for each marker (COI, Table S2; 18S, Table S3) and both markers together (Table S4), for subsequent analyses. Qualitative data of species distribution among taxonomic groups was displayed through bar graphs (GraphPad Software, Inc.). The proportion of overlapping and unique species detected among sampling times, for each substrate, and between substrates for each sampling time was determined for both sampling locations and displayed using Venn diagrams, using the web tool InteractiVenn [60].

Multivariate analyses were carried out considering presence/absence of the taxa due to the qualitative nature of the molecular data. Non-metric multidimensional scaling (nMDS) analyses were performed using PAleontological STatistics (PAST) version 4.03 [61] for Windows and based on Bray–Curtis resemblance coefficient between samples to visualize community distribution from the two sampled locations (Ría de Vigo and Ría de Ferrol), for all substrates and sampling times, for 18S and 18S + COI.

3. Results

A total of 190 species, representing 11 phyla (Annelida, Arthropoda, Bryozoa, Chordata, Cnidaria, Echinodermata, Entoprocta, Mollusca, Nemertea, Platyhelminthes, and Porifera; Table S4) were detected in this study combining data from ARMS and ASMS, and all sampling sites and times, in both locations. Through the observation of photographs, differences between sampling sites and times were patent within each substrate (Figure 2).

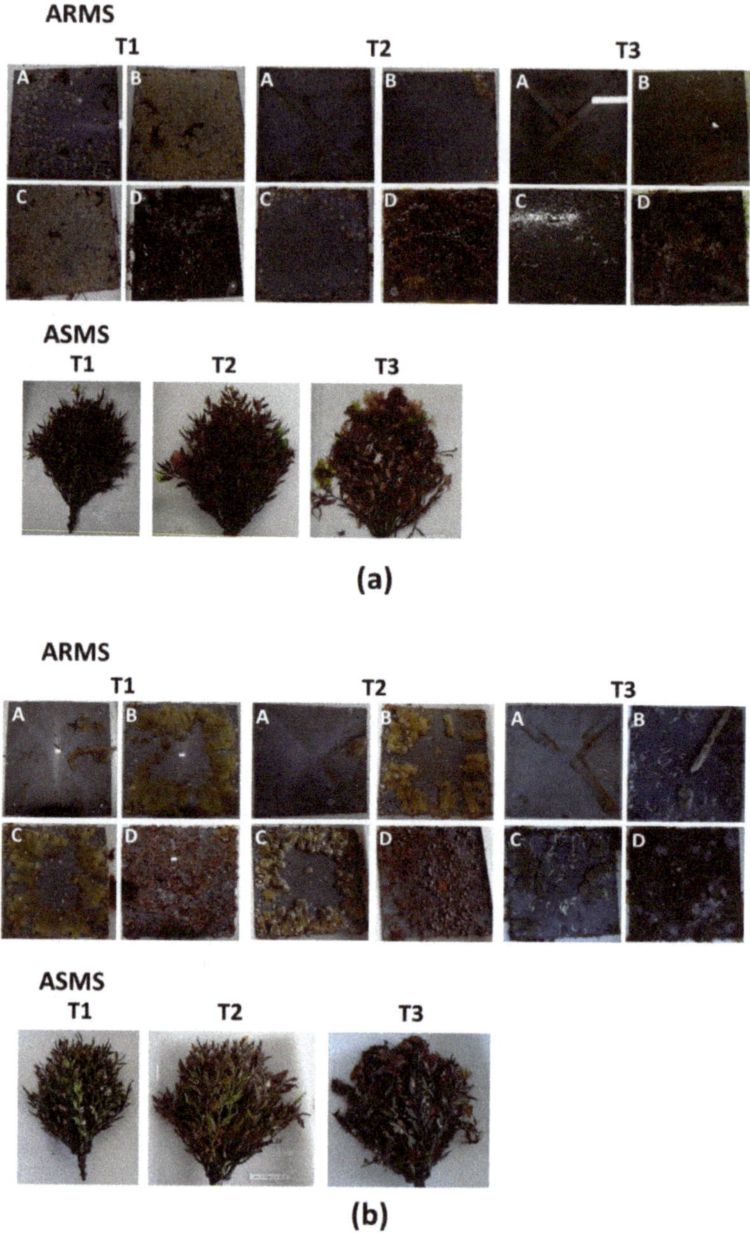

Figure 2. Sampled ARMS face plates and ASMS collected after 6 (T1), 9 (T2), and 12 (T3) months of deployment at (**a**) Ría de Ferrol and (**b**) Ría de Vigo. In ARMS: (**A**) Plate 1 top, (**B**) Plate 5 bottom, (**C**) Plate 9 top, and (**D**) Plate 9 bottom.

High-throughput sequencing from marine macrozoobenthic samples, for both markers and for the total of 24 samples, generated a total of 1,348,329 usable reads for both markers (Table S1) of these, 49% were assigned to marine macrozoobenthos species (30% using mICOIintF/LoboR1 and 19% with TAReuk454FWD1/TAReukREV3).

One-hundred and four species, distributed by 12 high-rank taxa (i.e., phylum, subphylum, and class level) were retrieved from ARMS samples, where 61 species were recovered from Ría de Ferrol and 76 species from Ría de Vigo (Figure 3, Table S5). At both locations, Annelida (15 and 19 species), Crustacea (14 and 12 species), and Echinodermata (10 and 15 species, in Ría de Ferrol and Ría de Vigo, respectively), were the most well represented. In addition, Cnidaria was also among the top rank contributors for the total number of species recovered in Ría de Vigo (15 species) (Figure 3, Table S5). On the other hand, a higher number of species (143) were recovered from ASMS samples, distributed by 14 high-rank taxa, with exclusive detection of Entoprocta and Pycnogonida, where 85 species were recovered from Ría de Ferrol and 100 species from Ría de Vigo (Table S5). The major contributors to ASMS community diversity were Crustacea (21 and 26 species), Annelida (13 species, in both locations), and Cnidaria (9 and 25 species, in Ría de Ferrol and Ría de Vigo, respectively) (Figure 3, Table S5). Furthermore, Gastropoda was also among the top rank contributors for the total number of species in Ría de Ferrol (nine species) (Figure 3, Table S5).

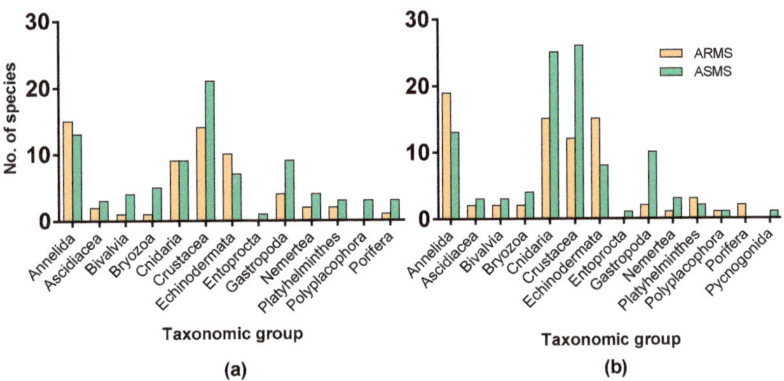

Figure 3. Taxonomic distribution of the cumulative species detected on each substrate and location: along all sampling times since substrate deployment. (**a**) Ría de Ferrol and (**b**) Ría de Vigo.

The lowest number of species was recovered in ARMS deployed in Ría de Ferrol for 6 months (17 species), but since we were not able to produce any amplicons for COI in this particular sampling time and location, this lower number might be the result of having data only for one genetic marker (Figure 4). On the other hand, the highest number of species (71 species) was recorded in ASMS deployed in Ría de Vigo, after 12 months (Figure 4). In general, the highest number of species recovered from ARMS in both locations was found after 12 months of deployment and when data from both genetic markers were combined together (34 and 43 species, for Ría de Ferrol and Ría de Vigo, respectively). On the other hand, in ASMS, the pattern was more variable; while the highest number of species was recovered after 12 months of deployment in Ría de Vigo (71 species), in Ría de Ferrol the highest diversity was attained after 6 months of deployment (51 species).

The contribution of dominant taxonomic groups changed over time of deployment of both substrates at both locations (Figure 4, Table S6). In Ría de Vigo, for ARMS, the contribution of Annelida (6 to 13 species) and Cnidaria (0 to 10 species) increased over time, while for Crustacea (3 to 8 species) and Echinodermata (9 to 13 species), reached the peak after 9 months, but decreased after 12 months. For ASMS, while a similar pattern was found for Cnidaria (9 to 21 species), which increased across time, an opposite pattern was found for Annelida, which decreased over time (10 to 6 species), and Echinodermata (2 to 7 species) and Crustacea (11 to 19 species), which increased over time of deployment (Figure 4).

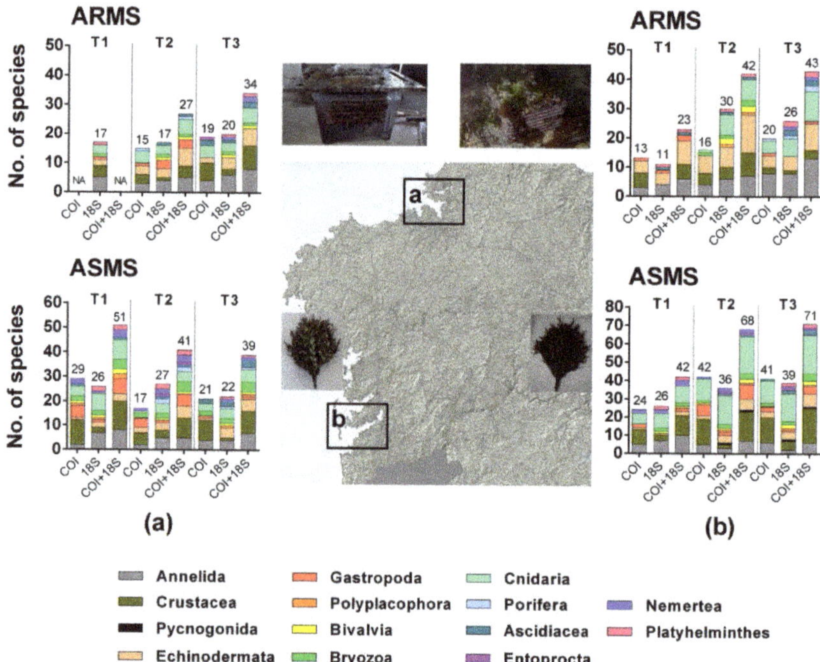

Figure 4. ARMS and ASMS substrates from the two sampling sites: (**a**) Ría de Ferrol and (**b**) Ría de Vigo. Bar charts represent the abundance and taxonomic distribution of species detected on each substrate among sampling times (T1, 6 months; T2, 9 months and T3, 12 months of deployment).

In Ría de Ferrol, a similar pattern was found in ARMS for Annelida (5 to 8 species) and Cnidaria (3 to 5 species), which increased over time, while an opposite trend to the found in Ría de Vígo was observed for Crustacea (4 to 8 species) and Echinodermata (2 to 6 species), which increased with time of deployment. In ASMS, for Annelida, Cnidaria, and Crustacea, the lowest species number was found after 9 months of deployment (5, 4, and 8 species, respectively), while contributions were slightly higher after 6 months of deployment (8, 8, and 12 species, respectively). Other groups displayed an increase in contribution across time, such as Echinodermata (3 to 5 species), while Gastropoda (6 to 1 species) decreased along the time of deployment (Figure 4, Table S6).

Comparing the species composition over time for each substrate individually, and considering the total species detected for each site (combining mobile and sessile fauna, and species recovered with both markers), the similarity of species occurrences was very low, especially for ARMS (5 and 9%, for Ría de Ferrol and Ría de Vigo, respectively), in comparison with ASMS (16 and 20%, for Ría de Ferrol and Ría de Vigo, respectively) (Figure 5, Table S7).

A great percentage of species (45 to 72% in Ría de Ferrol, 48% when joining data from all sampling times, and 53 to 62% in Ría de Vigo, 43% when joining data from all sampling times, respectively), were detected exclusively in ASMS (Figures 6 and S1, Table S8). Only 13 to 21% (25% for all data together) and 8 to 25% (32% for all data together), of the species, were detected on both substrates, in Ría de Ferrol and Ría de Vigo, respectively (Figure 6 and Figure S1, Table S8). Although from a quantitative perspective, ASMS retrieved more taxa in all sampling times and at both sampling sites; qualitatively the set of species observed in each substrate differed considerably (Figure 6, Table S8). When considering all exclusive species retrieved by each substrate in the overall experiment, a clear increase in Crustacea and Gastropoda and a decrease in the contribution of Echinodermata was observed for ASMS, deployed at both locations

(Figure 6, Table S8). Most of these exclusive species retrieved from ASMS at both locations were high mobility species, such as amphipods (e.g., *Ampithoe rubricata* (Montagu, 1808), *Caprella acanthifera* Leach, 1814 and *Jassa herdmani* (Walker, 1893)), but also gastropods such as *Calliostoma zizyphinum* (Linnaeus, 1758) or *Rissoa parva* (da Costa, 1778). In Ría de Vigo, an increase in the number of exclusive Cnidaria species, in particular of hydrozoans (e.g., *Abietinaria filicula* (Ellis and Solander, 1786), *Nemertesia antennina* (Linnaeus, 1758), *Halecium mediterraneum* Weismann, 1883), was also recovered from ASMS (Figure 6b, Table S8). In addition, exclusive groups such as Entoprocta (*Pedicellina cernua* (Pallas, 1774)) and Polyplacophora (e.g., *Acanthochitona fascicularis* (Linnaeus, 1767)) were exclusively detected in ASMS in Ría de Ferrol, while Entoprocta (*P. cernua*) and Pycnogonida (*Achelia echinata* Hodge, 1864) were exclusively detected in ASMS deployed in Ría de Vigo (Figure 6, Table S8). On the other hand, the most exclusive species detected on ARMS were Annelida and Echinodermata (e.g., *Sabellaria spinulosa* (Leuckart, 1849), detected in ARMS deployed at both locations or *Asterias rubens* Linnaeus, 1758, detected exclusively in ARMS deployed in Ría de Vigo) (Figure 6, Table S8).

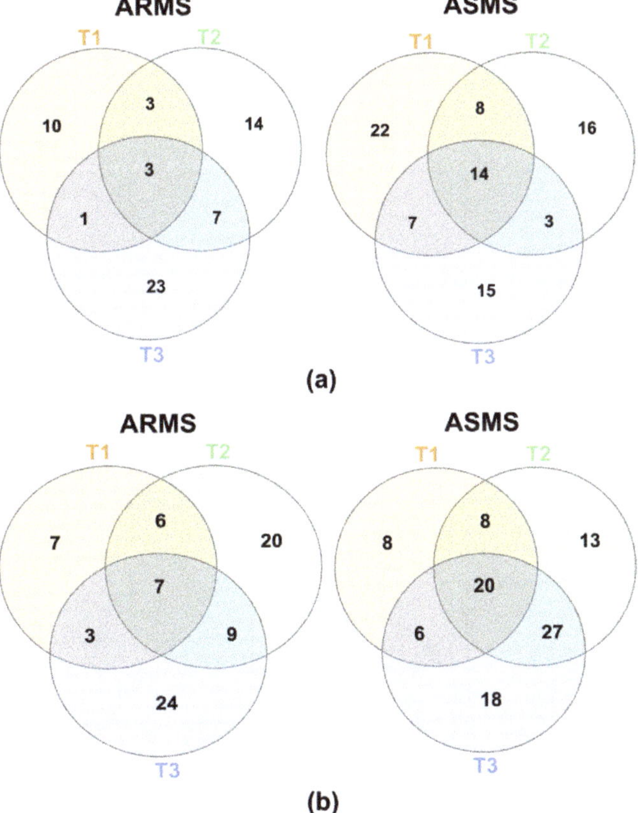

Figure 5. Partitioning of the marine macrozoobenthic species detected exclusively at each time of deployment (T1, 6 months; T2, 9 months and T3, 12 months) and shared by all sampling times (overlapping circles) for each substrate at each location: (**a**) Ría de Ferrol and (**b**) Ría de Vigo.

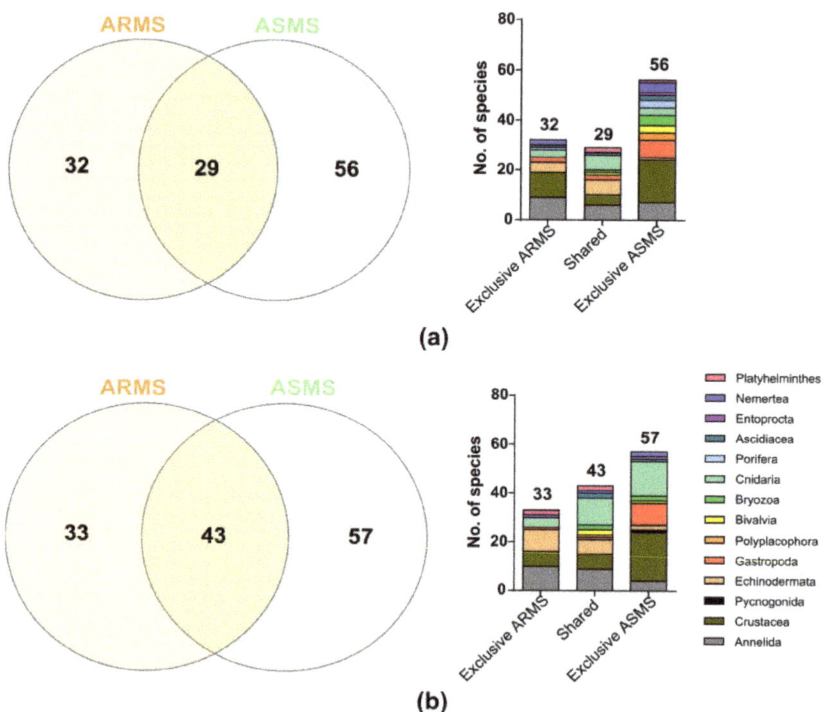

Figure 6. Partitioning of the marine macrozoobenthic species detected exclusively by ARMS, exclusively by ASMS, and shared by both substrates (overlapping circles), through all time of deployment in: (**a**) Ría de Ferrol and (**b**) Ría de Vigo, and respective taxonomic classification of each set of species.

Considering the substrates deployed at Ría de Ferrol (Figure 6a, Table S9), the majority of the species detected in one substrate only (32 in ARMS and 56 in ASMS) were exclusive species (i.e., species detected only in one substrate and one sampling time point; 61%). Furthermore, only 16% of the species in ASMS were pervasive (i.e., detected in all sampling times), while in ARMS only 5% of the species were detected over all sampling times. Among the species detected by both substrates, only two species were detected in all sampling times [the ophiuroid *Ophiothrix* (*Ophiothrix*) *oerstedii* Lütken, 1856, and the bryozoan *Tubulipora liliacea* (Pallas, 1766)].

The same pattern was found in the substrates collected from Ría de Vigo (Figure 6b, Table S9): 33 species were detected only in ARMS while 57 species were exclusively observed in ASMS, with 41% of the species being detected only in one substrate and one sampling time point. The percentage of pervasive species was low in both substrates (9% for ARMS and 20% for ASMS, respectively), whereas no species were detected by both substrates, for all sampling times combinations.

Comparing sampling locations, by combining the detected species by both substrates on each sampling location, a slightly higher number of species was detected in Ría de Vigo (133 species) than in Ría de Ferrol (117 species); Figure 7, Tables S9 and S10). However, differences in taxonomic groups were recorded (e.g., Pycnogonida was only detected in Ría de Vigo). Among the species detected only in Ría de Vigo, the majority were exclusive species (i.e., species detected only in one substrate/sampling time combination, 29%). The same pattern was observed in Ría de Ferrol, where 37% were exclusive species.

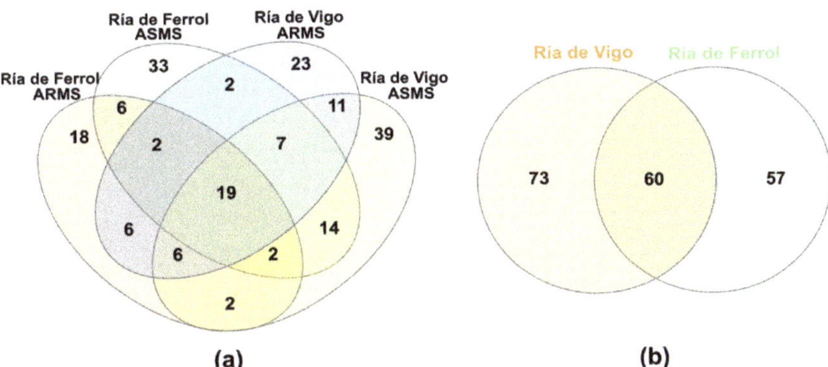

Figure 7. Partitioning of the marine macrozoobenthic species detected exclusively and shared (overlapping circles): (**a**) by each combination of substrate type (ARMS and ASMS) and location (Ría de Vigo and Ría de Ferrol) and (**b**) total number of species detected exclusively and shared (overlapping circles) in both locations.

Regarding the shared species among sampling sites, most of them had a variable occurrence independently of the substrate or sampling times (Table S9). For instance, *Pilumnus hirtellus*, was recovered in Ría de Vigo after 6 months (ARMS) and 9 months (ARMS and ASMS), and after 12 months of deployment in Ría de Ferrol (ARMS). The caprelid *Caprella acanthifera* detected in ASMS on Ría de Ferrol after 6 months of deployment, was then detected in Ría de Vigo at 9 and 12 months of deployment of the same substrate type. Another example was recorded for three decapods (*Eualus cranchii* in ASMS, *E. occultus* in ARMS, and *Hyppolyte varians* in ASMS) that were first detected after 9 months of deployment in Ría de Vigo, and then after 12 months of the same substrate type in Ría de Ferrol (Table S9). Overall, no species was recorded as completely pervasive (detected in all sampling times, substrates, and study sites) (Table S9), but 10% of the species were recovered from both substrates and at both sampled locations (19 species, Figure 7). The prevalence of exclusive species (i.e., species detected only in 1 substrate/sampling time/site combination) was observed for 43.2% of the total species (82 species) (Table S9).

Non-metric multidimensional scaling, based on species detected on each substrate and sampling time, at the two sampling locations and by using data recovered with the 18S marker (for which amplification was successful for all fractions, substrates, and sampling times, at both sites, Figure 8a) and using data from both markers together (18S + COI) (Figure 8b), revealed aggregation of the samples according to the sampling location, and within each sampling location according to the substrate, for both situations (18S alone and both markers together) (Figure 8). However, a higher variation between ARMS is evident by the lower similarity shared among time points, while a high similarity was observed among ASMS samples retrieved at the different time points, for both locations. The differences observed in the nMDS diagrams were supported by two-way PERMANOVA, which indicated that both location and substrate, but not the interaction between both factors, significantly affected the community structure of macrozoobenthic species retrieved either with 18S or with both markers (Table S11).

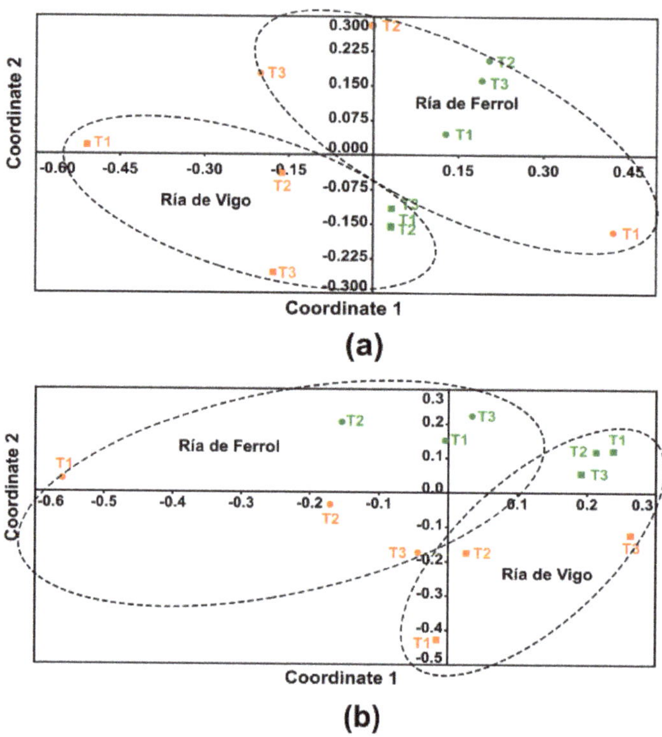

Figure 8. Non-metric multidimensional scaling (nMDS), based on Bray–Curtis similarity index, of the species detected in ASMS (green) and ARMS (orange), for all sampling times (T1, 6 months; T2, 9 months and T3, 12 months of deployment) on each sampled location (Ría de Vigo, squares, and Ría de Ferrol, circles) recovered: (**a**) with the 18S and (**b**) with both markers.

4. Discussion

Over the past decade, the coupling of ARMS with DNA metabarcoding has been increasingly employed and become a customary approach for the monitoring of marine benthic communities [23,27,32,62]. The success of this approach appears to be the combination of the benefits of substrate standardization using ARMS, supplemented with the high-throughput capacity of DNA metabarcoding, thereby providing a very efficient and comparable approach for monitoring these complex and taxonomically challenging communities [63,64]. In spite of the large benefits of substrate standardization, they may come with a cost, particularly if their widespread use tends to lessen, or reduce, the employment of other sampling strategies. In this study, using an alternative substrate with a different structural complexity that resembles macroalgae, we illustrate how a fair fraction of the marine benthic diversity may be missed by employing an ARMS-exclusive sampling strategy. Globally, our results showed: (i) differences in communities' composition and diversity as a result of the duration of the colonization period, in both substrates and sites; (ii) clear differentiation of the communities between Ría de Vigo and Ría de Ferrol, in both substrates; (iii) large and consistent differentiation between ARMS and ASMS communities within each site.

Changes in community composition during colonization periods of increasing duration would be expected as a result of ecological succession and seasonal fluctuation processes, e.g., [40]. In our previous study in Ría de Vigo, using artificial substrates made of different materials, and employing an identical metabarcoding approach [40], we observed similar fluctuations, although differences in the timing and duration of deployment do not allow direct comparisons. In the current study, we confirmed the occurrence of considerable over-time variations in two separate coastal areas and, importantly, that were also captured by two very distinct types of substrates. The pronounced fluctuations recorded over time indicate that sampling only after 12 months of deployment, which is the minimum deployment period normally used in ARMS [23,26,27,32,34,62,65] may fail to capture a fair diversity of taxa and species. Although more or less long periods of deployment may be required for ecological succession to be completed, and for the colonizing assemblage to reach a point of stability mirroring the natural community in that spot, our data indicates that maximum diversity can be reached under 12 months of deployment. Whilst comparisons with other studies are difficult, mostly due to the employment of different species identification approaches (i.e., morphology-based identifications) and different sampled locations, a previous study suggested that complete colonization of ASMS occurs within 3 months of deployment [66]. However, using ASMS in Ría de Vigo, we detected the maximum number of species after 12 months of deployment and after 6 months in Ría de Ferrol (consistent with Carreira-Flores et al. [29]). Furthermore, higher diversity at intermediate stages of succession has been reported in an earlier study, namely after 7 months of deployment of artificial substrates in Vigo, assessed using both morphology and DNA metabarcoding [40]. The overgrowth of mussels in the substrates that were observed after 10 and 15 months may have contributed to excluding some species and reduced the taxa diversity at later succession stages. The timing of initial deployment will also affect the speed of colonization given that availability of propagules of key species in different stages of succession will vary with season [23,30]. These different results highlighted the importance of seasonal sampling in long-term monitoring to learn when a species is expected to occur, to provide information about communities' changes over temporal scales, and to signal possible faulty detection of pervasive species, which could flag possible changes in the ecosystem. These findings reinforce the importance of considering the time and duration of the deployment of artificial substrates for the monitoring of coastal macrozoobenthic communities. On the other hand, ARMS plain surfaces may require longer colonization periods, particularly due to the lack of refuge and settlement spaces, whereas ASMS canopy may facilitate colonization thereby accelerating the stability of the macrozoobenthic community.

Globally, the variation over time also appeared to be considerably greater in ARMS compared to ASMS, a pattern that, if confirmed, could also have implications for future monitoring considerations. It should be noted that ARMS monitoring was originally developed for tropical reefs [21] where ecological succession may take long but once completed, may be less prone to intense seasonal fluctuations as the ones experienced by temperate communities such as the ones here studied.

In spite of the variations over time, the main differences in community composition were found between locations, with the macrozoobenthos communities from both Rías being clearly distinct from each other, independently of timing and substrate, and each recording a high number of exclusive species. These differences probably reflect particular features of each Ría, and possibly also their geographic location. However, the employment of DNA metabarcoding may have also contributed to more detailed taxonomic profiling, which in turn may have provided greater discrimination ability than using morphological approaches alone. Indeed, most studies to date comparing metabarcoding and morphological approaches have reported the recovery of a higher number of species and diversity with the former [11,15,31,33]. Whereas few ARMS-metabarcoding studies addressed overtime variation in macrozoobenthos, most of them report patterns of spatial variation at local or regional scales [15,23,27,32,34,65]. Interestingly, ARMS and ASMS appear to have consistent patterns of variation over time and between sites, with Ría de Vigo generally showing higher diversity compared to Ría de Ferrol, and comparable overtime variations within the site. Hence, both substrates appear to be similarly effective in capturing beta diversity.

Different communities' composition detected between the two substrates is likely related to the complexity of the shape of the substrates, particularly due to the complex branching pattern in ASMS, in contrast to the ARMS cavities made of flat surface PVC. Additionally, the three-dimensional structure of each substrate has different levels of exposure to light (even within each ARMS plate [28]), to predators, and differences in water flow. A fair number of studies have employed ARMS as substrates for species colonization [20,23,27] and recommend it as a prime tool for standardized monitoring of macrozoobenthic communities [27]. However, our results demonstrated that a fair portion of the macrozoobenthos diversity may fail to be captured by employing exclusively this artificial substrate, at least in the studied region. Hence, the efficiency of this monitoring tool can be partially compromised if there is the risk of systematic overlooking of fractions of diversity. The results showed that compared to ARMS, in ASMS we frequently detected more species for each taxonomic group (except for Annelida, and Echinodermata), and two additional high-rank taxa (i.e., Entoprocta and Pycnogonida) were exclusively detected. However, both substrates demonstrated to be complementary in their ability to be colonized by macrozoobenthic species, since a low proportion of species were recorded concurrently in both substrates. Although missing taxa may not have a great impact in studies aiming for bioassessments of the ecological status [9], these can be critical for studies aiming for long-term monitoring and assessing global change-induced alterations in species ranges and communities' composition, or detection of non-indigenous species (NIS). Results obtained for ASMS, for the three sampling times and for each location, consistently indicated more species and wider taxonomic diversity, as well as approx. up to 70% of exclusive species. Most of these exclusive species were high mobility species, such as crustaceans, in particular amphipods, which have colonized ASMS, at both locations (e.g., *A. rubricata*, *C. acanthifera* and *J. herdmani*), but also Gastropoda (e.g., *C. zizyphinum* and *R. parva*, were found exclusively in ASMS at both locations). Since the substrates were disposed side-by-side at each location, their tridimensional structure would have been key to favor differently the colonization of species. Contrary to ARMS, which has shaded areas, ASMS allowed greater algal growth, that attracted a particular set of species. Another factor could be the high abundance of hydrozoan species in ASMS (in particular for the ones deployed in Ría de Vigo), that serve as food to several gastropod species (e.g., *C. zizyphinum* feeds on hydrozoans). On the other hand, most exclusive species detected on ARMS were Annelida and Echinodermata (e.g., the annelid *S. spinulosa* was detected in ARMS at both locations), which can be frequently found in more shaded and sheltered areas, such as circalittoral bedrock, boulders, or cobbles, which display a greater resemblance with ARMS. Thus, the complementarity between substrates highlights the need to optimize sampling strategies, where employment of both substrates may provide a broader phylogenetic scope and detect sets of species that otherwise could be invariably overlooked.

The communities from each sampling site were well separated into two groups, where exclusive species were the drivers for those aggregations. No pervasive species were identified (all sampling times, substrate, and locations), and only 2% of the species and no species at all were recorded in all sampling times and substrates, in Ría de Ferrol and Ría de Vigo, respectively. In total, as much as 43.2% of the species recorded were exclusive, i.e., only detected in one substrate or sampling time or location. Differences in species detection in both sites and among sampling times could be a result of species distribution patterns. Habitat specificities and geographical distribution are the drivers for species trends and patterns. The new complexity of the habitat, a consequence of substrate colonization, as well as new spaces for shelter and settlement, can lead to shifts in species abundance and occurrence. In addition to the consequences of natural sampling variation, exclusive species appear to occur more randomly over time and space, suggesting that seasonality may have played a role in the observed patterns. Data from other studies combining ARMS with barcoding revealed patterns similar to ours, with high percentages of exclusive species detected (e.g., 44% of all species [67] and more than 50% of species [32]). Compositional changes in marine macrozoobenthic communities associated with ARMS and ASMS highlighted regional differences and suggest that a substrate with higher complexity (for example a combination between ARMS and ASMS) will result in more space for species settlement and improve colonization capacity.

5. Conclusions

Although the benefits of ARMS employment in marine macrozoobenthic monitoring appear to be well established [27], here we show how its exclusive use could introduce a recurrent bias, with sizeable fractions of the biodiversity being systematically overlooked. The obtained results indicate that no single substrate structure is able to capture comprehensively the diversity of a marine hard-bottom community, and the complementarity recorded between substrates highlighted the necessity to consider implications to sampling design. More intensive sampling strategies are now at reach thanks to the greater throughput provided by DNA metabarcoding. However, comprehensive monitoring of these communities would require ponderation on the use of complementary substrate structures, that encompass the natural structural complexity of coastal marine habitats. Reefs and rocky-bottom shores are structurally very elaborate, and it would be impractical to attempt to fully reproduce such complexity. Recent technological developments in material science and 3D printing systems [68] may offer some standardized solutions that could benefit the comprehensiveness and accuracy of marine benthic monitoring.

Supplementary Materials: The following supporting information can be downloaded at: https://www.mdpi.com/article/10.3390/d15050657/s1, A detailed description of all data used to support results is provided in Supplementary Materials. Figure S1: Partitioning of the marine macrozoobenthic species detected exclusively by ARMS, exclusively by ASMS, and shared by both substrates (overlapping circles), after 6 months (T1), 9 months (T2), and 12 months (T3) of deployment, in Ría de Ferrol (A) and Ría de Vigo (B); Table S1: No. of merged, usable reads (quality-filtered) and taxonomically assigned reads to species level (>97% for COI and >99% for 18S) and with equal or more than 8 reads obtained on each recovered fraction (M: mobile, S: sessile) from each substrate (ARMS and ASMS) and sampling time (6, 6 months; 9, 9 months and 12, 12 months of deployment), on each location (Ría de Ferrol and Ría de Vigo). NA, treatments where no amplicons were produced; Table S2: Taxonomic classification of all species detected in all substrates, sampling times, and locations with COI. RF, Ría de Ferrol; RV, Ría de Vigo; 6, 6 months; 9, 9 months and 12, 12 months of deployment; Table S3: Taxonomic classification of all species detected in all substrates, sampling times, and locations with 18S. RF, Ría de Ferrol; RV, Ría de Vigo; 6, 6 months; 9, 9 months and 12, 12 months of deployment; Table S4: Taxonomic classification of all species detected in all substrates, sampling times and locations with COI + 18S. RF, Ría de Ferrol; RV, Ría de Vigo; 6, 6 months; 9, 9 months and 12, 12 months of deployment; Table S5: Total no. of species, distributed by each taxonomic group, recovered by both markers from each substrate deployed in Ría de Ferrol, Ría de Vigo and in the total experiment; Table S6: No. of species distributed by each taxonomic group (Figures 3 and 4) and % of contribution of each taxonomic group for the total no. of species recovered by both genetic markers (with the exception of ARMS deployed in Ría de Ferrol for 6 months, where only data from 18S was available) for each substrate (ARMS and ASMS) and sampling point (6, 6 months; 9, 9 months and 12, 12 months of deployment), at both locations (Ría de Ferrol and Ría de Vigo); Table S7: No. of species recovered on each sampling time (6, 6 months; 9, 9 months and 12, 12 months of deployment), in the total of all sampling times and no. and % of species shared among all sampling times, for each substrate (ARMS and ASMS), on each location (Ría de Ferrol and Ría de Vigo) (Figure 5); Table S8: No. of species and % detected exclusively and shared between substrates (ARMS and ASMS), on each sampling time (6, 6 months; 9, 9 months and 12, 12 months of deployment) and though all experiment, for each location (Ría de Ferrol and Ría de Vigo), and respective lists and taxonomic classifications of species detected exclusively on each substrate and in both, for each location; Table S9: No. of detections for each substrate (ARMS and ASMS), on each location (RF, Ría de Ferrol; RV, Ría de Vigo) and in the totality of each sampling location and both locations. Additionally, 6, 6 months; 9, 9 months, and 12, 12 months of deployment. Numbers 1 to 12, indicate the number of detections, with 1 being one single combination of location, substrate, and sampling time and 12 the full combination of locations, substrates, and sampling time points; Table S10: No. of detections of each species on each sampling location (Ría de Ferrol and Ría de Vigo) and shared between both locations. *, 1 to 3, indicate the number of sampling time detections, for each location and #, 1, indicates exclusive species of each location, and 2, species shared by both locations; Table S11: Results from two-way PERMANOVA analyses testing the effect of location (Lo) and substrate (Su) and the interaction between both factors (LoxSu) on macrozoobenthos community structure recovered with 18S and with both markers (COI + 18S).

Author Contributions: Conceptualization, B.R.L., J.S.T. and F.O.C.; methodology, B.R.L., J.S.T. and F.O.C.; formal analysis, B.R.L., S.D. and J.S.T.; investigation, B.R.L., S.D., J.S.T. and F.O.C.; resources, J.S.T. and F.O.C.; data curation, B.R.L. and S.D.; writing—original draft preparation, B.R.L., S.D., J.S.T. and F.O.C.; writing—review and editing, B.R.L., S.D., J.S.T. and F.O.C.; visualization, B.R.L. and S.D.; supervision, J.S.T. and F.O.C.; project administration, J.S.T. and F.O.C.; funding acquisition, J.S.T. and F.O.C. All authors have read and agreed to the published version of the manuscript.

Funding: This study was supported by the project ATLANTIDA (NORTE-01-0145-FEDER-000040), funded by Programa Operacional Regional do Norte (NORTE2020) and by the "Contrato-Programa" UIDB/04050/2020 funded by national funds through the FCT I.P. (Foundation for Science and Technology). Financial support granted by the FCT to BRL (PD/BD/127994/2016) and SD (CEECIND/00667/2017) is also acknowledged.

Institutional Review Board Statement: Not applicable.

Data Availability Statement: All data used to produce all figures and results description are provided in Supplementary Materials. Metabarcoding datasets will be made available upon request.

Acknowledgments: We would like to thank all the members of the Marine Biology Station of A Graña and all the members of the ECIMAT Marine Station for providing the resources and support during sampling campaigns. We also acknowledge the two anonymous reviewers for their valuable comments.

Conflicts of Interest: The authors declare no conflict of interest. The funders had no role in the design of the study; in the collection, analyses, or interpretation of data; in the writing of the manuscript; or in the decision to publish the results.

References

1. Doney, S.C.; Ruckelshaus, M.; Emmett Duffy, J.; Barry, J.P.; Chan, F.; English, C.A.; Galindo, H.M.; Grebmeier, J.M.; Hollowed, A.B.; Knowlton, N.; et al. Climate Change Impacts on Marine Ecosystems. *Annu. Rev. Mar. Sci.* **2012**, *4*, 11–37. [CrossRef]
2. Niiranen, S.; Yletyinen, J.; Tomczak, M.T.; Blenckner, T.; Hjerne, O.; MacKenzie, B.R.; Müller-Karulis, B.; Neumann, T.; Meier, H.E.M. Combined Effects of Global Climate Change and Regional Ecosystem Drivers on an Exploited Marine Food Web. *Glob. Chang. Biol.* **2013**, *19*, 3327–3342. [CrossRef] [PubMed]
3. Burrows, M.T.; Schoeman, D.S.; Buckley, L.B.; Moore, P.; Poloczanska, E.S.; Brander, K.M.; Brown, C.; Bruno, J.F.; Duarte, C.M.; Halpern, B.S.; et al. The Pace of Shifting Climate in Marine and Terrestrial Ecosystems. *Science* **2011**, *334*, 652–655. [CrossRef] [PubMed]
4. Duarte, C.M. Global Change and the Future Ocean: A Grand Challenge for Marine Sciences. *Front. Mar. Sci.* **2014**, *1*, 63. [CrossRef]
5. Spalding, M.D.; Fox, H.E.; Allen, G.R.; Davidson, N.; Ferdaña, Z.A.; Finlayson, M.; Halpern, B.S.; Jorge, M.A.; Lombana, A.; Lourie, S.A.; et al. Marine Ecoregions of the World: A Bioregionalization of Coastal and Shelf Areas. *BioScience* **2007**, *57*, 573–583. [CrossRef]
6. Pereira, S.G.; Lima, F.P.; Queiroz, N.C.; Ribeiro, P.A.; Santos, A.M. Biogeographic Patterns of Intertidal Macroinvertebrates and Their Association with Macroalgae Distribution along the Portuguese Coast. *Hydrobiologia* **2006**, *555*, 185–192. [CrossRef]
7. Henriques, S.; Pais, M.P.; Batista, M.I.; Teixeira, C.M.; Costa, M.J.; Cabral, H. Can Different Biological Indicators Detect Similar Trends of Marine Ecosystem Degradation? *Ecol. Indic.* **2014**, *37*, 105–118. [CrossRef]
8. Bianchi, C.; Pronzato, R.; Cattaneo-Vietti, R.; Benedetti-Cecchi, L.; Morri, C.; Pansini, M.; Chemello, R.; Milazzo, M.; Fraschetti, S.; Terlizzi, A.; et al. Hard Bottoms. In *Mediterranean Marine Benthos: A Manual of Methods for Its Sampling and Study. Biologia Marina Mediterranea*; Gambi, M.C., Dappiano, M., Eds.; SIBM: Genova, Italy, 2004; Volume 11.
9. Borja, A.; Elliott, M.; Andersen, J.H.; Berg, T.; Carstensen, J.; Halpern, B.S.; Heiskanen, A.-S.; Korpinen, S.; Lowndes, J.S.S.; Martin, G.; et al. Overview of Integrative Assessment of Marine Systems: The Ecosystem Approach in Practice. *Front. Mar. Sci.* **2016**, *3*, 20. [CrossRef]
10. Danovaro, R.; Carugati, L.; Berzano, M.; Cahill, A.E.; Carvalho, S.; Chenuil, A.; Corinaldesi, C.; Cristina, S.; David, R.; Dell'Anno, A.; et al. Implementing and Innovating Marine Monitoring Approaches for Assessing Marine Environmental Status. *Front. Mar. Sci.* **2016**, *3*, 213. [CrossRef]
11. Duarte, S.; Leite, B.; Feio, M.; Costa, F.; Filipe, A. Integration of DNA-Based Approaches in Aquatic Ecological Assessment Using Benthic Macroinvertebrates. *Water* **2021**, *13*, 331. [CrossRef]
12. Tanhua, T.; Pouliquen, S.; Hausman, J.; O'Brien, K.; Bricher, P.; de Bruin, T.; Buck, J.J.H.; Burger, E.F.; Carval, T.; Casey, K.S.; et al. Ocean FAIR Data Services. *Front. Mar. Sci.* **2019**, *6*, 440. [CrossRef]
13. Steyaert, M.; Priestley, V.; Osborne, O.; Herraiz, A.; Arnold, R.; Savolainen, V. Advances in Metabarcoding Techniques Bring Us Closer to Reliable Monitoring of the Marine Benthos. *J. Appl. Ecol.* **2020**, *57*, 2234–2245. [CrossRef]

14. Bourlat, S.J.; Borja, A.; Gilbert, J.; Taylor, M.I.; Davies, N.; Weisberg, S.B.; Griffith, J.F.; Lettieri, T.; Field, D.; Benzie, J.; et al. Genomics in Marine Monitoring: New Opportunities for Assessing Marine Health Status. *Mar. Pollut. Bull.* **2013**, *74*, 19–31. [CrossRef]

15. Cahill, A.E.; Pearman, J.K.; Borja, A.; Carugati, L.; Carvalho, S.; Danovaro, R.; Dashfield, S.; David, R.; Féral, J.-P.; Olenin, S.; et al. A Comparative Analysis of Metabarcoding and Morphology-Based Identification of Benthic Communities across Different Regional Seas. *Ecol. Evol.* **2018**, *8*, 8908–8920. [CrossRef]

16. Aylagas, E.; Borja, A.; Muxika, I.; Rodriguez-Ezpeleta, N. Adapting metabarcoding-based benthic biomonitoring into routine marine ecological status assessment networks. *Ecol. Ind.* **2018**, *95*, 194–202. [CrossRef]

17. Ros, M.; Vázquez-Luis, M.; Guerra-García, J.M. The Role of Marinas and Recreational Boating in the Occurrence and Distribution of Exotic Caprellids (Crustacea: Amphipoda) in the Western Mediterranean: Mallorca Island as a Case Study. *J. Sea Res.* **2013**, *83*, 94–103. [CrossRef]

18. Pearman, J.K.; Aylagas, E.; Voolstra, C.R.; Anlauf, H.; Villalobos, R.; Carvalho, S. Disentangling the Complex Microbial Community of Coral Reefs Using Standardized Autonomous Reef Monitoring Structures (ARMS). *Mol. Ecol.* **2019**, *28*, 3496–3507. [CrossRef]

19. Sedano, F.; Tierno de Figueroa, J.M.; Navarro-Barranco, C.; Ortega, E.; Guerra-García, J.M.; Espinosa, F. Do Artificial Structures Cause Shifts in Epifaunal Communities and Trophic Guilds across Different Spatial Scales? *Mar. Environ. Res.* **2020**, *158*, 104998. [CrossRef]

20. Pavloudi, C.; Yperifanou, E.; Kristoffersen, J.; Dailianis, T.; Gerovasileiou, V. Artificial Reef Monitoring Structures (ARMS) Providing Insights on Hard Substrate Biodiversity and Community Structure of the Eastern Mediterranean Sea. *ACA* **2021**, *4*, e64760. [CrossRef]

21. Zimmerman, T.L.; Martin, J.W. Artificial Reef Matrix Structures (Arms): An Inexpensive and Effective Method for Collecting Coral Reef-Associated Invertebrates. *GCR* **2004**, *16*, 59–64. [CrossRef]

22. Leray, M.; Knowlton, N. DNA Barcoding and Metabarcoding of Standardized Samples Reveal Patterns of Marine Benthic Diversity. *Proc. Natl. Acad. Sci. USA* **2015**, *112*, 2076–2081. [CrossRef] [PubMed]

23. Ransome, E.; Geller, J.B.; Timmers, M.; Leray, M.; Mahardini, A.; Sembiring, A.; Collins, A.G.; Meyer, C.P. The Importance of Standardization for Biodiversity Comparisons: A Case Study Using Autonomous Reef Monitoring Structures (ARMS) and Metabarcoding to Measure Cryptic Diversity on Mo'orea Coral Reefs, French Polynesia. *PLoS ONE* **2017**, *12*, e0175066. [CrossRef] [PubMed]

24. Templado, J.; Paulay, G.; Gittenberger, A.; Meyer, C. Chapter 11. Sampling the Marine Realm. *ABC Taxa* **2010**, *8*, 273–307.

25. Plaisance, L.; Caley, M.J.; Brainard, R.E.; Knowlton, N. The Diversity of Coral Reefs: What Are We Missing? *PLoS ONE* **2011**, *6*, e25026. [CrossRef] [PubMed]

26. Chang, J.J.M.; Ip, Y.C.A.; Bauman, A.G.; Huang, D. MinION-in-ARMS: Nanopore Sequencing to Expedite Barcoding of Specimen-Rich Macrofaunal Samples From Autonomous Reef Monitoring Structures. *Front. Mar. Sci.* **2020**, *7*, 448. [CrossRef]

27. Obst, M.; Exter, K.; Allcock, A.L.; Arvanitidis, C.; Axberg, A.; Bustamante, M.; Cancio, I.; Carreira-Flores, D.; Chatzinikolaou, E.; Chatzigeorgiou, G.; et al. A Marine Biodiversity Observation Network for Genetic Monitoring of Hard-Bottom Communities (ARMS-MBON). *Front. Mar. Sci.* **2020**, *7*, 572680. [CrossRef]

28. David, R.; Uyarra, M.C.; Carvalho, S.; Anlauf, H.; Borja, A.; Cahill, A.E.; Carugati, L.; Danovaro, R.; De Jode, A.; Feral, J.-P.; et al. Lessons from Photo Analyses of Autonomous Reef Monitoring Structures as Tools to Detect (Bio-)Geographical, Spatial, and Environmental Effects. *Mar. Pollut. Bull.* **2019**, *141*, 420–429. [CrossRef]

29. Carreira-Flores, D.; Neto, R.; Ferreira, H.; Cabecinha, E.; Díaz-Agras, G.; Gomes, P.T. Artificial Substrates as Sampling Devices for Marine Epibenthic Fauna: A Quest for Standardization. *Reg. Stud. Mar. Sci.* **2020**, *37*, 101331. [CrossRef]

30. Carreira-Flores, D.; Neto, R.; Ferreira, H.; Cabecinha, E.; Díaz-Agras, G.; Gomes, P.T. Two Better than One: The Complementary of Different Types of Artificial Substrates on Benthic Marine Macrofauna Studies. *Mar. Environ. Res.* **2021**, *171*, 105449. [CrossRef]

31. Lobo, J.; Shokralla, S.; Costa, M.H.; Hajibabaei, M.; Costa, F.O. DNA Metabarcoding for High-Throughput Monitoring of Estuarine Macrobenthic Communities. *Sci. Rep.* **2017**, *7*, 15618. [CrossRef]

32. Carvalho, S.; Aylagas, E.; Villalobos, R.; Kattan, Y.; Berumen, M.; Pearman, J.K. Beyond the Visual: Using Metabarcoding to Characterize the Hidden Reef Cryptobiome. *Proc. R. Soc. B* **2019**, *286*, 20182697. [CrossRef]

33. Duarte, S.; Vieira, P.E.; Leite, B.R.; Teixeira, M.A.L.; Neto, J.M.; Costa, F.O. Macrozoobenthos Monitoring in Portuguese Transitional Waters in the Scope of the Water Framework Directive Using Morphology and DNA Metabarcoding. *Estuar. Coast. Shelf Sci.* **2023**, *281*, 108207. [CrossRef]

34. Pearman, J.K.; Chust, G.; Aylagas, E.; Villarino, E.; Watson, J.R.; Chenuil, A.; Borja, A.; Cahill, A.E.; Carugati, L.; Danovaro, R.; et al. Pan-Regional Marine Benthic Cryptobiome Biodiversity Patterns Revealed by Metabarcoding Autonomous Reef Monitoring Structures. *Mol. Ecol.* **2020**, *29*, 4882–4897. [CrossRef]

35. Moreira, J.; Díaz-Agras, G.; Candás, M.; Señarís, M.P.; Urgorri, V. Leptostracans (Crustacea: Phyllocarida) from the Ría de Ferrol (Galicia, NW Iberian Peninsula), with Description of a New Species of Nebalia Leach, 1814. *Sci. Mar.* **2009**, *73*, 269–285. [CrossRef]

36. Urgorri, V.; Troncoso, J.; Dobarro, J. Malacofauna asociada a una biocenosis de maerl en la ría de Ferrol (Galicia, NO España). *An. Biol.* **1992**, *18*, 161–174.

37. Prego, R.; Fraga, F. A Simple Model to Calculate the Residual Flows in a Spanish Ria. Hydrographic Consequences in the Ria of Vigo. *Estuar. Coast. Shelf Sci.* **1992**, *34*, 603–615. [CrossRef]

38. Barroso, C.; Moreira, M.; Gibbs, P. Comparison of Imposex and Intersex Development in Four Prosobranch Species for TBT Monitoring of a Southern European Estuarine System (Ria de Aveiro, NW Portugal). *Mar. Ecol. Prog. Ser.* **2000**, *201*, 221–232. [CrossRef]

39. Prego, R.; Filgueiras, A.V.; Santos-Echeandía, J. Temporal and Spatial Changes of Total and Labile Metal Concentration in the Surface Sediments of the Vigo Ria (NW Iberian Peninsula): Influence of Anthropogenic Sources. *Mar. Pollut. Bull.* **2008**, *56*, 1031–1042. [CrossRef]

40. Leite, B.R.; Vieira, P.E.; Troncoso, J.S.; Costa, F.O. Comparing species detection success between molecular markers in DNA metabarcoding of coastal macroinvertebrates. *Metabarcoding Metagenomics* **2021**, *5*, e70063. [CrossRef]

41. Ivanova, N.V.; Dewaard, J.R.; Hebert, P.D.N. An Inexpensive, Automation-Friendly Protocol for Recovering High-Quality DNA. *Mol. Ecol. Notes* **2006**, *6*, 998–1002. [CrossRef]

42. Steinke, D.; deWaard, S.L.; Sones, J.E.; Ivanova, N.V.; Prosser, S.W.J.; Perez, K.; Braukmann, T.W.A.; Milton, M.; Zakharov, E.V.; deWaard, J.R.; et al. Message in a Bottle—Metabarcoding Enables Biodiversity Comparisons across Ecoregions. *GigaScience* **2022**, *11*, giac040. [CrossRef] [PubMed]

43. Leray, M.; Yang, J.Y.; Meyer, C.P.; Mills, S.C.; Agudelo, N.; Ranwez, V.; Boehm, J.T.; Machida, R.J. A New Versatile Primer Set Targeting a Short Fragment of the Mitochondrial COI Region for Metabarcoding Metazoan Diversity: Application for Characterizing Coral Reef Fish Gut Contents. *Front. Zool.* **2013**, *10*, 34. [CrossRef] [PubMed]

44. Lobo, J.; Costa, P.M.; Teixeira, M.A.; Ferreira, M.S.; Costa, M.H.; Costa, F.O. Enhanced Primers for Amplification of DNA Barcodes from a Broad Range of Marine Metazoans. *BMC Ecol.* **2013**, *13*, 34. [CrossRef] [PubMed]

45. Stoeck, T.; Bass, D.; Nebel, M.; Christen, R.; Jones, M.D.M.; Breiner, H.-W.; Richards, T.A. Multiple Marker Parallel Tag Environmental DNA Sequencing Reveals a Highly Complex Eukaryotic Community in Marine Anoxic Water. *Mol. Ecol.* **2010**, *19*, 21–31. [CrossRef]

46. Lejzerowicz, F.; Esling, P.; Pillet, L.; Wilding, T.A.; Black, K.D.; Pawlowski, J. High-Throughput Sequencing and Morphology Perform Equally Well for Benthic Monitoring of Marine Ecosystems. *Sci. Rep.* **2015**, *5*, 13932. [CrossRef]

47. Illumina 16S. Metagenomic Sequencing Library Preparation. In *Preparing 16S Ribosomal RNA Gene Amplicons for the Illumina MiSeq System*; Illumina: San Diego, CA, USA, 2013.

48. Comeau, A.M.; Douglas, G.M.; Langille, M.G.I. Microbiome Helper: A Custom and Streamlined Workflow for Microbiome Research. *MSystems* **2017**, *2*, e00127-16. [CrossRef]

49. Schmieder, R.; Edwards, R. Quality Control and Preprocessing of Metagenomic Datasets. *Bioinformatics* **2011**, *27*, 863–864. [CrossRef]

50. Schloss, P.D.; Westcott, S.L.; Ryabin, T.; Hall, J.R.; Hartmann, M.; Hollister, E.B.; Lesniewski, R.A.; Oakley, B.B.; Parks, D.H.; Robinson, C.J.; et al. Introducing Mothur: Open-Source, Platform-Independent, Community-Supported Software for Describing and Comparing Microbial Communities. *Appl. Environ. Microbiol.* **2009**, *75*, 7537–7541. [CrossRef]

51. Kozich, J.J.; Westcott, S.L.; Baxter, N.T.; Highlander, S.K.; Schloss, P.D. Development of a Dual-Index Sequencing Strategy and Curation Pipeline for Analyzing Amplicon Sequence Data on the MiSeq Illumina Sequencing Platform. *Appl. Environ. Microbiol.* **2013**, *79*, 5112–5120. [CrossRef]

52. Ratnasingham, S. MBRAVE: The Multiplex Barcode Research And Visualization Environment. *BISS* **2019**, *3*, e37986. [CrossRef]

53. Ratnasingham, S.; Hebert, P.D.N. BOLD: The Barcode of Life Data System (http://www.barcodinglife.org). *Mol. Ecol. Notes* **2007**, *7*, 355–364. [CrossRef]

54. Leite, B.R.; Vieira, P.E.; Teixeira, M.A.L.; Lobo-Arteaga, J.; Hollatz, C.; Borges, L.M.S.; Duarte, S.; Troncoso, J.S.; Costa, F.O. Gap-analysis and annotated reference library for supporting macroinvertebrate metabarcoding in Atlantic Iberia. *Reg. Stud. Mar. Sci.* **2020**, *36*, 101307. [CrossRef]

55. Lavrador, A.S.; Fontes, J.T.; Vieira, P.E.; Costa, F.O.; Duarte, S. Compilation, Revision, and Annotation of DNA Barcodes of Marine Invertebrate Non-Indigenous Species (NIS) Occurring in European Coastal Regions. *Diversity* **2023**, *15*, 174. [CrossRef]

56. Pruesse, E.; Peplies, J.; Glöckner, F.O. SINA: Accurate High-Throughput Multiple Sequence Alignment of Ribosomal RNA Genes. *Bioinformatics* **2012**, *28*, 1823–1829. [CrossRef]

57. Quast, C.; Pruesse, E.; Yilmaz, P.; Gerken, J.; Schweer, T.; Yarza, P.; Peplies, J.; Glöckner, F.O. The SILVA Ribosomal RNA Gene Database Project: Improved Data Processing and Web-Based Tools. *Nucleic Acids Res.* **2012**, *41*, D590–D596. [CrossRef]

58. Rognes, T.; Flouri, T.; Nichols, B.; Quince, C.; Mahé, F. VSEARCH: A Versatile Open Source Tool for Metagenomics. *PeerJ* **2016**, *4*, e2584. [CrossRef]

59. Camacho, C.; Coulouris, G.; Avagyan, V.; Ma, N.; Papadopoulos, J.; Bealer, K.; Madden, T.L. BLAST+: Architecture and Applications. *BMC Bioinform.* **2009**, *10*, 421. [CrossRef]

60. Heberle, H.; Meirelles, G.V.; da Silva, F.R.; Telles, G.P.; Minghim, R. InteractiVenn: A Web-Based Tool for the Analysis of Sets through Venn Diagrams. *BMC Bioinform.* **2015**, *16*, 169. [CrossRef]

61. Hammer, O.; Harper, D.; Ryan, P. PAST: Paleontological Statistics Software Package for Education and Data Analysis. *Palaeontol. Electron.* **2001**, *4*, 9.

62. Al-Rshaidat, M.M.D.; Snider, A.; Rosebraugh, S.; Devine, A.M.; Devine, T.D.; Plaisance, L.; Knowlton, N.; Leray, M. Deep COI Sequencing of Standardized Benthic Samples Unveils Overlooked Diversity of Jordanian Coral Reefs in the Northern Red Sea. *Genome* **2016**, *59*, 724–737. [CrossRef]

63. Gaither, M.R.; DiBattista, J.D.; Leray, M.; Heyden, S. Metabarcoding the Marine Environment: From Single Species to Biogeographic Patterns. *Environ. DNA* **2022**, *4*, 3–8. [CrossRef]
64. Ip, Y.C.A.; Chang, J.J.M.; Oh, R.M.; Quek, Z.B.R.; Chan, Y.K.S.; Bauman, A.G.; Huang, D. Seq' and ARMS Shall Find: DNA (Meta)Barcoding of Autonomous Reef Monitoring Structures across the Tree of Life Uncovers Hidden Cryptobiome of Tropical Urban Coral Reefs. *Mol. Ecol.* **2022**. [CrossRef] [PubMed]
65. Pearman, J.K.; Anlauf, H.; Irigoien, X.; Carvalho, S. Please Mind the Gap—Visual Census and Cryptic Biodiversity Assessment at Central Red Sea Coral Reefs. *Mar. Environ. Res.* **2016**, *118*, 20–30. [CrossRef] [PubMed]
66. Alves, A.C.M. Macroalgas Formadoras de Habitat: Espaço Neutro para Macrofauna Marinha? Master's Thesis, Universidade do Minho, Braga, Portugal, 2017.
67. Plaisance, L.; Knowlton, N.; Paulay, G.; Meyer, C. Reef-Associated Crustacean Fauna: Biodiversity Estimates Using Semi-Quantitative Sampling and DNA Barcoding. *Coral Reefs* **2009**, *28*, 977–986. [CrossRef]
68. Wolfe, K.; Mumby, P.J. RUbble Biodiversity Samplers: 3D-printed Coral Models to Standardize Biodiversity Censuses. *Methods Ecol. Evol.* **2020**, *11*, 1395–1400. [CrossRef]

 *diversity*

Article

Metabarcoding Extends the Distribution of *Porphyra corallicola* (Bangiales) into the Arctic While Revealing Novel Species and Patterns for Conchocelis Stages in the Canadian Flora

Gary W. Saunders [1,*] and **Cody M. Brooks** [1,2]

1   Department of Biology, Centre for Environmental & Molecular Algal Research, University of New Brunswick, Fredericton, NB E3B 5A3, Canada; cody.brooks@dfo-mpo.gc.ca
2   Bedford Institute of Oceanography, Department of Fisheries and Oceans, Dartmouth, NS B2Y 4A2, Canada
*   Correspondence: gws@unb.ca

Abstract: *Porphyra corallicola* was described based on a filamentous red alga inadvertently introduced into culture from a crustose coralline alga. This species is known only in its sporophyte (Conchocelis) stage, being possibly asexual and lacking the charismatic and "collectable" gametophyte stage. Consequently, little is known of its range and distribution. Taxon-targeted metabarcoding was explored as a pathway to gain insights into the vertical (intertidal versus subtidal) and biogeographical distribution of this species, as well as to assess host diversity. We also wanted to ascertain if other species occur in only the Conchocelis stage in the Canadian flora. Primers targeting a short (521 bp) region of the plastid *rbc*L gene in the Bangiales were used to screen DNA from 285 coralline crusts collected throughout Canada and adjacent waters. In addition to confirming the presence of *P. corallicola* in the Bay of Fundy, this species was recovered from coralline crusts along the coast of Nova Scotia ($n = 1$) and in the low Arctic (Labrador; $n = 2$), greatly extending its range and suggesting it is a cold-water taxon. We have confirmed its presence in both the low intertidal and subtidal (to 10 m), and its occurrence in three different coralline species, suggesting that it lacks host specificity. In total, nine genetic groups of Bangiales were uncovered in our survey, six matching entries currently in GenBank and three apparently novel genetic groups—two from the northeast Pacific and one from the low Arctic. Notable host and ecological patterns are discussed. This method, when further developed, will facilitate the study of Conchocelis stages in nature, which will greatly enhance ecological knowledge of bangialean species.

Keywords: *Bangia*; Bangiales; Conchocelis; *Fuscifolium*; *Porphyra*; *Porphyra corallicola*; *Pyropia*; taxon-targeted metabarcoding; *Wildemania*

Citation: Saunders, G.W.; Brooks, C.M. Metabarcoding Extends the Distribution of *Porphyra corallicola* (Bangiales) into the Arctic While Revealing Novel Species and Patterns for Conchocelis Stages in the Canadian Flora. *Diversity* **2023**, *15*, 677. https://doi.org/10.3390/d15050677

Academic Editors: Thomas J. Trott and Michael Wink

Received: 23 March 2023
Revised: 4 May 2023
Accepted: 15 May 2023
Published: 18 May 2023

1. Introduction

For many, Bangiales conjures images of beautiful filmy rose to purple blades, while others may be inclined to think of the typically dark red to golden multiseriate filaments assigned to the form genus *Bangia* [1]; however, these represent only the gametophyte stage. In 1949, Drew [2] published results linking the endolithic filaments of *Conchocelis rosea* Batters to the leafy *Porphyra umbilicalis* Kützing, adding to previous studies that linked germination of spores from other *Porphyra* spp. with a filamentous stage now known to be the sporophytic stage in a life history with an alternation of heteromorphic generations [1]. Despite the significance of these and subsequent discoveries, especially for the aquaculture of nori, there have been few studies on the actual ecology of Conchocelis stages in situ including aspects of host range, as well as vertical and biogeographical distribution. Although there is a fair level of interest in Conchocelis stages inhabiting corals (e.g., [3] and references therein), our knowledge of cold-temperate systems is more limited.

For European waters, Drew [4] provided the first ecological insights. Although not the focus of her manuscript, what little was known on the ecology of the sporophyte relative to the gametophyte was discussed. In terms of habitat, Drew [4] notes that in addition to calcareous invertebrate shells, the Conchocelis stage grows in calcareous stone and crustose coralline algae (notably *Lithothamnion laevigatum* Foslie, now *Phymatolithon laevigatum* (Foslie) Foslie, although there has been considerable confusion in the identification of crustose corallines e.g., [5]) (Figure 1). With so few studies, the view of Conchocelis as a shell-boring filament has become the default in general texts, which is only partially compatible with the work of Drew [4] and others, notably their occurrence in crustose coralline algae [4,6]. Although subsequent works have looked at thermal tolerances for growth and reproduction (summarized in [6]), little is known of the distribution, diversity and habitat specificity of the many species in nature.

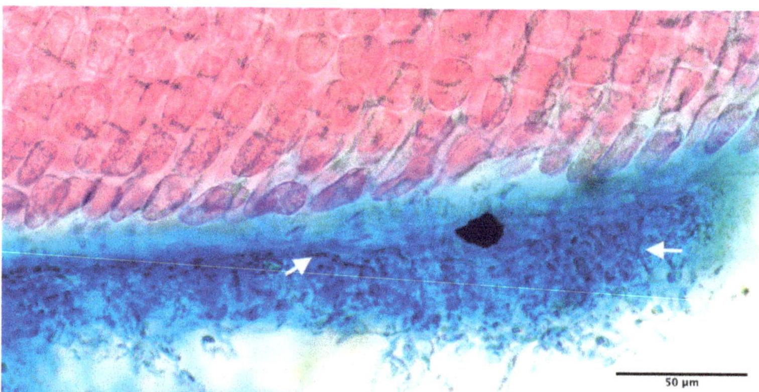

Figure 1. The original voucher from the collection that yielded the culture of *Porphyra corallicola* H. Kucera & G. W. Saunders; the filaments (white arrows) growing among cells of a coralline crust overgrown by *Peyssonnelia rosenvingei* F. Schmitz (aniline blue stained).

In terms of distribution, Drew [4] reports that although the Conchocelis stage "occurs occasionally in the intertidal belt, it is usually found by dredging in water up to 32 m in depth". Indeed, whereas *Porphyra umbilicalis* is widely regarded as an intertidal species [6,7], *Conchocelis rosea* is considered largely subtidal in distribution (although the unequivocal linking of these two species remains uncertain given the cryptic habit of the Conchocelis stages [6]). For the NW Atlantic, detailed phenological observations for *Porphyra linearis* Greville reported that the Conchocelis stage was absent from the intertidal band of the gametophyte stage but found subtidally at ~9 m depth [8]. In short, Conchocelis stages are typically considered subtidal in distribution (see [6] for a summary), which has caused some to ponder how the spores contribute to the recruitment of the intertidal gametophytic stage. In considering this conundrum, Drew [4] posited that the Conchocelis stage may be more prevalent in the intertidal than realized, considering that "minute pieces" of shell (washed up by storms into the intertidal) and barnacles were likely hosts for Conchocelis filaments and sources for the recruitment of the intertidal gametophyte stages.

A complicating factor is the difficulty, perhaps even inability, to identify various Conchocelis stages in the field to their respective species (i.e., link to a known gametophyte stage) as these filaments are largely cryptic [4,6,9]. As well, although asexual species of Bangiales are reported for the erect (gametophyte) stage, this does not appear to have been considered as a possibility for the Conchocelis stage. The description of *Porphyra corallicola* H.Kucera & G.W.Saunders (Figure 1) provided a departure in that it was possibly the only species of the Bangiales intentionally described based on its Conchocelis stage in the absence of knowledge of the gametophyte, if one even exists [7]. This species was accidentally introduced into culture when one of the authors (GWS) attempted to culture the red crust *Peyssonnelia rosenvingei* F.Schmitz. Finding it odd that the culture was filamentous and subsequently that molecular data indicated that it was a Conchocelis, a re-examination of the voucher revealed bangialean filaments growing within the tissue of a calcified coralline crust that had been overgrown by *P. rosenvingei* (Figure 1). Thus *P. corallicola* was described based on a single low intertidal collection growing in a crustose coralline alga from the lower Bay of Fundy [7].

The primary aim of this study was to assess *P. corallicola* considering host preference, vertical distribution (intertidal versus subtidal) and biogeography (especially in light of the connectivity of the Canadian flora through the Arctic e.g., [10,11]) by screening archived crustose coralline DNA at UNB. The secondary aims included assessing how many other species of Conchocelis were living in coralline crusts, to identify any other putatively asexual filament-only species and to look for any other trends of host specificity, vertical distribution and/or geographical distribution of Conchocelis stages relative to their gametophyte stages.

To accomplish these aims while appreciating the cryptic nature of Conchocelis stages, a marker system specific to the Bangiales but excluding coralline algae (and as many other epi-endophytic organisms that inhabit them as possible) was designed. Although not the best barcode marker for species discrimination among red algae, the *rbc*L provides reasonable species resolution among Bangiales [7] and excludes a wide variety of potential non-photosynthetic contaminants [12]. Primers were developed to amplify 521 bp of this marker for Bangiales, but attempts at Sanger sequencing revealed multiple bangialean taxa in some hosts (ambiguities in the data consistent with two or more bangialean taxa being present), which resulted in the application of NextGen sequencing—in essence taxon-targeted metabarcoding—to further reveal Conchocelis' diversity and ecology in coralline crusts. Although a preliminary survey, we have uncovered some interesting patterns as well as four putative new species of Bangiales in Canadian waters.

2. Materials and Methods

A total of 285 collections were selected for this study from archival crustose coralline DNAs (Table S1). In total, 237 collections were from the northwest Atlantic (ranging from Connecticut to Newfoundland and Labrador in the low Arctic), 41 from British Columbia (plus an additional collection was from Washington), three from Hudson Bay, two from Nunavut and a single collection from Norway (Table S1). The previously extracted DNA followed published protocols [13]. Amplification targeted a 521 bp region of the *rbc*L gene using the reverse primer TLR6 (5′ GTATAACCAATWACAAGRTC 3′ [12]) and the novel forward primer ConcF3 (5′ GWGTIGATCCAGTTCCRAAYGTTG 3′) and a published PCR profile for red algal *rbc*L [12]. ConcF3 was designed to amplify Bangiales to the exclusion of other red algae by aligning 35 and 441 *rbc*L sequences, respectively, to identify a suitable primer. Successful amplicons were sent to the Integrated Microbiome Resource (IMR) at Dalhousie University for short-read amplicon sequencing on an Illumina MiSeq machine following [14].

Raw data were processed using QIIME2 [15] and DADA2 software [16], generally following the Microbiome Helper standard operating procedures [14,17], with a relaxed expected error rate of three during read trimming. Current reference libraries used for metabarcoding analyses are generally lacking in red algal coverage and were found unsuitable for adequately identifying species within the Bangiales, so a custom reference library was created of 65 sequences (Table S2) amalgamated from publicly available data in GenBank and supplemented with newly generated sequences following established protocols [12]. All sequences used were generated at UNB and thus unequivocally linked to a voucher. To visualize the genetic groups obtained through metabarcoding in context of the reference library, all sequences were aligned by eye and subsequently subjected to UPGMA cluster analyses (Jukes and Cantor corrected distances) in Geneious Prime 2023.1.1 (https://www.geneious.com accessed on 7 March 2023).

3. Results

Of the original 285 specimens, 56 were successfully amplified and sequenced resulting in a total of 47,847 raw reads. Following denoising, dereplication and chimera filtering steps in DADA2, these raw reads were reduced to just 13,510 (28.2%), representing 49 unique operational taxonomic units (OTUs), which were assigned names by the Microbiome Helper pipeline using the custom reference database. Novel OTUs (not in the reference database) were compared against publicly available data in GenBank to search for their best match. Forward reads were truncated at a length of 294 bp with a median Phred score of 35, and reverse reads at a length of 283 bp with a median Phred score of 24 to allow adequate overlap for the merging of paired-end reads. A review by eye uncovered that seven of these 49 OTUs were chimeric, despite previous filtering. Using a ~0.5% threshold (allowing for 2–3 substitutions owing to variation within a species and/or PCR and sequencing errors), the remaining 42 sequences resolved into nine genetic species groups (Figure 2). Five of the previous matched known species for Canadian waters, one matched an unnamed species of *Fuscifolium* in GenBank (KP781730) while the remaining three were newly encountered species tentatively assigned to *Bangia* sensu lato (Figure 2 and Table 1). These nine genetic species groups were distributed among 28 positive Conchocelis identification events (CIEs) (Table 1 and Table S3).

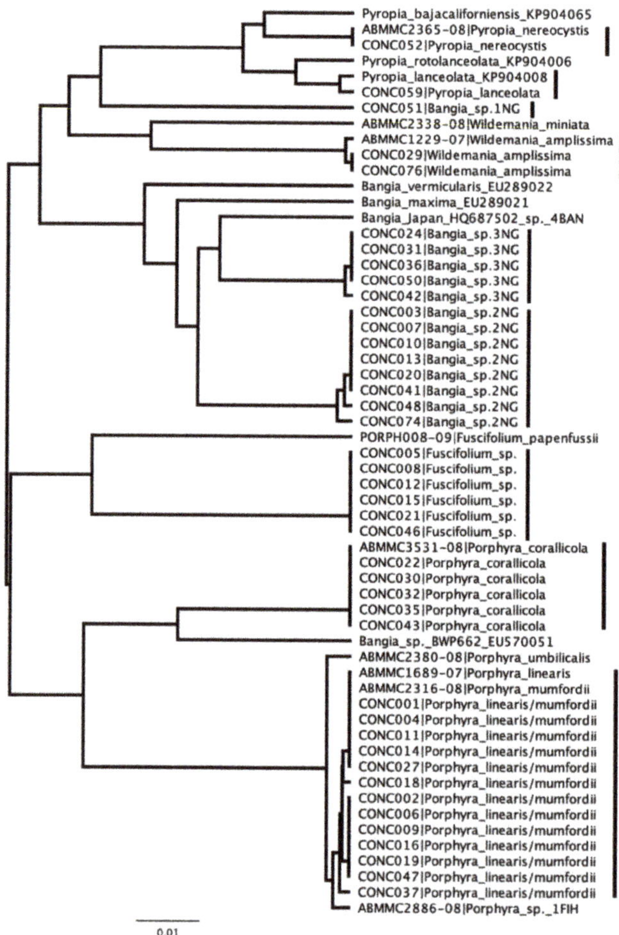

Figure 2. UPGMA clustering of our 42 unique Conchocelis OTUs (CONC###) with their closest matches from the reference database (Table S2) resolved as nine genetic species groups (vertical lines). Note that *Fuscifolium* sp. matched (100%) a GenBank entry from Chile (KP781730), which was not included in our reference database.

Table 1. List of positive Conchocelis identification events (CIEs) acquired, including species assignment, DNA match and host and distributional data. For details see Table S3.

Species Assignment	DNA Match	Hab.	Location
Bangia sp. 1NG	*Bangia* sp. 2Ban (93.61%)	Subtidal (10 m) in *Clathromorphum* sp. (GWS040344)	Labrador
Bangia sp. 2NG	*Bangia* Japan HQ687502 (~95%)	Subtidal (6 m) in *Lithophyllum* sp. 2BCcrust (GWS021028)	British Columbia
Bangia sp. 2NG	*Bangia* Japan HQ687502 (~95%)	Subtidal (10 m) in inverts or *Leptophytum* sp. 1SanJuan (GWS014430)	British Columbia
Bangia sp. 2NG	*Bangia* Japan HQ687502 (~95%)	Subtidal (7 m) in shell or *Leptophytum* sp. 1SanJuan (GWS036253)	Washington
Bangia sp. 3NG	*Bangia* Japan HQ687502 (~96%)	Low intertidal pool in *Lithothamnion* sp. 6BCcrust (GWS009940)	British Columbia
Bangia sp. 3NG	*Bangia* Japan HQ687502 (~96%)	Subtidal (10 m) in shell or *Lithothamnion* sp. 2glaciale (GWS014308)	British Columbia
Bangia sp. 3NG	*Bangia* Japan HQ687502 (~96%)	Subtidal (10 m) in *Lithophyllum* sp. 2BCcrust (GWS019653)	British Columbia
Fuscifolium sp.	*Fuscifolium* sp. CHa Chile (100%)	Subtidal (10 m) in inverts or *Leptophytum* sp. 1SanJuan (GWS014430)	British Columbia
Porphyra corallicola	*Porphyra corallicola* (99.47%)	Subtidal (4 m) in *Lithothamnion glaciale* (GWS011835)	Nova Scotia
Porphyra corallicola	*Porphyra corallicola* (99.47%)	Subtidal (10 m) in *Clathromorphum* sp. (GWS040344)	Labrador
Porphyra corallicola	*Porphyra corallicola* (99.65%)	Subtidal (10 m) in *Lithothamnion lemoineae* (overgrowing a dead crust) (GWS040346)	Labrador
Porphyra corallicola	*Porphyra corallicola* (99.65%)	Low intertidal pool in *Lithothamnion lemoineae* (GWS046571)	New Brunswick
Porphyra linearis	*Porphyra mumfordii/linearis* (99.29–99.82%)	Subtidal (5 m) in *Lithothamnion glaciale* (GWS003728)	New Brunswick
Porphyra linearis	*Porphyra mumfordii/linearis* (99.29–99.82%)	Subtidal (10 m) in *Phymatolithon* sp. 6ATcrust (GWS008908)	New Brunswick
Porphyra linearis	*Porphyra mumfordii/linearis* (99.29–99.82%)	Subtidal (10 m) in *Lithothamnion glaciale* (GWS011765)	New Brunswick
Porphyra linearis	*Porphyra mumfordii/linearis* (99.29–99.82%)	Subtidal (3 m) in mussel or *Lithothamnion glaciale* (GWS018163)	Maine
Porphyra linearis	*Porphyra mumfordii/linearis* (99.29–99.82%)	Low intertidal in *Phymatolithon laevigatum* (GWS039831)	Norway
Porphyra linearis	*Porphyra mumfordii/linearis* (99.29–99.82%)	Low intertidal in *Phymatolithon laevigatum* (GWS045271)	New Brunswick
Porphyra linearis	*Porphyra mumfordii/linearis* (99.29–99.82%)	Low intertidal pool in *Lithothamnion lemoineae* (GWS046571)	New Brunswick
Porphyra mumfordii	*Porphyra mumfordii/linearis* (99.29–99.82%)	Low intertidal pool in *Lithothamnion* sp. 6BCcrust (GWS009940)	British Columbia
Porphyra mumfordii	*Porphyra mumfordii/linearis* (99.29–99.82%)	Subtidal (10 m) in shell or *Lithothamnion* sp. 2glaciale (GWS014308)	British Columbia
Porphyra mumfordii	*Porphyra mumfordii/linearis* (99.29–99.82%)	Subtidal (6 m) in invert or *Leptophytum* sp. 1SanJuan (GWS020757)	British Columbia
Porphyra mumfordii	*Porphyra mumfordii/linearis* (99.29–99.82%)	Subtidal (6 m) in invert or *Leptophytum* sp. 1SanJuan (GWS020843)	British Columbia
Porphyra mumfordii	*Porphyra mumfordii/linearis* (99.29–99.82%)	Subtidal (3 m) in *Crusticorallina adhaerens* (GWS030995)	British Columbia
Pyropia lanceolata	*Pyropia lanceolata* (99.11%)	Subtidal (5 m) in snail or *Lithothamnion* sp. 2glaciale (GWS028036)	British Columbia
Pyropia nereocystis	*Pyropia nereocystis* (99.47%)	Subtidal (6 m) in *Lithophyllum* sp. 2BCcrust (GWS021028)	British Columbia
Wildemania amplissima	*Wildemania amplissima* (99.47%)	Lowest intertidal in *Lithothamnion glaciale* (GWS008885)	New Brunswick
Wildemania amplissima	*Wildemania amplissima* (99.64%)	Low intertidal in periwinkle or *Clathromorphum* sp. 1circumscriptum (GWS039527)	New Brunswick

Of primary consideration here is that four of the CIEs were for *Porphyra corallicola*, extending the range from the Bay of Fundy into the low Arctic (Labrador; Table 1 and Figure 3). The Arctic also returned a new *Bangia* sp. (sp. 1NG; Table 1 and Figure 3). Although *Wildemania amplissima* grows in both the Pacific and Atlantic, our two positive CIEs were from New Brunswick (Table 1 and Figure 3). Our marker region could not distinguish between *Porphyra linearis* and *Porphyra mumfordii*; however, the former is considered an Atlantic species and the latter a Pacific species, accounting for seven and five of the twelve CIEs, respectively (Table 1 and Figure 3). *Pyropia lanceolata* and *Pyropia nereocystis* were recovered from NE Pacific crusts, consistent with the expected range of these species, as were two of the new *Bangia* spp. (sp. 2NG and sp. 3NG) and a range extension for *Fuscifolium* sp., which was previously reported from Chile (KP781730) (Table 1 and Figure 3).

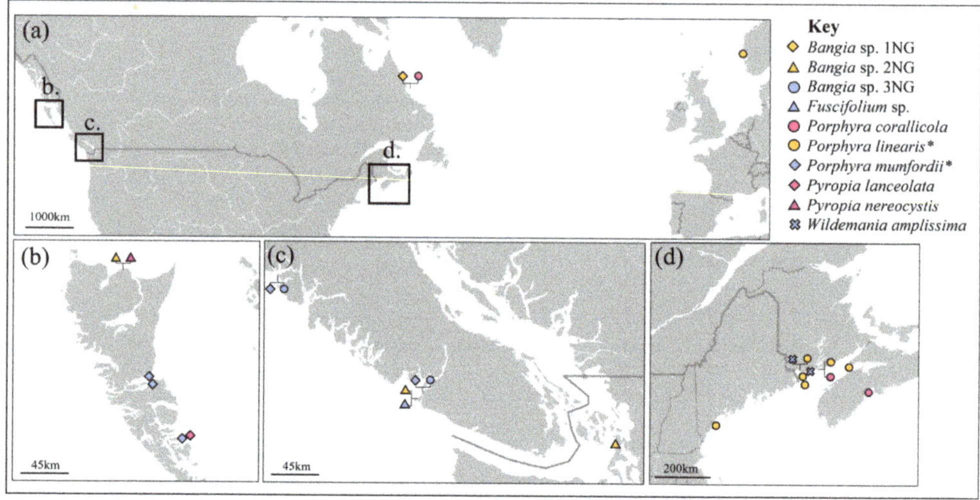

Figure 3. Maps showing the locations of our CIEs: entire study area (**a**) with inserts for Haida Gwaii (**b**), central to southern Vancouver Island (**c**) and the Maritime Provinces/Maine (**d**). Asterisks (*) indicate that *Porphyra linearis* and *Porphyra mumfordii* sequences are identical throughout the target region, so these species have been delineated based on their respective biogeographic ranges.

Of the crusts tested, 19% (54 of 283 (2 of the 285 lacked distributional data; Table S1)) were collected from the intertidal while the CIEs returned were 29% intertidal (8 of 28; Table 1). In total, 85% of the specimens tested were Atlantic/Arctic in distribution (243 of 285), but only 50% of the CIEs were from this region (14 of 28; Table 1). Ten of the twenty-eight CIEs were recovered from coralline crusts that were growing on shells or invertebrates, opening the possibility that the actual host may not have been the coralline crust but the latter's host (Table 1). At least 11 host coralline species in six divergent genera were uncovered with no obvious patterns of host specificity (Table 1).

4. Discussion

The primary objective of this study was to assess the utility of taxon-targeted metabarcoding in extending knowledge on the range of *Porphyra corallicola*, which is currently known only from the type culture isolated from the low intertidal zone at Maces Bay along the lower coast of the Bay of Fundy in New Brunswick [7]. Our new data identified a second low intertidal collection from near the type location at Musquash Head (host GWS046571), but also three subtidal records: one from Nova Scotia (host GWS011835) and two from the low Arctic in Labrador (hosts GWS040344 and GWS040346) (Table 1 and Figure 3). These CIEs were recovered from three hosts—an unidentified *Clathromorphum* sp., *Lithothamnion glaciale* and *L. lemoineae*—all relatively robust species and all growing directly on rock (GWS040346 was partially overgrowing a dead crustose coralline, Table 1), consistent with the crusts being the host for *Porphyra corallicola*. Our results have thus extended the vertical, host and biogeographical range of this species.

In the NW Atlantic, we also uncovered two CIEs for *Wildemania amplissima*, both intertidal in the Bay of Fundy. We have collected the gametophyte stage of this species widely in Canadian waters from the intertidal to shallow subtidal and it is common in the lower Bay of Fundy during spring and summer based on records in our database [18]. The most encountered CIEs in this area matched *Porphyra mumfordii/linearis* (our marker region cannot distinguish between these two species), which were found from the low intertidal to subtidal in a variety of hosts (Table 1). Interestingly, this was also the most common CIE in British Columbia, suggesting that these two closely related species may prefer crustose coralline algae as habitat for their Conchocelis stages. As *Porphyra mumfordii* is a Pacific species (although see [19]) and *P. linearis* an Atlantic species, we have used this biogeographical pattern to make tentative species assignments (Table 1 and Figure 3). While we have collected the gametophyte stage of *P. mumfordii* widely in British Columbia from the intertidal, we have only collected *Porphyra linearis* during winter in the intertidal at a few exposed locations [18]. There are two notable exceptions, but both are presumptive Conchocelis stages. A single subtidal collection from Massachusetts was reported as growing in a coralline crust (MK185874), and our only previous Bay of Fundy collection for this species was also subtidal (GWS041780; OQ706563), growing in undetermined "calcified substrata". Despite the limited vertical, ecological and seasonal distribution of the gametophyte, the Conchocelis stage appears to be more widely distributed in all categories. Notably, we are yet to encounter the gametophyte stage in the Bay of Fundy, but the previous record (GWS041780) and five of the seven Conchocelis stages detected here were from this area (Table 1 and Figure 3). Thus, there may be a generalization that Conchocelis stages have broader biogeographical and ecological ranges than their gametophytic counterparts as has been noted in other red algae with alternations of heteromorphic generations (e.g., [11]).

In the NE Pacific, in addition to the known species *Pyropia lanceolata* and *Pyropia nereocystis*, we uncovered three novel species: what appear to be newly discovered species *Bangia* sp. 2NG and *Bangia* sp. 3NG, as well as a range extension for *Fuscifolium* sp. currently reported from Chile (Table 1). To determine if these species represent stages in a sexual life history with an alternation of generations (e.g., *Porphyra linearis*) or novel asexual Conchocelis-only species (e.g., *Porphyra corallicola*) will require more study. We can note that through our work [18] and that of colleagues (e.g., [1]), hundreds of specimens of bangialean gametophytes have been genetically screened from British Columbia and have not turned up matches to these species. We do have a gametophytic *Bangia* sp. (GWS008341; UPA, JN029024) from northern British Columbia that currently lacks *rbc*L data, which may be a match to one of these species. Prior to this study we had only encountered a single genetic group for *Fuscifolium* in British Columbia, *Fuscifolium papenfussii* (V.Krishnamurthy) S.C.Lindstrom (e.g., JN028940), and indeed only one other species is included in this genus (*Fuscifolium tasa* (Yendo) S.C.Lindstrom) for which there are *rbc*L data in GenBank [1]. The novel *Fuscifolium* sp. may or may not have an erect gametophytic stage in British Columbia but is reported as a blade (gametophyte) in Chile [20]. More collections are needed from British Columbia to determine if this species undergoes an alternation of heteromorphic generations in that region or simply persists as an asexual Conchocelis stage.

It is also notable that while only 15% of the crusts screened were from the Pacific (Table S1), 50% of the CIEs were from this region (14 of 28; Table 1). In looking at our own records for gametophyte stages, there are more species of Bangiales in the NE Pacific than NW Atlantic (~25 vs. 17 [18]), but this slight difference does not appear to account for the discrepancy. It could be that more Pacific species are sexual and/or prefer coralline algae as hosts for their Conchocelis stages (there is considerably more crustose coralline diversity in the NE Pacific, perhaps providing more opportunity; compare [21,22]). More study is needed.

The novel entity designated here as *Bangia* sp. 1NG was also from the low Arctic. It was only a distant match (93.61%) to a genetic group that we call *Bangia* sp. 2Ban, that unequivocally has a gametophyte with the *Bangia* morphology and is widely distributed in the North Atlantic with genetic matches from Rhode Island to Norway [18]. Thus, if *Bangia* sp. 1NG does have an alternation of generations, the gametophyte is likely to have this morphology and again more collections are needed from northern waters. Although there is a general notion that the Conchocelis stages are subtidal in distribution (e.g., [8]), Drew [4] suggested that they may be more common in the intertidal than realized. Only 19% of the crusts we tested were collected from the intertidal, while the CIEs were 29% intertidal (Table 1). This supports Drew's assertion, but it is important to recognize that they are also abundant subtidally with some of our records from as deep as 10 m (Table 1).

The three novel taxa uncovered here are all in the genus *Bangia*. Currently in Canada, floristic guides recognize only *Bangia atropurpurea* (Roth) C.Agardh [21] or have defaulted to *Bangia* spp., reporting three genetic groups in need of taxonomic study [22]. Our own lab has uncovered two genetic groups in BC and two in the NW Atlantic [18], with those numbers increasing to three (or four) and three, respectively, following this survey (Figure 2). None of the genetic groups discussed here from Canada are a genetic match for bona fide *B. atropurpurea*, which is a freshwater species [1]. Considerable taxonomic research remains for this genus in our waters and indeed the genus itself is not monophyletic [1].

This study has limitations as it was opportunistic in the use of existing archived crustose coralline DNA. Nonetheless, it has merit in joining the few studies that use archival DNA for purposes other than taxonomy (for examples, see [23,24]), and it serves as a proof of concept that this technique can be used to identify Conchocelis stages in the field, which sets the foundation for a more structured survey of coralline crusts (and other calcareous substrata) based on targeted sampling. In carrying out such a project, care must be taken in acquiring the crusts to ensure that any Conchocelis stages identified are unequivocally growing in the coralline alga. For example, 10 of the 28 CIEs that we recovered here were from coralline crusts that were growing on shells or invertebrates, opening the possibility that the host may not have been the actual coralline crust (Table 1). Although we always sample with care to avoid contamination from underlying or host material, it can be difficult to obtain "clean" samples from nature. However, this last-mentioned caveat raises the potential of under sampling as we intentionally avoided old or what appeared to be infested pieces of material during sampling to facilitate acquiring a clean target sequence for coralline crusts during our routine DNA barcode surveys. A survey for Conchocelis stages would do the opposite, focusing on old and infested pieces of hosts. It is also notable that the Conchocelis stages have been cultured without the use of calcareous substrata (e.g., [4,9]), raising the possibility that they may grow in noncalcareous substrata. Hence, the screening of fleshy macroalgae may return unexpected results.

The 521 bp region of the *rbc*L used as a marker here failed to resolve all the target species. The *rbc*L-3P is considered a suitable secondary barcode marker for red algae [12], but that region is 800 bp in length and outside the range for the Illumina technology used here. Further, it is more variable than the region used here, which obstructed efforts at primer design for a shorter fragment from this region of the gene. For red algae, the COI-5P and ITS are recognized as better barcode markers in terms of species resolution [25], but they come with other shortfalls with regard to the current study. Being highly variable, COI-5P primer design has been a challenge for red algae [12], and we were unable to design primers to include all Bangiales to the exclusion of other red algae. The ITS is potentially more amenable to primer design but can be highly variable in length in red algae, which could result in taxa being missed with Illumina. Both of these markers are also found in a wider variety of taxa (e.g., animals, fungi, etc.) than the *rbc*L, which would further complicate primer design and could invite further PCR bias and lead to reduced Conchocelis read counts and underestimating species richness [26]. In future, a longer fragment of the *rbc*L could be used (if primers can be developed), but this would require a different sequencing technology. In the end, the marker used here separated all but two of the taxa included in our reference database (Table S2), which is good resolution by metabarcoding standards.

Metabarcoding can struggle with high false-negative rates [27], owing in part to the necessary discarding of large quantities of "junk data" by the denoising program (DADA2 in this case, although this is a universal feature). Typically, "junk data" are rare reads, particularly singletons and doubletons, which are notoriously difficult to discern from sequencing artefacts [16]. In this study, old stocks of DNA were screened for minute traces of microscopic endophytes belonging to a single order, so unsurprisingly our quantity of raw reads was low (~43,000) and the quantity of rare reads was high. This paucity of data was exacerbated when denoising removed the singletons, doubletons and rare reads (71.8% of raw reads), many of which could be true, rare species [28]. The authors in [29] observed (albeit in fungal systems) that up to 44% of the discarded singletons and doubletons alone could be true rare species rather than sequencing artefacts, to say nothing of other low-abundance reads also discarded by DADA2. Although the results of this study include several novel groups, the true diversity of Conchocelis present in these crustose hosts is likely under-represented here. In contrast, had we not carefully examined the resulting alignment for chimeric sequences post bioinformatics pipeline, we would have had seven additional OTUs and would be reporting on the remarkable levels of undiscovered bangialean diversity in Canadian waters. Clearly, better methods for analyzing these types of data are needed for metabarcoding surveys to reach their full potential.

As a final consideration, without detailed microscopy it can be difficult to confirm that the sequences are from an endophytic Conchocelis stage and not simply from juveniles of the gametophytic stage. We note that we have never collected the gametophyte of *Porphyra linearis* in the Bay of Fundy, and yet five of the seven positive CIEs were from this region (Table 1 and Figure 3). Further, the *P. linearis* gametophyte is a mid to upper intertidal winter annual [8], while all of our CIEs (including the two discussed above as being encountered during routine DNA barcode screening) were low intertidal ($n = 3$) or subtidal ($n = 6$) and collected from April to September (Table S3). Finally, in contrast to the gametophyte stage of *Porphyra linearis*, which is seemingly rare in the Bay of Fundy and confined to winter and the mid to upper intertidal, the closely allied *Porphyra umbilicalis* Kützing (Figure 2) is common in the Bay of Fundy (we have 78 archived collections [18]), and occurs from the upper intertidal to the shallow subtidal and in all seasons (we have collections from March to December; we rarely collect in December to March) in this area [18,30]). However, in our experience, *Porphyra umbilicalis* is asexual and confined to the blade morphology in this region, which may account for this species not being encountered in our study, consistent with our CIEs (Table 1) being legitimately from Conchocelis stages. On the other hand, we did not encounter the Conchocelis for the less closely related and presumably sexual *Porphyra purpurea* (Roth) C.Agardh, which is also common in this region [7,18]. Does its Conchocelis grow exclusively in calcareous substrata of animal origin, or is it actually asexual in the study region? Unlike *Porphyra linearis*, we have not unexpectedly encountered the sporophyte of *P. purpurea* in our routine barcode surveys (which include abundant coralline crusts, but not animals), which is consistent with both the previous hypotheses. With the tools developed here we can begin to resolve these uncertainties and further shed light on the ecology of the Conchocelis stages and species in the Bangiales.

Supplementary Materials: The following supporting information can be downloaded at https://www.mdpi.com/article/10.3390/d15050677/s1, Table S1: Listing of coralline DNAs screened for Conchocelis stages with their associated collection data; Table S2: List of all species included in the custom Bangiales reference database and their associated GenBank accession numbers; Table S3: Resulting CIEs with their assigned name, match and associated collection data.

Author Contributions: G.W.S. conceptualized the project and developed the marker system used to generate sequence data, as well as completed the original writing of a draft preparation. C.M.B. managed the short-read amplicon sequencing, including pipeline assembly and data analyses. Both authors contributed to project administration and data management, preparation of the final document for submission and the interpretation of the data. All authors have read and agreed to the published version of the manuscript.

Funding: This research was funded by the Natural Sciences and Engineering Research Council of Canada, through a Discovery Grant to G.W. Saunders (170151-2013), with infrastructure support from the Canada Foundation for Innovation and the New Brunswick Innovation Foundation.

Institutional Review Board Statement: Not applicable.

Data Availability Statement: Sequences generated for the *rbc*L reference database are available through the Barcode of Life Data System published dataset DS-BANGIAL1 Bangiales DNA Barcode Survey Data release [18], as well as GenBank (Table S2). The Illumina data are available in GenBank PRJNA951650.

Acknowledgments: This work was only possible through access to a number of archived DNA extractions from coralline crusts collected by numerous members of the Saunders Laboratory over the years, all of whom are thanked for their contribution. T. Moore generated some of the reference bangialean *rbc*L sequences used in this study (Table S2) and is thanked for her assistance in other technical aspects of this project.

Conflicts of Interest: The authors declare no conflict of interest. The funders had no role in the design of the study; in the collection, analyses, or interpretation of data; in the writing of the manuscript; or in the decision to publish the results.

References

1. Sutherland, J.E.; Lindstrom, S.C.; Nelson, W.A.; Brodie, J.; Lynch, M.D.J.; Hwang, M.S.; Choi, H.G.; Miyata, M.; Kikuchi, N.; Oliveira, M.C.; et al. A new look at an ancient order: Generic revision of the Bangiales (Rhodophyta). *J. Phycol.* **2011**, *47*, 1131–1151. [CrossRef] [PubMed]
2. Drew, K.M. *Conchocelis*-phase in the life-history of *Porphyra umbilicalis* (L.) Kütz. *Nature* **1949**, *164*, 748–749. [CrossRef]
3. Tribollet, A.; Pica, D.; Puce, S.; Radtke, G.; Campbell, S.E.; Golubic, S. Euendolithic *Conchocelis* stage (*Bangiales*, Rhodophyta) in the skeletons of live stylasterid reef corals. *Mar. Biodiv.* **2018**, *48*, 1855–1862. [CrossRef]
4. Drew, K.M. Studies in the Bangioideae III. The life-history of *Porphyra umbilicalis* (L.) Kütz. var. *laciniata* (Lightf.) J. Ag.: A. The Conchocelis-phase in culture. *Ann. Bot.* **1954**, *18*, 183–211.
5. Peña, V.; Bélanger, D.; Gagnon, P.; Richards, J.L.; Le Gall, L.; Hughey, J.R.; Saunders, G.W.; Lindstrom, S.C.; Rinde, E.; Husa, V.; et al. *Lithothamnion* (Hapalidiales, Rhodophyta) in the Arctic and Subarctic: DNA sequencing of type and recent specimens provides a systematic foundation. *Eur. J. Phycol.* **2021**, *56*, 468–493. [CrossRef]
6. Brodie, J.A.; Irvine, L.M. *Seaweeds of the British Isles. Volume 1. Rhodophyta. Part 3B. Bangiophycidae*; Intercept: Andover, UK, 2003; pp. i–xiii, 1–167.
7. Kucera, H.; Saunders, G.W. A survey of Bangiales (Rhodophyta) based on multiple molecular markers reveals cryptic diversity. *J. Phycol.* **2012**, *48*, 869–882. [CrossRef]
8. Bird, C.J. Aspects of the life history and ecology of *Porphyra linearis*. (Bangiales, Rhodophyceae) in nature. *Can. J. Bot.* **1973**, *51*, 2371–2379. [CrossRef]
9. Bird, C.J.; Chen, L.C.-M.; McLachlan, J. The culture of *Porphyra linearis* (Bangiales, Rhodophyceae). *Can. J. Bot.* **1972**, *50*, 1859–1863. [CrossRef]
10. Bringloe, T.T.; Saunders, G.W. DNA barcoding of the marine macroalgae from Nome, Alaska (Northern Bering Sea) reveals many trans-Arctic species. *Polar Biol.* **2019**, *42*, 851–864. [CrossRef]
11. Bringloe, T.T.; Verbruggen, H.; Saunders, G.W. Unique diversity in Arctic marine forests is shaped by diverse recolonisation pathways and far northern glacial refugia. *Proc. Natl. Acad. Sci. USA* **2020**, *117*, 22590–22596. [CrossRef] [PubMed]
12. Saunders, G.W.; Moore, T.E. Refinements for the amplification and sequencing of red algal DNA barcode and RedToL phylogenetic markers: A summary of current primers, profiles and strategies. *Algae* **2013**, *28*, 31–43. [CrossRef]
13. Saunders, G.W.; McDevit, D.C. Methods for DNA barcoding photosynthetic protists emphasizing the macroalgae and diatoms. *Methods Molec. Biol.* **2012**, *858*, 207–222.
14. Comeau, A.M.; Douglas, G.M.; Langille, G.I. Microbiome helper: A custom and streamlined workflow for microbiome research. *Methods Protoc.* **2017**, *2*, e00127-16. [CrossRef]
15. Bolyen, E.; Rideout, J.R.; Dillon, M.R.; Bokulich, N.A.; Abnet, C.C.; Al-Ghalith, G.A.; Alexander, H.; Alm, E.J.; Arumugam, M.; Asnicar, F.; et al. Reproducible, interactive, scalable and extensible microbiome data science using QIIME2. *Nat. Biotechnol.* **2019**, *37*, 852–857. [CrossRef] [PubMed]

16. Callahan, B.J.; McMurdie, P.J.; Rosen, M.J.; Han, A.W.; Johnson, A.J.A.; Holmes, S.P. DADA2: High-resolution sample inference from Illumina amplicon data. *Nat. Methods* **2016**, *13*, 581–583. [CrossRef] [PubMed]

17. Martin, M. Cutadapt removes adapter sequences from high-throughput sequencing reads. *Embnet. J.* **2011**, *17*, 10–12. [CrossRef]

18. Saunders, G.W.; Moore, T.E. Dataset—DS-BANGIAL1 Bangiales DNA Barcode Survey Data Release. 2023. Available online: http://www.boldsystems.org/index.php/Public_SearchTerms?query=DS-BANGIAL1 (accessed on 7 March 2023).

19. Saunders, G.W.; Brooks, C.M.; Scott, S. Preliminary DNA barcode report on the marine red algae (Rhodophyta) from the British Overseas Territory of Tristan da Cunha. *Cryptogam. Algol.* **2019**, *40*, 105–117. [CrossRef]

20. Guillemin, M.-L.; Contreras-Porcia, L.; Ramírez, M.E.; Macaya, E.C.; Contador, C.B.; Woods, H.; Wyatt, C.; Brodie, J. The bladed Bangiales (Rhodophyta) of the South Eastern Pacific: Molecular species delimitation reveals extensive diversity. *Mol. Phylogenet. Evol.* **2016**, *94*, 814–826. [CrossRef]

21. Sears, J.R. *NEAS Keys to Benthic Marine Algae of the Northeastern Coast of North America from Long Island Sound to the Strait of Belle Isle*; Northeast Algal Society Contr. No. 2. Univ. Mass. Dartmouth Campus Book Store: Dartmouth, MA, USA, 2002; pp. i–xvii, 1–161.

22. Gabrielson, P.W.; Lindstrom, S.C. *Keys to the Seaweeds and Seagrasses of Southeast Alaska, British Columbia, Washington, and Oregon*; Phycological Contribution Number 9; Island Blue/Printorium Bookworks: Victoria, BC, USA, 2018; pp. i–iv, 1–180.

23. Datlof, E.M.; Amend, A.S.; Earl, K.; Hayward, J.; Morden, C.W.; Wade, R.; Zahn, G.; Hynson, N.A. Uncovering unseen fungal diversity from plant DNA banks. *PeerJ* **2017**, *5*, e3730. [CrossRef]

24. Heberling, J.M.; Burke, D.J. Utilizing herbarium specimens to quantify historical mycorrhizal communities. *Appl. Plant Sci.* **2019**, *7*, e01223. [CrossRef]

25. Saunders, G.W. Applying DNA barcoding to red macroalgae: A preliminary appraisal holds promise for future applications. *Phil. Trans. R. Soc. B.* **2005**, *360*, 1879–1888. [CrossRef]

26. Tedersoo, L.; Nilsson, R.H.; Abarenkov, K.; Jairus, T.; Sadam, A.; Saar, I.; Bahram, M.; Bechem, E.; Chuyong, G.; Kõljalg, U. 454 Pyrosequencing and Sanger sequencing of tropical mycorrhizal fungi provide similar results but reveal substantial methodological biases. *New Phytol.* **2011**, *188*, 291–301. [CrossRef]

27. Harper, L.R.; Handley, L.L.; Hahn, C.; Boonham, N.; Rees, H.C.; Gough, K.C.; Lewis, E.; Adams, I.P.; Brotherton, P.; Phillips, S.; et al. Needle in a Haystack? A comparison of eDNA metabarcoding and targeted qPCR for detection of the great crested newt (*Triturus cristatus*). *Ecol. Evol.* **2017**, *8*, 6330–6341. [CrossRef] [PubMed]

28. Zhan, A.; Hulák, M.; Sylvester, F.; Huang, X.; Adebayo, A.A.; Abbot, C.L.; Adamowicz, S.J.; Heath, D.D.; Cristescu, M.E.; MacIsaac, H.J. High-sensitivity pyrosequencing for detection of rare species in aquatic communities. *Methods Ecol. Evol.* **2013**, *4*, 558–565. [CrossRef]

29. Brown, S.P.; Veach, A.M.; Rigdon-Huss, A.R.; Grond, K.; Lickteig, S.K.; Lothamer, K.; Oliver, A.K.; Jumpponen, A. Scraping the bottom of the barrel: Are rare high throughput sequences artifacts? *Fungal Ecol.* **2019**, *13*, 221–225. [CrossRef]

30. Saunders, G.W. Seaweed of Canada. *Porphyra umbilicalis* Kützing (A). Available online: https://seaweedcanada.wordpress.com/pophyra-umbilicalis-linnaeus-kutzing-a/ (accessed on 7 March 2023).

Article

Spatiotemporal Variability in Subarctic *Lithothamnion glaciale* Rhodolith Bed Structural Complexity and Macrofaunal Diversity

David Bélanger [1] and Patrick Gagnon [2,*]

[1] Department of Biology, Memorial University of Newfoundland, St. John's, NL A1B 3X9, Canada; david.belanger@mun.ca

[2] Department of Ocean Sciences, Ocean Sciences Centre, Memorial University of Newfoundland, St. John's, NL A1C 5S7, Canada

* Correspondence: pgagnon@mun.ca

Abstract: Rhodoliths are non-geniculate, free-living coralline red algae that can accumulate on the seafloor and form structurally complex benthic habitats supporting diverse communities known as rhodolith beds. We combined in situ rhodolith collections and imagery to quantify variability, over 9 months and at two sites, in the structural complexity and biodiversity of a subarctic *Lithothamnion glaciale* rhodolith bed. We show that the unconsolidated rhodolith framework is spatially heterogeneous, yet provides a temporally stable habitat to an abundant and highly diverse macrofauna encompassing 108 taxa dominated by brittle stars, chitons, bivalves, gastropods, polychaetes, sea urchins, and sea stars. Specific habitat components, including large bivalve shells, affect rhodolith morphology and resident macrofauna, with increasingly large, non-nucleated rhodoliths hosting higher macrofaunal density, biomass, and diversity than increasingly large, shell-nucleated rhodoliths. The present study's fine taxonomic resolution results strongly support the notion that rhodolith beds are biodiversity hotspots. Their spatial and temporal domains provide clear quantitative evidence that rhodolith beds provide a stable framework under the main influence of biological forcing until sporadic and unusually intense physical forcing reworks it. Our findings suggest that shallow (<20 m depth) rhodolith beds are vulnerable to ongoing and predicted increases in the frequency and severity of wave storms.

Keywords: coralline algae; ecosystem engineering; benthic habitat; biodiversity; cryptofauna; ecological stability; climate forcing; generalized linear models; multivariate statistics; Newfoundland

Citation: Bélanger, D.; Gagnon, P. Spatiotemporal Variability in Subarctic *Lithothamnion glaciale* Rhodolith Bed Structural Complexity and Macrofaunal Diversity. *Diversity* **2023**, *15*, 774. https://doi.org/10.3390/d15060774

Academic Editors: Thomas J. Trott and Michael Wink

Received: 30 March 2023
Revised: 1 June 2023
Accepted: 7 June 2023
Published: 14 June 2023

1. Introduction

Habitat structure primarily refers to the composition and spatial arrangement of physical matter at a location [1,2]. Positive relationships between habitat structural complexity and biodiversity have been described in a variety of terrestrial [3–5] and aquatic [6–8] ecosystems, which build on early investigations relating the availability of inhabitable surfaces to faunal density and diversity [9–11]. More recent work suggests habitat complexity can indirectly affect microhabitat and niche availability by shaping foraging activities, intra- and interspecific competition, and predator–prey interactions [12–15]. Modern definitions of the concept of habitat complexity also include ecosystem engineers (sensu Jones 1994 [16]), i.e., species that create, significantly modify, or maintain habitat structure [17–19].

Biodiversity in marine benthic systems generally scales positively with the presence of habitat-forming species such as macroalgae [20–22] and marine calcifiers, including bivalves [23] and corals [7]. Rhodoliths are free-living nodules of primarily (>50% in composition) non-geniculate coralline red algae [24,25]. Under favorable conditions, rhodoliths aggregate over extensive areas of the seafloor to form complex benthic habitats supporting diverse communities called rhodolith beds [26]. With their calcium carbonate skeleton, rhodoliths function as autogenic ecosystem engineers [16], providing a three-dimensional, unconsolidated, branching framework (sensu [27]) that macrobenthic organisms can colonize [28]. Rhodoliths exhibit various sizes, shapes, and growth forms, thus significantly contributing to benthic habitat structural complexity [29,30]. They are long-lived (up to 100 years) [31,32] and slow-growing (generally < 1 mm y^{-1}) [33–35], and presumably provide stable habitats since water motion is insufficient to move (transport or overturn) rhodoliths [36].

Rhodoliths mainly reproduce through fragmentation but can also originate from spore settlement on hard organic or inorganic particles the coralline tissue eventually overgrows [37]. While in the first case, the resulting new rhodoliths are composed exclusively of algal tissue, the latter condition produces rhodoliths that contain an exogenous core (nucleus). Nucleus size and shape may strongly influence rhodolith morphology and reduce the inner space available for colonization by macrofauna [38,39].

A few studies have linked the high biodiversity generally associated with rhodolith beds to the structural complexity of the habitat they provide [40–42]. This functional aspect of rhodolith beds is particularly important given that they normally form over comparatively featureless sedimentary bottoms. Rhodolith beds also act as nursery grounds for several ecologically and economically important species by enhancing the larval settlement of mollusks [43,44], echinoderms [45], corals [46], and sponges [47].

Reports of rhodolith beds along the eastern Canadian seaboard date back more than 70 y [27,48]. Newfoundland rhodoliths are essentially composed of *Lithothamnion glaciale*, a dominant coralline red alga within the photic zone of subarctic marine environments [49]. A growing number of studies acknowledge the ecological importance of rhodolith beds as biodiversity hotspots [26,28,39]. However, few studies have characterized the spatiotemporal variability of rhodolith bed macrofaunal diversity and structural complexity and their relationship.

The present study builds on pioneering work by Gagnon et al. [30], who provided a general description of rhodolith abundance and morphology, as well as of macrofaunal diversity, in two subarctic rhodolith beds in southeastern Newfoundland. One of these beds, located along the shore of the town of St. Philip's and hereafter termed the "St. Philip's bed", is relatively large (~25,000 m^2) and highly biodiverse ([30] and present study). This bed has since been studied for its sedimentary processes [36], calcium carbonate ($CaCO_3$) production [32], resilience to thermal shifts and eutrophication [33–35], and trophodynamics [50]. This knowledge base offers an excellent opportunity to further investigate and explain spatial and temporal variability in macrofaunal biodiversity patterns.

We combined in situ rhodolith collections and imagery to quantify seasonal variability in the structural complexity and biodiversity of the St. Philip's bed. Specifically, we sampled the bed in winter, spring, summer, and fall of the same year at two different locations (sites), to test the hypotheses that (1) the bed's structure differs between sites, yet is temporally stable within each site; (2) the abundance and diversity of rhodolith macrofauna scale positively with rhodolith size; (3) rhodolith macrofauna differs between sites; and (4) rhodolith macrofauna varies seasonally. These hypotheses stem from prior knowledge (per studies of the St. Philip's bed cited above) about the biology and ecology of *L. glaciale* rhodoliths, thermal and hydrodynamic conditions in the St. Philip's rhodolith bed, and the ontogeny of dominant rhodolith macrofauna.

2. Materials and Methods

2.1. Study Site

The study system was a rhodolith bed located off the town of St. Philip's on the eastern shore of Conception Bay, Newfoundland, Canada (Figure 1A). There, the relatively steep rocky seabed extends to a depth of ~10 m, where it grades into a gently sloping sedimentary bottom interspersed with bedrock outcrops rising to up to ~2 m above the seabed (Figure 1B). At depths of 12 to 20 m, large patches of branching (i.e., fruticose, sensu [29]) *Lithothamnion glaciale* rhodoliths cover a sedimentary bottom primarily composed of coarse biogenic sands; lithic pebbles (hereafter "pebble"); large (~5–10 cm across) disarticulated shells of dead horse mussels (*Modiolus modiolus*); scattered lithic cobbles/boulders; and loose fragments of brown seaweeds (mainly kelp *Agarum clathratum*) originating from individuals growing on surrounding bedrock or boulders (Figure 1B,C). We selected two study sites in this bed based on marked differences in seafloor composition and rhodolith size and structure. The first site, SP15, was located at depths of 15–17 m. It was partially enclosed by short (≤50 cm in height) bedrock outcrops and densely covered with mainly praline rhodoliths (sensu [51]), with no external signs of nucleation. The second site, SP18, was at a distance of ~50 m from SP15, at depths of 18–20 m. It was almost entirely enclosed by tall (~2 m) bedrock outcrops (Figure 1B) and differed from SP15 by a noticeably greater occurrence of (1) disarticulated shells of dead horse mussels (*Modiolus modiolus*) variably covered in coralline red algae; and (2) larger boxwork rhodoliths (sensu [51]), with clear signs of internal nucleation (Figure 1D,E).

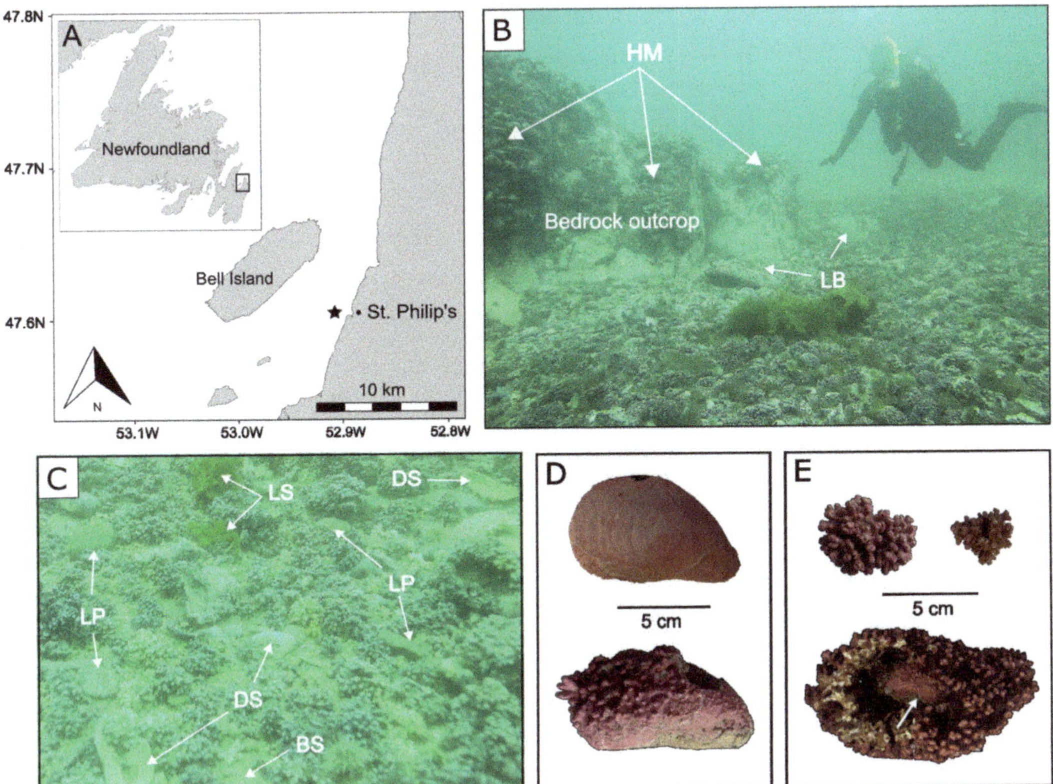

Figure 1. (**A**) Location of the present study's focal *Lithothamnion glaciale* rhodolith bed (star), off the town of St. Philip's in Conception Bay, Newfoundland (eastern Canada). (**B**) Portion of the rhodolith bed at site SP18 showing lithic boulders (LB) and one of the bedrock outcrops colonized by live horse mussels (HM, *Modiololus modiolus*). (**C**) Close-up view of the rhodolith bed surface showing live and dead rhodoliths; lithic pebbles (LP); disarticulated horse mussel shells (DS), biogenic sands (BS), and loose seaweed (LS) fragments. (**D**) Disarticulated horse mussel shells at early (top) and advanced (bottom) stages of coralline algal encrustation. (**E**) Non-nucleated praline rhodoliths (top) and large (~12 cm across), partially broken/eroded mussel-shell (white arrow) nucleated boxwork rhodolith (bottom).

2.2. Rhodolith Bed Structure

In 2013, we surveyed both study sites at ~3-month intervals to measure spatial and temporal variations in rhodolith bed structure and rhodolith-associated macrofauna. Sampling was carried out when sea temperature was nearing annual minimal (winter), maximal (summer), and intermediate (spring and fall) values (see Table 1 for sampling dates). On each sampling event, scuba divers acquired video transects of the bed and hand-collected rhodoliths for laboratory analysis.

Table 1. Schedule of video transects acquisition and rhodolith collections at the two study sites (SP15 and SP18). Temperature is the sea temperature averaged from hourly recordings at the surface of the rhodolith bed at SP15 over a 2-week period encompassing the sampling dates listed for both sites.

Survey	Site	Sampling Date (2013)	Temperature °C (±SD)
Winter	SP15	14 Mar	−0.1 (0.2)
	SP18	4 Mar	
Spring	SP15	17 Jun	5.5 (1.9)
	SP18	10 Jun	
Summer	SP15	17 Sep	11.7 (1.5)
	SP18	1 Oct	
Fall	SP15 and SP18	15 Dec	4.4 (0.6)

2.2.1. Bed Composition

On each survey, scuba divers laid on the surface of the bed three 20 m long transects, ~4 m apart from each other. Divers filmed, at a fixed distance of 1.5 m above the bed, both sides of each transect with a submersible video camera system (Sony HDV 1080i/MiniDV with an Amphibico Endeavor housing) equipped with a scaling bar. We converted videos to non-overlapping image frames with VLC Media Player 2.2.3 (VideoLan Organization, Paris, France) and randomly selected 20 frames from each transect. A 0.5×1 m digital quadrat was overlaid on each frame with ImageJ version 1.51 (National Institutes of Health, U.S.A., Bethesda, MD) and filled with 50 evenly spaced (10 cm apart) points. We classified the substratum under each point as either of six categories: (1) live rhodolith, (2) dead rhodolith, (3) shell, (4) sediment, (5) pebble, and (6) cobble/boulder. Live or dead rhodoliths consisted of, respectively, pigmented (showing pinkish) or unpigmented (showing whitish) individuals. Shell consisted of large (≥ 5 cm across) disarticulated shells (whole or fragmented) of horse mussel, *Modiolus modiolus*. Biogenic sands were coarse grains dominated by small (<0.4 cm) fragments of dead rhodoliths. Per the Wentworth scale for grain size classification [52], we categorized lithic fragments with a maximum diameter of 4–64 mm and >64 mm, as pebble and cobble/boulder, respectively. Given their rare occurrence, we merged cobbles and boulders into a single category (category 6 above). Loose seaweed fragments were ephemeral components of the rhodolith bed and hence were not considered a substratum type. Bed composition for each transect was estimated by summing the occurrences of substratum types under each of the 50 points within each digital quadrat. We repeated this procedure for each of the 20 image frames of each transect, for a total of 1000 occurrences per transect.

2.2.2. Rhodolith Abundance and Morphology

Upon completing transect videoing, divers hand-collected all rhodoliths within three 30×30 cm (0.09 m^2) quadrats placed on the right side of each transect at 2, 10, and 18 m marks, for a total of 72 quadrats (3 quadrats $\times$ 3 transects $\times$ 2 sites $\times$ 4 surveys). The systematic placement of quadrats at predetermined distances along transects ensured unbiased sampling. Rhodoliths from each quadrat were brought to the sea surface in numbered, seawater-filled, sealed plastic bags, and transported to the Ocean Sciences Centre (OSC) of Memorial University of Newfoundland. Upon arrival at the OSC, bags with their seawater and rhodoliths content were placed in large flow-through tanks supplied with ambient seawater pumped in from a depth of ~5 m in the adjacent embayment, Logy Bay. Water in each bag was replaced with new seawater from the tanks three times daily until the completion of the measurements described below.

For each quadrat sample, we counted all rhodoliths and measured their longest (L), intermediate (I), and shortest (S) axes to determine their size (volume) and shape. The volume of each rhodolith was estimated with the following equation:

$$V = \frac{4}{3}\pi abc \tag{1}$$

where V is the volume of an ellipsoid, a denotes the radius of the longest axis (L/2), b denotes the radius of the intermediate axis (I/2), and c is the radius of the shortest axis (S/2). We summed the volume of all rhodoliths in each quadrat, hereafter referred to as "rhodolith volume" (per 0.09 m^2 quadrat), to investigate the relationship between rhodolith volume and macrofaunal density, biomass, and diversity (see Sections 2.3 and 2.4).

We used ternary diagrams created with the open source software TRIPLOT developed by Graham and Midgley [53] and based on work by Sneed and Folk [54] on particle shapes to plot variation in rhodolith shape distribution: (1) within and between the two sites, and (2) among non-nucleated, pebble-nucleated, and shell-nucleated rhodoliths. The software uses mathematical relationships between the three rhodolith axes (S/L, I/L, and (L-)/L-S)) to assign each rhodolith to 1 of the 10 categories lying in a continuum of shapes between spheroidal, discoidal, and ellipsoidal [25]. Compact, compact–platy, compact–bladed, and compact–elongate rhodoliths are largely spheroidal, whereas platy and very-platy rhodoliths, or elongate and very elongate rhodoliths, are comparatively more discoidal and ellipsoidal, respectively. Bladed and very bladed rhodoliths are in between discoidal and ellipsoidal. We analyzed rhodolith shapes according to Sneed and Folk's [54] four main shape categories (compact, platy, bladed, and elongate) because the majority of rhodoliths at our two study sites were between discoidal and ellipsoidal (i.e., bladed or very bladed). We broke all rhodoliths into ~2 cm^3 fragments to determine the presence and type of nucleus (pebble or shell). We determined total rhodolith biomass per quadrat after removing all nuclei and macrofauna (see Section 2.3) and oven drying the rhodolith fragments at 40 °C to constant weight.

2.3. Macrofaunal Density, Biomass, and Diversity

For each quadrat sample, we inspected rhodolith fragments and carefully extracted all visible macrofauna with tweezers and forceps. We transferred the fragments into a 5 L bucket filled with filtered seawater and manually stirred for ~1 min to dislodge the remaining macrofauna. The content was further poured onto a 5 mm mesh sieve, which retained rhodolith fragments, and was placed atop a 500 μm mesh sieve, which retained all the remaining macrofauna. Macrofauna was preserved in a solution of 4% formaldehyde mixed with seawater.

We sorted macrofauna from each sample into 12 taxonomic groups (Ophiuroidea, Asteroidea, Echinoidea, Holothuroidea, Polyplacophora, Gastropoda, Bivalvia, Polychaeta, Crustacea, Porifera, Nemertea, and Sipuncula) and measured, using a balance with a precision of ±0.001 g (PB3002-S/FACT; Mettler Toledo, Columbus, OH, USA), the total wet weight for each group after gently blotting the samples. We chose to express macrofaunal biomass in wet weight rather than dry weight to preserve the samples for future biodiversity analyses. We identified and tallied all organisms to the lowest possible taxonomic rank. Because of the time constraints associated with the identification of high numbers of small and often juvenile specimens, we limited the identification of polychaetes to family. However, using one randomly chosen sample per site and per season (n = 8 samples), we produced a list of all the identifiable polychaete species. Sponges were highly fragmented and could only be tallied as present or absent. For each sample, we calculated (1) macrofaunal density by dividing the total number of the collected organisms by the quadrat surface area (0.09 m^2); and (2) the Shannon diversity index (*H*).

We used *American Seashells* [55] as the main guide for mollusk identification. We used detailed identification keys by Pettitbone [56], Fauchald [57], Appy et al. [58], and Pocklington [59] to identify polychaetes families and species. Amphipods were identified by professional taxonomists. We used field guides by Gosner [60], Pollock [61], Abbott and Morris [62], and Squires [63] to complete invertebrate identification.

2.4. Statistical Analysis

2.4.1. Rhodolith Bed Structure

We used a two-way permutational multivariate analysis of variance (PERMANOVA) (Anderson [64]) with the fixed factors of Site (SP15 and SP18) and Season (winter, spring, summer, and fall) to compare spatial and seasonal variability in (1) seafloor composition based on six substratum types, namely live rhodolith, dead rhodolith, shell, sand, pebble, and cobble/boulder (n = 24; 3 transects × 2 sites × 4 seasons); and (2) rhodolith shape based on four shape categories derived from Sneed and Folk's [54] ternary diagram: compact (i.e., compact, compact–platy, compact–bladed, and compact–elongate), platy (platy and very platy), bladed (bladed and very bladed), and elongate (elongate and very elongate) (n = 72; 9 quadrats × 2 sites × 4 seasons). We used a negative binomial regression [65] with the fixed factor of Site (SP15 and SP18) to compare rhodolith density (counts per quadrat) between the two sites with a negative binomial distribution, which best accounted for overdispersion of rhodolith count data. We used two two-way ANOVAs, each with the fixed factors of Site (SP15 and SP18) and Season (winter, spring, summer, and fall) to compare (1) rhodolith biomass and (2) rhodolith volume between the sites (n = 72 for each ANOVA; 2 sites × 9 quadrats per site × 4 collections). Two binomial regressions [65] with the fixed factor of Site, enabled the comparison of (1) the proportion of nucleated rhodoliths (nucleated or non-nucleated) and (2) the proportion of each nucleus type (pebble or shell) between the sites (n = 72 each; 2 sites × 9 quadrats per sites × 4 collections).

2.4.2. Macrofaunal Density, Biomass, and Diversity

As rhodolith volume increases, so does the space available for colonization by macrofauna between rhodolith branches. Therefore, we included rhodolith volume as an explanatory variable in our analyses of macrofaunal density, biomass, and diversity. We used two analyses of covariance (ANCOVA), each with the fixed factors of Site (SP15 and SP18) and Season (winter, spring, summer, and fall) and covariate Volume (total rhodolith volume per quadrat) to compare macrofaunal biomass and diversity (the Shannon diversity index) between sites and among seasons. We used negative binomial regressions to model overdispersed invertebrate counts with the fixed factors of Site (SP15 and SP18) and Season (winter, spring, summer, and fall) to compare macrofaunal density between sites and among seasons.

All analyses (except PERMANOVAs) were performed on untransformed data; however, the densities and biomass values of rhodolith and macrofauna are reported in the text and figures as values per meter square (rather than values per 0.09 m^2 quadrat) to facilitate the comparison with data in the literature. For all ANOVAs and ANCOVAs, we verified the homogeneity of the variance and normality of the residuals by examining the distribution and normal probability plots of the residuals, respectively [66]. Tukey HSD multiple comparison tests based on least-square means were used to detect the differences among levels within a factor. All statistical analyses were carried out with R 3.6.1 R (R Core Team 2019) [67]. We used R packages MASS [68] and VEGAN [69] to fit generalized linear models (binomial and negative binomial regressions) and PERMANOVA, respectively. PERMANOVA only included species/taxa that accounted for at least 0.1% of total macrofaunal abundance (52 taxa) and were based on Bray–Curtis dissimilarity for square-root-transformed data. We used pairwise comparisons to detect the differences among levels within factors and similarity percentage (SIMPER) analyses, with 9999 permutations to quantify the contribution of each taxon to overall between-group dissimilarity. All analyses used a 5% significance level (α = 0.05). All means are presented with a 95% confidence interval (mean ± 95% CI) unless stated otherwise. Analysis of variance/deviance tables and detailed SIMPER outputs are presented in Supplement S1 and Supplement S3, respectively.

3. Results

3.1. Rhodolith Bed Habitat Structure

3.1.1. Seafloor Composition

Live rhodolith and sediment were the most frequent substratum types, with mean percent cover of 63% and 23%, respectively, followed by dead rhodolith, with 9% surface cover, pebble (3.6%), shell (2.4%), and cobble/boulder (<1%). PERMANOVA showed spatial variation in substratum type (Table S1; Figure 2A), whereas SIMPER analysis revealed that the seven-times lower percent cover of shells at SP15 compared with SP18 explained 30% of the dissimilarity between the two sites (Table S8). PERMANOVA also showed seasonal variation in seafloor composition. However, post hoc pairwise comparison detected differences at a 10% significance level ($p = 0.096$) only, and SIMPER analysis revealed that the higher percent cover of sediment, pebbles, and shells in winter, compared with fall, explained 64% of the dissimilarity between these seasons (Tables S1A and S8; Figure 2A).

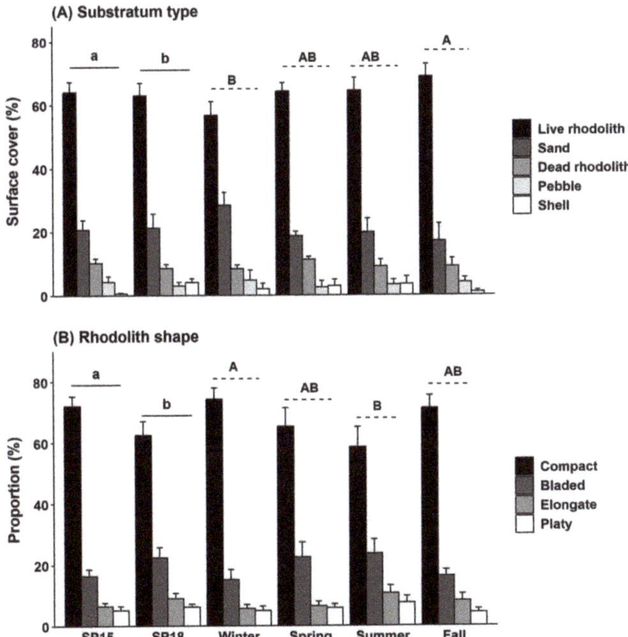

Figure 2. Variation in mean (**A**) surface cover (±95% CI) of substratum types, and (**B**) proportion of rhodolith shapes, between the two study sites (SP15 and SP18) and among the four sampling seasons. PERMANOVA analyses showed no Site–Season interaction for substratum type or rhodolith shape (see Table S1 in Supplement S1), and hence data were pooled across seasons (n = 12 and 36 per site for substratum types and rhodolith shapes, respectively) and across sites (n = 6 and 18 per season for substratum types and rhodolith shapes, respectively). Error bars are 95% CIs. Sites (seasons) not sharing the same lowercase (uppercase) letters differ statistically at a significance level of 5% (solid lines) or 10% (dashed lines). Cobble/boulder surface cover was ≤1% at both sites and in all seasons, and hence is not shown in (**A**) for simplicity (see Section 3.1).

3.1.2. Rhodolith Abundance and Morphology

Rhodolith density and biomass did not differ between sites or among seasons (Tables S2A and S3A), averaging 822 ± 44 individuals m^{-2} and 11.54 ± 0.47 kg rhodoliths m^{-2}, respectively. Rhodoliths ranged in size from 0.6 to 527.2 cm^3 at SP15, and from 0.2 to 392.0 cm^3 at SP18, for a mean rhodolith size 1.4 times greater at SP15 (25.6 ± 0.87 cm^3) than at SP18 (18.8 ± 0.54 cm^3). Rhodolith distributions for the 5–50 cm^3 size classes were similar at both sites, but the proportion of smaller (<5 cm^3) and larger (>50 cm^3) rhodoliths was 7% lower and 5% higher at SP15 than at SP18, respectively (Figure 3A). Rhodolith volume was 1.4 times higher at SP15 ($25,600 \pm 870$ cm^3) than at SP18 ($18,800 \pm 540$ cm^3) but did not vary seasonally (Table S3B).

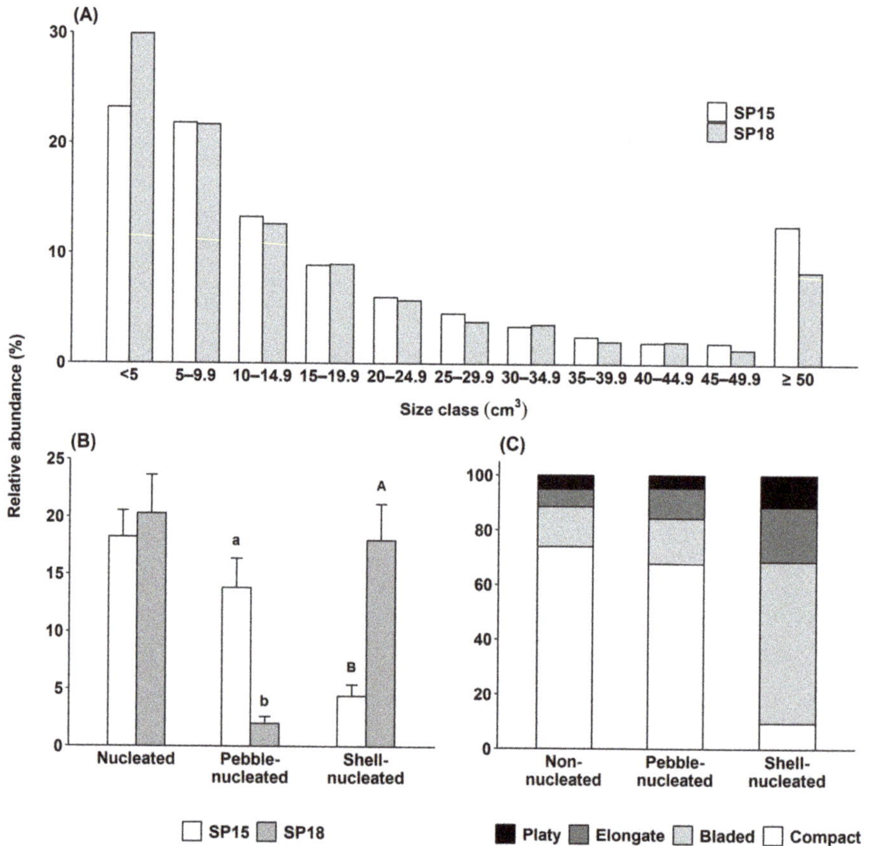

Figure 3. Relative abundance of (**A**) rhodoliths per 5 cm^3 size class intervals at the two study sites. Relative abundances were calculated using all rhodoliths collected during the four seasonal surveys at SP15 (n = 2849) and SP18 (n = 2805); (**B**) nucleated rhodoliths and main nucleus type at the two study sites (n = 36 for each site, error bars = 95% CI). Bars with different letters differ statistically ($p < 0.05$); and (**C**) each of four rhodolith shape categories for non-nucleated (n = 4371), pebble-nucleated (n = 383), and shell-nucleated (n = 538) rhodoliths (data were pooled across sites (SP15 and SP18) and seasons (winter, spring, summer, and fall)).

The proportion of nucleated rhodoliths was similar at both sites (Table S2B), averaging $18 \pm 1\%$ at SP15 and $20 \pm 2\%$ at SP18 (Figure 3B). Nuclei consisted of either small pebbles or entire or fragmented horse mussel (*Modiolus modiolus*) shells, except for one gastropod shell and one small (~2 cm) piece of wood. The proportion of the two main nucleus types (pebble and shell), however, differed between sites (Table S2C), with seven times more pebble-nucleated, and four times fewer shell-nucleated rhodoliths at SP15 than SP18 (Figure 3B). Large ($\geq$50 cm^3) rhodoliths were 47% nucleated at SP15, including 34% pebble-nucleated rhodoliths. At SP18, the proportion of rhodoliths increased to 68%, including 61% shell-nucleated rhodoliths.

Compact shapes (i.e., rhodoliths in any one of the four compact shape classes) dominated SP15 (73%) and SP18 (62%) rhodoliths, followed by bladed (bladed and very-bladed; 16% and 22%), elongate (elongate or very elongate; 6% and 9%), and platy (platy or very platy; 5% and 6%) shapes (Figure 4A,B). PERMANOVA showed a significant spatial variation in rhodolith shape, whereas SIMPER analysis revealed that the higher proportion of compact rhodoliths and lower proportion of bladed, elongate, and platy rhodoliths at SP15, compared with SP18, explained 16% of the dissimilarity between the two sites (Tables S1B and S9; Figure 2B). Rhodolith shapes also varied seasonally, with a higher proportion of compact rhodoliths and a lower proportion of three other shapes, which explained 19% of the dissimilarity between winter and fall (Tables S1B and S9; Figure 2B).

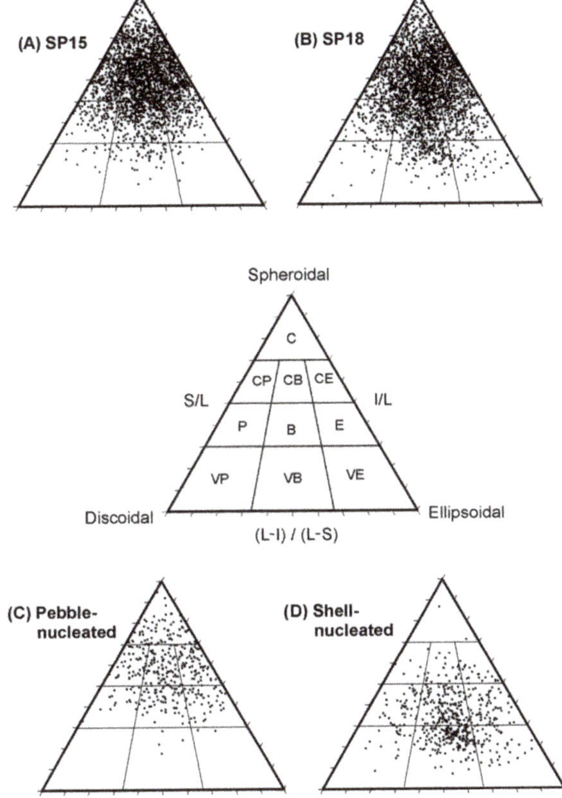

Figure 4. Ternary diagrams showing the distribution of rhodoliths among the ten shape classes (compact (C), compact–platy (CP), compact–bladed (CB), compact–elongate (CE), platy (P), bladed (B), elongate (E), very platy (VP), very bladed (VB), very elongate (VE)) as defined by Sneed and Folk [54] associated with the spheroidal, discoidal, and ellipsoidal rhodolith morphologies at (**A**) site SP15 (n = 2489) and (**B**) site SP18 (n = 2802), and for (**C**) pebble-nucleated rhodoliths (n = 383), and (**D**) shell-nucleated rhodoliths (n = 538). Rhodoliths at SP15 and SP18 were pooled across seasons (winter, spring, summer, and fall). Pebble- and shell-nucleated rhodoliths were pooled across sites and seasons.

Rhodolith shape distribution was similar in non-nucleated and pebble-nucleated rhodoliths, with compact shapes dominating, followed by bladed, elongate, and platy shapes (Figure 3C). Shell-nucleated rhodoliths were predominantly bladed or elongate (79%), whereas compact shapes were the least represented (9%) (Figures 3C and 4D). The proportions of bladed and elongate shapes were, on average, 3.8 and 2.5 times higher in shell-nucleated than in non-nucleated or pebble-nucleated rhodoliths, respectively (Figures 3C and 4).

3.2. Rhodolith Macrofauna

We counted and identified 50,775 specimens from 108 taxa (Table S7). Brittle stars and chitons numerically dominated macrofaunal abundance, accounting for 36% and 18% of total specimens, respectively. Polychaetes, gastropods, and bivalves accounted for 14%, 11%, and 7% of total abundance, respectively, followed by urchins (4%), amphipods (3%), sea stars and isopods (2% each), and nemerteans and sipunculids (1% each) (Table S7). Sea cucumbers, decapods, ostracods, tunicates, platyhelminthes, and cnidarians also occurred in low (<1%) abundance (Table S7). Small fragments of sponges (*Sycon* sp.) and colonial tunicates (*Didemnum* sp.) were present at both sites and in all seasons (Table S7). Nearly 50% of the polychaetes belonged to the families Sabellidae (mainly *Pseudopotamilla reniformis*, 33%) and Terrebellidae (15%), while the wrinkled rock borer (*Hiatella arctica*) made up 82% of all bivalves. Macrofaunal density ranged from 911 to 25,240 individuals m^{-2} and increased with rhodolith volume at a similar rate at both sites, averaging 7833 $\pm$ 444 (SE) individuals m^{-2} (Table S4; Figure 5A). Density also scaled positively with rhodolith volume in all seasons, but it increased at a lower rate and was on average 25% higher in spring (8878 $\pm$ 689 individuals m^{-2}) than in the other seasons (7193 $\pm$ 796 individuals m^{-2}; Table S4; Figure 5A).

Macrofaunal biomass ranged from 38.9 g m^{-2} to 1.153 kg m^{-2} and was dominated by brittle stars (64%) and sea urchins (13%), followed by chitons and bivalves (8% each), polychaetes (4%), sea stars (2%), and gastropods and nemerteans (1% each). Sponges, crustaceans, sipunculids, and sea cucumbers each accounted for <1% of the biomass. Mean biomass did not vary significantly with rhodolith volume, was 1.6 times higher at SP15 (346.7 $\pm$ 31.1 g m^{-2}) than at SP18 (216.7 $\pm$ 30.0 g m^{-2}), and 2 times higher in spring (32.4 $\pm$ 3.8 g) (360.0 $\pm$ 42.2 g m^{-2}) than in summer (16.4 $\pm$ 3.7 g) (182.2 $\pm$ 41.1 g m^{-2}) (Table S5, Figure 5B). Macrofaunal diversity did not vary significantly among seasons but increased with the rhodolith volume at SP15 and was negatively related to rhodolith volume at SP18 (Table S6, Figure 5C).

PERMANOVA analysis indicated that macrofaunal assemblages varied spatially. SIMPER analysis showed that the higher abundance of daisy brittle stars (*Ophiura aculeata*), mottled chitons (*Tonicella marmorea*), wrinkled rock borer clams (*Hiatella arctica*), sabellid polychaetes, and keyhole limpets (*Puncturella noachina*), together with lower abundance of brittle stars (*Ophiopholis robusta*) and northern white chitons (*Stenosemus albus*), explained 62% of the dissimilarity between the two sites (Tables S1C and S10, Figure 6A). Macrofaunal assemblages also varied seasonally, with SIMPER analysis indicating that a higher abundance of the same seven taxa noted above accounted for ~65% of the dissimilarity between spring and the three other seasons (Tables S1C and S10; Figure 6B). The only exception was the lower abundance of brittle stars (*O. aculeata*) in spring than in fall (Table S10). All 52 taxa included in the PERMANOVA analysis (see Section 2.4) were present at both sites and in all seasons. The seven taxa that contributed most to the spatial and seasonal dissimilarities mentioned above accounted for 69% of all the specimens identified.

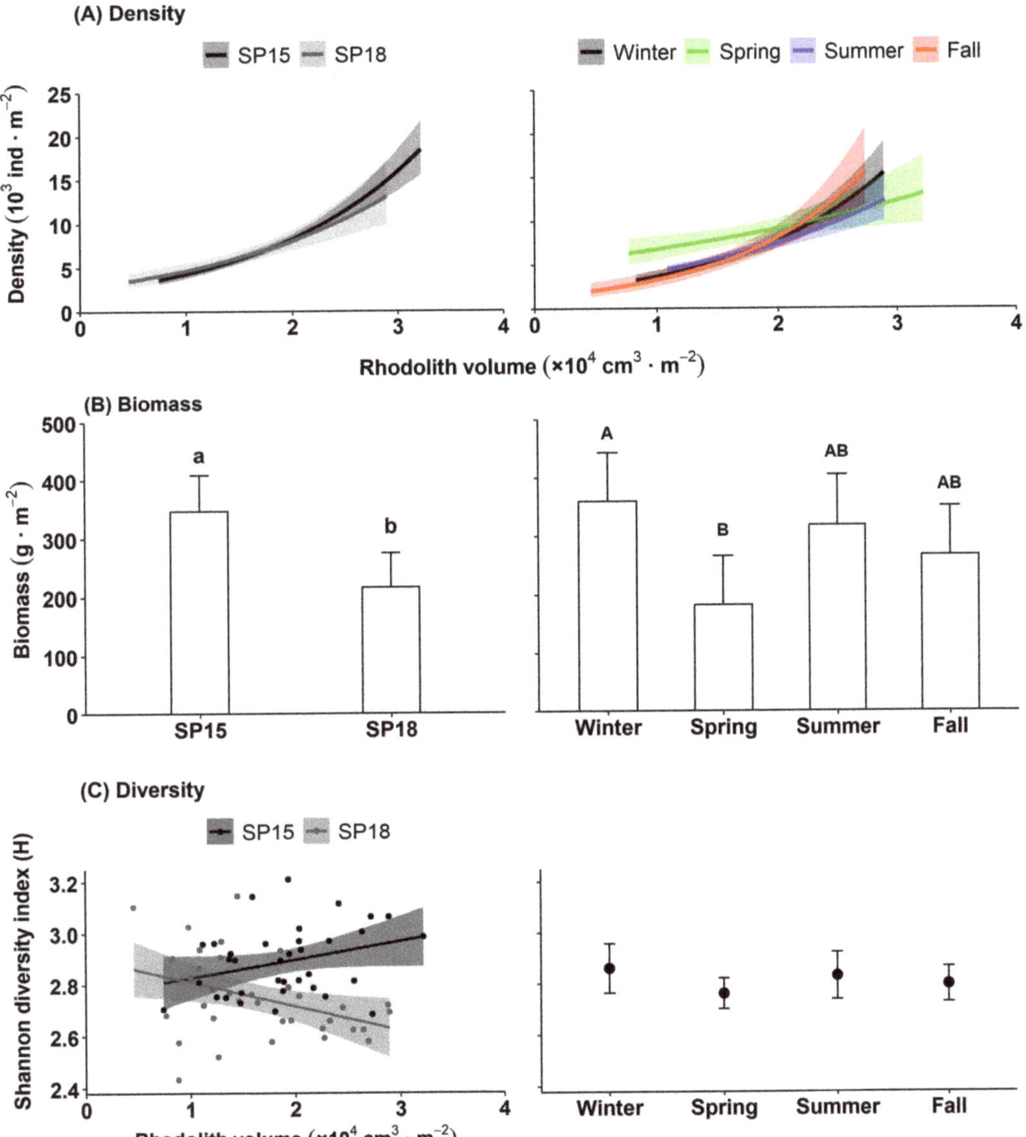

Figure 5. Mean macrofaunal (**A**) density, (**B**) biomass, and (**C**) diversity at the two study sites (**left panels**) and within each season (**right panels**) (n = 36 each). Regression lines indicate significant relationships between (**A**) macrofaunal density (negative binomial regression applied to macrofaunal counts) or (**B**) diversity (linear regression applied to Shannon diversity index) and rhodolith volume. Errors are 95% confidence intervals (±95% CIs). Bars not sharing the same letters differ statistically ($p < 0.05$).

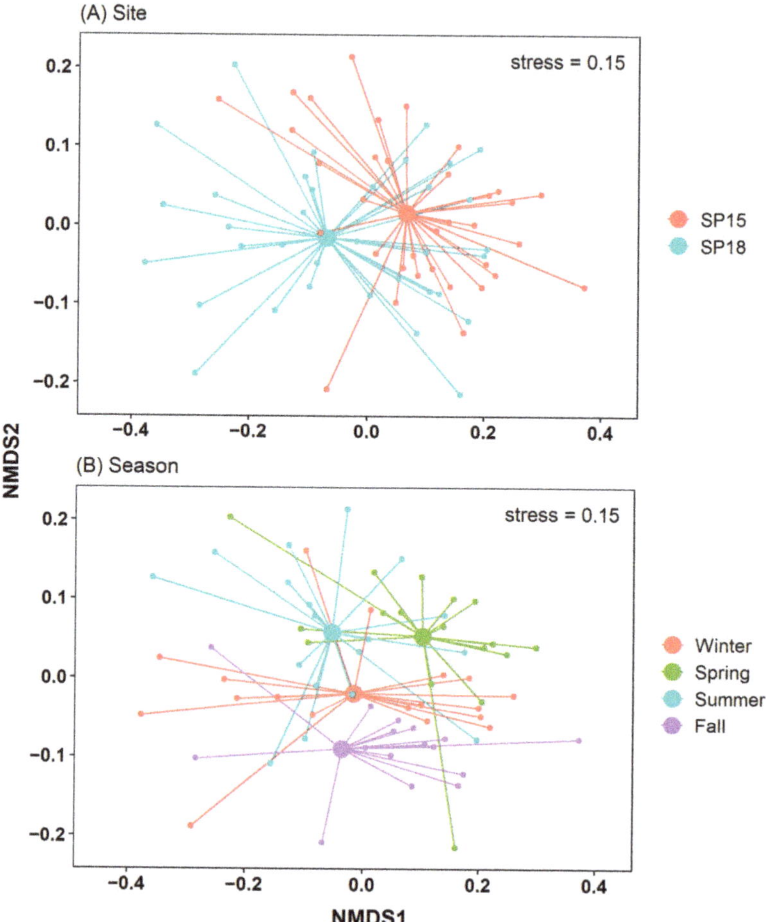

Figure 6. Non-metric multidimensional scaling (NMDS) using Bray–Curtis dissimilarity for macro-faunal counts showing dissimilarity of macrofaunal assemblages (53 taxa) between (**A**) the two study sites (SP15 and SP18), and (**B**) the four sampling seasons (n = 72 for each). Smaller and larger solid dots indicate individual sample and group (site or season) centroids, respectively.

4. Discussion

Our nine-month survey of a subarctic rhodolith (*Lithothamnion glaciale*) bed revealed a stable, unconsolidated rhodolith framework containing a highly diverse macrofaunal community of at least 108 taxa. We found that (1) substratum types varied seasonally and between the shallow (SP15; 15 to 17 m deep) and deeper (SP18; 18 to 20 m deep) study sites but that rhodolith density and biomass remained stable; (2) macrofaunal density scaled positively with rhodolith volume; (3) macroinvertebrate diversity showed contrasting relationships with rhodolith volume at SP15 (positive) and SP18 (negative); and (4) differences in macroinvertebrate assemblages between the two study sites and between spring and the other three seasons were mainly driven by variation in the density of up to the seven most abundant taxa.

4.1. Rhodolith Bed Structure

We showed that the rhodolith bed at both sites was structurally stable throughout the survey, containing >90% of live and dead rhodoliths and coarse biogenic sediments, as well as similar rhodolith densities (~822 individuals m^{-2}) and biomasses (~11.5 kg rhodoliths m^{-2}) in any given season. Such rhodolith densities and biomasses within the 15–20 m depth range (1) parallel the ~858 rhodoliths m^{-2} reported by Gagnon et al. [30] for a shallower (8 to 10 m deep) portion of the same bed sampled four years before the present study, and (2) are up to twice lower than the peak 19.5 kg rhodoliths m^{-2} measured in other portions of the bed within the 8–15 m depth range four years before [30] and four years after [50] the present study. That rhodolith biomass varies spatially within the bed despite a homogenous rhodolith density, indicates that rhodolith size varies across the bed. In a study of the growth response of *L. glaciale* rhodoliths to seasonal changes in sea temperature and light in the same bed, Bélanger and Gagnon [33] showed that rhodolith tissue growth did not vary within the 8–25 m depth range despite a 6× decrease in photosynthetically active radiation (PAR) with depth. Growth, therefore, is an unlikely driver of the uneven distribution of variably sized rhodoliths across the St. Philip's bed. We propose that differences in habitat characteristics between both sites, including the greater abundance of larger and taller rocky outcrops surrounding SP18, may help create hydrodynamic conditions more favorable for rhodolith growth or long-term preservation of rhodolith size and shape, even during occasional, yet damaging wind and wave storms (see below).

The rhodolith bed framework differed between the two study sites, primarily because of a greater abundance of empty and disarticulated shells of horse mussels (*Modiolus modiolus*) at SP18 than at SP15. Live horse mussels in the rhodolith bed predominantly form clumps attached to bedrock outcrops, with shells often covered in thick and branching coralline crusts. The relatively tall and steep bedrock outcrops enclosing SP18 host larger populations of *M. modiolus* than those on the shorter and more dispersed outcrops surrounding SP15 (authors' scuba observations). When they die, mussels, and their continuously expanding and thickening cover of coralline algae, presumably detach and fall down the outcrops, and water currents and mobile organisms move the empty shells across the bed. Such active and passive transportation and displacement of empty, disarticulated mussel shells, which would eventually become rhodoliths as new *L. glaciale* tissue develops and would accumulate over time, may well explain our observed 4x greater abundance of mussel-shell-nucleated rhodoliths at SP18 than at SP15.

The size and shape of the nuclei likely influenced the overall morphology of nucleated rhodoliths [38,70,71]. For example, shell-nucleated rhodoliths were predominantly large ($\geq$50 cm^3) with a bladed or elongated shape, characteristic of bivalve shells, whereas pebble nuclei were much smaller than shell nuclei, accordingly yielding 2.5x fewer large rhodoliths. We also showed that rhodolith shape distribution was similar among pebble-nucleated and non-nucleated rhodoliths, with ~70% of compact shapes, while shell-nucleated rhodoliths were mostly bladed (59%) or elongate (20%) but seldom compact. Our finding that the majority of rhodoliths at both study sites were nonetheless non-nucleated and compact in shape parallels observations in other sections of the St. Philip's bed since 2012 [30,36,50]. Predominantly compact shapes have also been documented in arctic [72,73], temperate [71,74], and tropical [75,76] rhodolith beds. Compact shapes presumably result from uniform, radial rhodolith growth facilitated by occasional overturning by hydrodynamic forces [74,77] or bioturbators [36]. Nucleation type and bioturbation are likely the main drivers of rhodolith size and morphology in the St. Philip's bed since hydrodynamic forces are generally insufficient to move rhodoliths [36].

Our noted differences in the rhodolith bed framework suggest mild seasonal reworking of the bed. Although invertebrates and fish can transport rhodoliths over distances of up to several meters per day [36,78,79], it is unlikely that bioturbation solely accounted for the 13% increase in rhodolith cover from winter to fall, and the 16% decrease in the abundance of compact rhodoliths from winter to summer. This is especially true of SP18, with the high bedrock outcrops that almost entirely enclosed the bed. Interestingly, a major wave storm swept our study sites during the first week of March 2013, (i.e., ~1 wk before the first rhodolith survey (winter)). Extraordinarily, strong surge and bottom currents completely ripped epibenthic enclosures while moving heavy (15 kg) cinder blocks over several meters across the bed we had installed several 10 s of meters away from SP15. This major disturbance pushed large quantities of rhodoliths into areas with lower hydrodynamic forces while exposing large patches of the sedimentary bottom normally overtopped by rhodoliths. Our casual observations of the rhodolith bed after the storm suggested sections of the rhodolith bed framework were gradually redistributing, presumably by the combined effects of water motion and bioturbation. Observed seasonal differences in rhodolith shape distribution were unexpected given that the slow growth rate of <1 mm y^{-1} in *L. glaciale* rhodoliths [32–35,80,81] is unlikely to yield perceptible changes in the shape of the majority of rhodoliths over the course of ~6 months. The marginally significant statistical differences of our post hoc comparisons suggested a mild seasonal effect on rhodolith shape distribution that may have been caused by the highly heterogeneous distribution of variably shaped rhodoliths across the bed.

Our findings, therefore, generally supported our hypothesis that the bed's structure differs between sites, yet is temporally stable within each site. Accordingly, we conclude that the rhodolith bed framework, particularly in terms of rhodolith size and shape, varies spatially across the bed and that local stability prevails under the main influence of biological forcing until sporadic and unusually intense physical forcing reworks the framework.

4.2. Rhodolith Macrofauna

We showed that rhodolith macrofaunal richness in the St. Philip's bed was >2 times higher than that reported for arctic *L. glaciale* beds [39,73]. Macrofaunal abundance, however, was dominated by a few species with brittle stars (*Ophiopholis aculeata* and *Ophiura robusta*) and chitons (*Tonicella marmorea* and *Stenosemus albus*), accounting for >50% of all the specimens identified, and ~70% of total macrofaunal biomass. Studies of other portions of the St. Philip's bed carried out 4 y before [30], as well as 1 y [36] and 4 y [50] after the present study, also highlighted the dominance of these and other abundant taxa, including green sea urchin (*Strongylocentrotus droebachiensis*), common sea star (*Asterias rubens*), and wrinkled rock borer (*Hiatella arctica*), therefore highlighting high temporal stability in the main composition of the macrofaunal community. The present study's finer taxonomic resolution unveiled the important contribution of gastropods (*Puncturella noachina* and *Moelleria costulata*) and polychaetes (Sabellidae and Terebellidae) to total macrofaunal abundance and diversity.

Unlike biomass, macrofaunal density scaled positively with rhodolith volume. Positive relationships between macrofaunal density and rhodolith size have been described in other rhodolith beds [82–84]. We found that most macrofauna was located within spaces between rhodolith branches, or deeper inside cavities running within the rhodoliths' calcified skeleton, which explains why larger rhodoliths housed higher macrofaunal density than smaller ones. Brittle stars and sea urchins were among the largest organisms sampled, contributing mostly to total macrofaunal biomass. Large brittle stars often had their central disc recessed in between distal portions of rhodolith branches, with their arms extending into the water column to collect food particles. Sea urchins, with their rigid spherical test, were located mainly on the outer surface of rhodoliths, with a small proportion of tiny individuals concealed in between branches. The largely spheroidal shape of most rhodoliths sampled implies that the surface-to-volume ratio decreased nearly logarithmically with increasing rhodolith size. Accordingly, the amount of surface and space available at the periphery of rhodoliths, mainly occupied by larger brittles stars and sea urchins, increased at a lower rate than the internal space available for colonization by smaller macrofauna. This imbalance could explain the lack of a significant relationship between macrofaunal biomass and rhodolith volume.

Macrofaunal diversity (*H*) increased at SP15 but decreased at SP18, with increasing rhodolith volume. Positive relationships between macrofaunal diversity and rhodolith size were linked to higher heterogeneity of microhabitats associated with larger rhodoliths in other beds [83,84]. The majority of large ($\geq$50 cm^3) rhodoliths at SP15 were non-nucleated or had a small pebble nucleus that made only a small volume of space unavailable to macrofauna. In contrast, ~60% of the largest rhodoliths at SP18 had a horse mussel-shell nucleus, creating a considerable amount of inner, open space that could potentially be colonized by macrofauna. Our results, therefore, only partially supported our hypothesis that the abundance and diversity of rhodolith macrofauna scale positively with rhodolith size. We conclude that the relationship between rhodolith size and rhodolith macrofaunal abundance or diversity varies ontogenetically, being largely allometric in non-nucleated rhodoliths and isometric in rhodoliths with proportionally large nuclei.

Significant differences in macrofaunal biomass and diversity between SP15 and SP18 supported our hypothesis that rhodolith macrofauna differs between sites. Rhodolith volume was ~20% higher, and large ($\geq$50 cm^3) rhodoliths were ~20% less nucleated at SP15 than SP18, providing additional space and microhabitat to a more abundant and diverse macrofaunal assemblage. PERMANOVA and SIMPER analyses of macrofaunal data revealed that 87% of all the taxa sampled were common to both sites, with most of the between-site dissimilarity explained by the high abundances at SP15 of a few dominant taxa, including *Ophiopholis aculeata*, *Tonicella marmorea*, *Hiatella arctica*, *Puncturella noachina*, and sabellid polychaetes. NMDS analysis corroborated the above-noted considerable overlap in macrofauna between the two study sites. Our results, therefore, suggest that the bed in St. Philip's is characterized by a relatively homogenous and biodiverse macrofaunal assemblage whose spatial variability is largely explained by differences in the distribution and abundance of a few numerically dominant taxa.

Our findings that macrofaunal density in spring was 25% higher than that in any of the three other seasons sampled, while macrofaunal biomass dropped by ~50 between winter and spring, attest to considerable temporal variability in the size structure of the macrofaunal community. Mass spawning in brittle stars is common in the northwest Atlantic [85,86]. Our observed 3$\times$ lower biomass-to-density ratio of brittle stars in spring compared to winter suggests that massive recruitment of juveniles occurred between the two seasons. Because brittle stars accounted for 36% and 64% of the total macrofaunal density and biomass, respectively, such major recruitment events could largely explain seasonal variability in the St. Philip's bed. Nevertheless, we showed that over 75% of all taxa were present in all seasons, whereas the taxa that were not present in all seasons accounted for <1% of the total macrofaunal abundance. PERMANOVA, SIMPER, and NMDS analyses revealed that the higher densities of the up-to-seven most abundant taxa (*Ophiura robusta*, *Ophiopholis aculeata*, *Tonicella marmorea*, *Stenosemus albus*, *Hiatella arctica*, *Punturella noachina*, and sabellid polychaetes) in spring were responsible for 50–60% of the dissimilarity in macrofaunal assemblages between this and the three other seasons. Our results therefore strongly support our hypothesis that rhodolith macrofauna varies seasonally.

5. Conclusions

Our nine-month field survey of the St. Philip's bed revealed that the rhodolith bed framework is spatially heterogeneous, yet provides a temporally stable habitat to an abundant and highly diverse (at least 108 taxa) macrofaunal assemblage dominated by a few taxa (including brittle stars, chitons, bivalves, gastropods, polychaetes, sea urchins, and sea stars). We showed that certain habitat components, such as bivalve shells, can have a profound effect on rhodolith morphology and associated macrofauna, with larger, non-nucleated rhodoliths generally hosting higher macrofaunal density, biomass, and diversity. Spatiotemporal variability in rhodolith macrofauna was mainly driven by seasonal fluctuations in the relative abundance of up-to-seven numerically dominant taxa, with little changes in otherwise homogenous species composition.

The present study, which was carried out at an unprecedentedly fine taxonomic resolution, strongly supports the notion that rhodolith beds are key biodiversity hotspots. Its spatial and temporal domains, also largely unparalleled for this kind of ecosystem, provide clear, quantitative evidence that rhodolith beds provide a stable framework under the main influence of biological forcing until sporadic and unusually intense physical forcing reworks it. These findings are worrisome in the face of observed and predicted changes in ocean climate because they suggest that rhodolith beds are particularly vulnerable to ongoing increases in the frequency and severity of wind and wave storms [87,88]. In other studies [33,35], we concluded that ongoing ocean warming will benefit subarctic *L. glaciale* rhodoliths by shortening the yearly period over which sea temperatures near 0 °C prevent growth. Further studies should address the resilience of rhodolith beds to wave storms and other types of physical or chemical forcing (e.g., ocean warming, acidification, and eutrophication) to better understand and predict the trajectory of rhodolith-based ecosystems.

Supplementary Materials: The following supporting information can be downloaded at: https://www.mdpi.com/article/10.3390/d15060774/s1, Table S1: Summary of permutational multivariate analysis of variance (PERMANOVA), based on Bray-Curtis dissimilarity for square-root transformed count data, examining the effect of the fixed factors Site (SP15 and SP18) and Season (winter, spring, summer, and fall) on (A) substratum type percent cover based on six categories: live rhodolith, dead rhodolith, sediment, shell, pebble, and cobble/boulder), (B) rhodolith shape distribution based on ten shape classes: compact, compact-platy, platy, very-platy, compact-bladed, bladed, very-bladed, compact-elongate, elongate, very-elongate, and (C) rhodolith macrofaunal community based on 52 taxa. All analyses were performed on balanced design using Type I sums of squares and a 5% significance level ($\alpha = 0.05$).; Table S2: Summary of (A) negative binomial regression examining the effect of fixed factors Site (SP15 and SP18) and Season (winter, spring, summer, and fall) on rhodolith density, and for (B) and (C) binomial regressions examining the effect of the fixed factor Site (SP15 and SP18) on rhodolith nucleation (nucleated and non-nucleated) and nucleus type (pebble-nucleated and shell-nucleated), respectively. All analyses were performed on balanced design using Type I sums of squares and a 5% significance level ($\alpha = 0.05$).; Table S3: Summary of two-way ANOVAs examining the effect of the fixed factors Site (SP15 and SP18) and Seasons (winter, spring, summer, and fall) on (A) rhodolith biomass, and (B) total rhodolith volume per 30×30-cm quadrat (0.09 m^2). All analyses were performed on balanced design using Type I sums of squares and a 5% significance level ($\alpha = 0.05$).; Table S4: Summary of negative binomial linear models (applied to non-transformed count data) examining the effect of the fixed factors Site (SP15 and SP18) and Season (winter, spring, summer, and fall), and covariate Volume (total rhodolith volume per 30×30-cm quadrat) on macrofaunal density. Analysis was performed on balanced design using Type I sums of squares and a 5% significance level ($\alpha = 0.05$).; Table S5: Summary of ANCOVAs examining the effect of the fixed factors Site (SP15 and SP18) and Season (winter, spring, summer, and fall), and covariate Volume (total rhodolith volume per 30×30-cm quadrat) on macrofaunal biomass (wet weight). Analysis was performed using Type I sums of squares and a 5% significance level ($\alpha = 0.05$).; Table S6: Summary of ANCOVAs examining the effect of the fixed factors Site (SP15 and SP18) and Season (winter, spring, summer, and fall), and covariate Volume (total rhodolith volume per 30×30-cm quadrat) on rhodolith macrofaunal community Shannon diversity index (H). Analysis was performed using Type I sums of squares and a 5% significance level ($\alpha = 0.05$).; Table S7: Macrofauna taxa identified in the St. Philip's rhodolith bed. Numbers indicate mean density ($\pm$SE) per 30×30-cm quadrat (0.09 m^2) at the two study sites (SP15 and SP18; data pooled across seasons), and in each sampling season (data pooled across sites) (see Table 1 for collection dates). Zero (0) values indicate absence for a given collection. Crosses (x) indicate sponge taxa that were present but not tallied because of fragmentation. Polychaete species listed under family names were identified from subsamples but not tallied for quantitative analyses.; Table S8: Summary of SIMPER analysis for the substratum type assemblages. Dissim is the average ($\pm$SD) contribution of each species/Taxon to the overall dissimilarity between contrasted sites (SP15, SP18) or seasons. CumSum is the ordered cumulative contribution to overall dissimilarity. avgA and avgB are the average substratum type occurrence (square root transformed) for the first and second contrasted site/season, respectively.; Tale S9: Summary of SIMPER analysis for the rhodolith shape assemblages. Dissim is the average ($\pm$SD) contribution of each species/Taxon to the overall dissimilarity between contrasted sites (SP15, SP18) or seasons. CumSum is the ordered cumulative contribution to overall dissimilarity. avgA and avgB are the average substratum type occurrence (square root transformed) for the first and second contrasted site/season, respectively.; Table S10: Summary of SIMPER analysis for macrofaunal community. Dissim is the average (±SD) contribution of each species/taxon to the overall dissimilarity between contrasted sites (SP15, SP18) or seasons. CumSum is the ordered cumulative contribution to overall dissimilarity. avgA and avgB are the average species/taxon abundances (square root transformed) for the first and second contrasted site/season, respectively.

Author Contributions: Conceptualization, D.B. and P.G.; methodology, D.B. and P.G.; validation, D.B. and P.G.; formal analysis, D.B.; investigation, D.B. and P.G.; resources, P.G.; writing—original draft preparation, D.B. and P.G.; writing—review and editing, P.G.; visualization, D.B.; supervision, P.G.; project administration, P.G.; funding acquisition, P.G. All authors have read and agreed to the published version of the manuscript.

Funding: This research was funded by the Natural Sciences and Engineering Research Council (NSERC Discovery Grant, Grant number RGPIN-355875-2008), Canada Foundation for Innovation (CFI Leaders Opportunity Funds, Grant number 16940), Research and Development Corporation of Newfoundland and Labrador (Ignite R&D, Grant number 5003.070.002), and Department of Fisheries and Aquaculture of Newfoundland and Labrador (DFA, Grant number BS-48456) grants to P.G.; D.B. was supported by the Memorial University of Newfoundland President's Doctoral Student Investment Fund Program.

Institutional Review Board Statement: Not applicable.

Data Availability Statement: The data presented in this study are available in the Supplementary Materials.

Acknowledgments: We are grateful to A. P. St-Pierre, D. Frey, and I. Lima for assistance with field and laboratory work, and Kent Gilkinson and Laure de Montety for help with the taxonomic identification of invertebrates. We also thank P. Snelgrove, R. Gregory, and two anonymous reviewers for constructive comments that helped improve the manuscript.

Conflicts of Interest: The authors declare no conflict of interest. The funders had no role in the design of the study; in the collection, analyses, or interpretation of data; in the writing of the manuscript; or in the decision to publish the results.

References

1. McCoy, E.D.; Bell, S.S. Habitat structure: The evolution and diversification of a complex topic. In *Habitat Structure*; Springer: Dordrecht, The Netherlands, 1991; pp. 3–27.
2. Byrne, L.B. Habitat Structure: A fundamental concept and framework for urban soil ecology. *Urban Ecosyst.* **2007**, *10*, 255–274. [CrossRef]
3. de Camargo, N.F.; Sano, N.Y.; Vieira, E.M. Forest vertical complexity affects alpha and beta diversity of small mammals. *J. Mammal.* **2018**, *99*, 1444–1454. [CrossRef]
4. Garden, J.G.; Mcalpine, C.A.; Possingham, H.P.; Jones, D.N. Habitat structure is more important than vegetation composition for local-level management of native terrestrial reptile and small mammal species living in urban remnants: A case study from Brisbane, Australia. *Austral. Ecol.* **2007**, *32*, 669–685. [CrossRef] [PubMed]
5. Zimbres, B.; Peres, C.A.; Machado, R.B. Terrestrial mammal responses to habitat structure and quality of remnant riparian forests in an Amazonian cattle-ranching landscape. *Biol. Conserv.* **2017**, *206*, 283–292. [CrossRef]
6. Beck, M.W. Comparison of the measurement and effects of habitat structure on gastropods in rocky intertidal and mangrove habitats. *Mar. Ecol. Prog. Ser.* **1998**, *169*, 165–178. [CrossRef]
7. Buhl-Mortensen, L.; Buhl-Mortensen, P.; Dolan, M.F.J.; Dannheim, J.; Bellec, V.; Holte, B. Habitat complexity and bottom fauna composition at different scales on the continental shelf and slope of northern Norway. *Hydrobiologia* **2012**, *685*, 191–219. [CrossRef]
8. Carvalho, L.R.S.; Loiola, M.; Barros, F. Manipulating habitat complexity to understand its influence on benthic macrofauna. *J. Exp. Mar. Biol. Ecol.* **2017**, *489*, 48–57. [CrossRef]
9. Preston, F.W. Time and space and the variation of species. *Ecology* **1960**, *41*, 612–627. [CrossRef]
10. MacArthur, R.H.; Wilson, E.O. (Eds.) *The Theory of Island Biogeography*; Princeton University Press: Princeton, NJ, USA, 1967.
11. Connor, E.F.; McCoy, E.D. The statistics and biology of the species-area relationship. *Am. Nat.* **1979**, *113*, 791–833. [CrossRef]
12. Dean, R.L.; Connell, J.H. Marine invertebrates in an algal succession. III. Mechanisms linking habitat complexity with diversity. *J. Exp. Mar. Biol. Ecol.* **1987**, *109*, 249–273. [CrossRef]
13. Hixon, M.A.; Menge, B.A. Species diversity: Prey refuges modify the interactive effects of predation and competition. *Theor. Popul. Biol.* **1991**, *39*, 178–200. [CrossRef]
14. Myhre, L.C.; Forsgren, E.; Amundsen, T. Effects of habitat complexity on mating behavior and mating success in a marine fish. *Behav. Ecol.* **2013**, *24*, 553–563. [CrossRef]

15. Bell, S.S.; McCoy, E.D.; Mushinsky, H.R. (Eds.) *Habitat Structure: The Physical Arrangement of Objects in Space*; Population and Community Biology Series; Springer: Dordrecht, The Netherlands, 1991; Volume 8.

16. Jones, C.G.; Lawton, J.H.; Shachak, M. Organisms as ecosystem engineers. *Oikos* **1994**, *69*, 373. [CrossRef]

17. Buhl-Mortensen, L.; Vanreusel, A.; Gooday, A.J.; Levin, L.A.; Priede, I.G.; Buhl-Mortensen, P.; Gheerardyn, H.; King, N.J.; Raes, M. Biological structures as a source of habitat heterogeneity and biodiversity on the deep ocean margins. *Mar. Ecol.* **2010**, *31*, 21–50. [CrossRef]

18. Coggan, N.V.; Hayward, M.W.; Gibb, H. A global database and "state of the field" review of research into ecosystem engineering by land animals. *J. Anim. Ecol.* **2018**, *87*, 974–994. [CrossRef] [PubMed]

19. Curd, A.; Pernet, F.; Corporeau, C.; Delisle, L.; Firth, L.B.; Nunes, F.L.D.; Dubois, S.F. Connecting Organic to Mineral: How the physiological state of an ecosystem-engineer is linked to its habitat structure. *Ecol. Indic.* **2019**, *98*, 49–60. [CrossRef]

20. Steneck, R.S.; Graham, M.H.; Bourque, B.J.; Corbett, D.; Erlandson, J.M.; Estes, J.A.; Tegner, M.J. Kelp forest ecosystems: Biodiversity, stability, resilience and future. *Environ. Conserv.* **2002**, *29*, 436–459. [CrossRef]

21. Eriksson, B.K.; Rubach, A.; Hillebrand, H. Biotic habitat complexity controls species diversity and nutrient effects on net biomass production. *Ecology* **2006**, *87*, 246–254. [CrossRef]

22. Hauser, A.; Attrill, M.J.; Cotton, P.A. Effects of habitat complexity on the diversity and abundance of macrofauna colonising artificial kelp holdfasts. *Mar. Ecol. Prog. Ser.* **2006**, *325*, 93–100. [CrossRef]

23. Koivisto, M.E.; Westerbom, M. Habitat structure and complexity as determinants of biodiversity in blue mussel beds on sublittoral rocky shores. *Mar. Biol.* **2010**, *157*, 1463–1474. [CrossRef]

24. Adey, W.H.; Macintyre, I.G. Crustose coralline algae: A re-evaluation in the geological sciences. *Geol. Soc. Am. Bull.* **1973**, *84*, 883–904. [CrossRef]

25. Bosence, D.W. Description and Classification of Rhodoliths (Rhodoids, Rhodolites). In *Coated Grains*; Peryt, T.M., Ed.; Springer: Berlin/Heidelberg, Germany, 1983; pp. 217–224. [CrossRef]

26. Foster, M.S. Rhodoliths: Between rocks and soft places. *J. Phycol.* **2001**, *37*, 659–667. [CrossRef]

27. Bosence, D.W.J. Coralline Algal Reef Frameworks. *J. Geol. Soc.* **1983**, *140*, 365–376. [CrossRef]

28. Nelson, W.A. Calcified macroalgae critical to coastal ecosystems and vulnerable to change: A review. *Mar. Freshw. Res.* **2009**, *60*, 787–801. [CrossRef]

29. Woelkerling, W.J.; Irvine, L.M.; Harvey, A.S. Growth-forms in non-geniculate coralline red algae (Corallinales, Rhodophyta). *Aust. Syst. Bot.* **1993**, *6*, 277–293. [CrossRef]

30. Gagnon, P.; Matheson, K.; Stapleton, M. Variation in rhodolith morphology and biogenic potential of newly discovered rhodolith beds in Newfoundland and Labrador (Canada). *Bot. Mar.* **2012**, *55*, 85–99. [CrossRef]

31. Frantz, B.R.; Kashgarian, M.; Coale, K.H.; Foster, M.S. Growth rate and potential climate record from a rhodolith using 14C accelerator mass spectrometry. *Limnol. Oceanogr.* **2000**, *45*, 1773–1777. [CrossRef]

32. Teed, L.; Bélanger, D.; Gagnon, P.; Edinger, E. Calcium carbonate ($CaCO_3$) production of a subpolar rhodolith bed: Methods of estimation, effect of bioturbators, and global comparisons. *Estuar. Coast. Shelf Sci.* **2020**, *242*, 106822. [CrossRef]

33. Bélanger, D.; Gagnon, P. High growth resilience of subarctic rhodoliths (*Lithothamnion glaciale*) to ocean warming and chronic low irradiance. *Mar. Ecol. Prog. Ser.* **2021**, *663*, 77–97. [CrossRef]

34. Bélanger, D.; Gagnon, P. Low growth resilience of subarctic rhodoliths (*Lithothamnion glaciale*) to coastal eutrophication. *Mar. Ecol. Prog. Ser.* **2020**, *642*, 117–132. [CrossRef]

35. Arnold, C.L.; Bélanger, D.; Gagnon, P. Growth resilience of subarctic rhodoliths (*Lithothamnion glaciale*, Rhodophyta) to chronic low sea temperature and irradiance. *J. Phycol.* **2022**, *58*, 251–266. [CrossRef] [PubMed]

36. Millar, K.R.; Gagnon, P. Mechanisms of stability of rhodolith beds: Sedimentological aspects. *Mar. Ecol. Prog. Ser.* **2018**, *594*, 65–83. [CrossRef]

37. Freiwald, A.; Henrich, R. Reefal coralline algal build-ups within the Arctic Circle: Morphology and sedimentary dynamics under extreme environmental seasonality. *Sedimentology* **1994**, *41*, 963–984. [CrossRef]

38. Ballantine, D.L.; Bowden-Kerby, A.; Aponte, N.E. Cruoriella rhodoliths from shallow-water back reef environments in La Parguera, Puerto Rico (Caribbean Sea). *Coral Reefs* **2000**, *19*, 75–81. [CrossRef]

39. Teichert, S. Hollow rhodoliths increase svalbard's shelf biodiversity. *Sci. Rep.* **2014**, *4*, 6972. [CrossRef]

40. Hinojosa-Arango, G.; Riosmena-Rodríguez, R. Influence of rhodolith-forming species and growth-form on associated fauna of rhodolith beds in the central-west Gulf of California, México. *Mar. Ecol.* **2004**, *25*, 109–127. [CrossRef]

41. Gabara, S.S.; Hamilton, S.L.; Edwards, M.S.; Steller, D.L. Rhodolith structural loss decreases abundance, diversity, and stability of benthic communities at Santa Catalina Island, CA. *Mar. Ecol. Prog. Ser.* **2018**, *595*, 71–88. [CrossRef]

42. Bracchi, V.A.; Meroni, A.N.; Epis, V.; Basso, D. Mollusk thanatocoenoses unravel the diversity of heterogeneous rhodolith beds (Italy, Tyrrhenian Sea). *Diversity* **2023**, *15*, 526. [CrossRef]

43. Kamenos, N.A.; Moore, P.G.; Hall-Spencer, J.M. Attachment of the juvenile queen scallop (*Aequipecten opercularis* (L.)) to maerl in mesocosm conditions; juvenile habitat selection. *J. Exp. Mar. Biol. Ecol.* **2004**, *306*, 139–155. [CrossRef]

44. Steller, D.L.; Cáceres-Martínez, C. Coralline algal rhodoliths enhance larval settlement and early growth of the Pacific calico scallop *Argopecten ventricosus*. *Mar. Ecol. Prog. Ser.* **2009**, *396*, 49–60. [CrossRef]

45. Pearce, C.M.; Scheibling, R.E. Induction of metamorphosis of larvae of the green sea urchin, *Strongylocentrotus droebachiensis*, by coralline red algae. *Biol. Bull.* **1990**, *179*, 304–311. [CrossRef] [PubMed]

46. Heyward, A.J.; Negri, A.P. Natural inducers for coral larval metamorphosis. *Coral Reefs* **1999**, *18*, 273–279. [CrossRef]

47. Whalan, S.; Webster, N.S.; Negri, A.P. Crustose coralline algae and a cnidarian neuropeptide trigger larval settlement in two coral reef sponges. *PLoS ONE* **2012**, *7*, e30386. [CrossRef] [PubMed]

48. Adey, W.H. Distribution of saxicolous crustose corallines in the northwestern North Atlantic. *J. Phycol.* **1966**, *2*, 49–54. [CrossRef]

49. Adey, W.H.; Hayek, L.-A.C. Elucidating marine biogeography with macrophytes: Quantitative analysis of the North Atlantic supports the thermogeographic model and demonstrates a distinct subarctic region in the northwestern Atlantic. *Northeast. Nat.* **2011**, *18*, 1–128. [CrossRef]

50. Hacker Teper, S.; Parrish, C.C.; Gagnon, P. Multiple trophic tracer analyses of subarctic rhodolith (*Lithothamnion glaciale*) bed trophodynamics uncover bottom-up forcing and benthic-pelagic coupling. *Front. Mar. Sci.* **2022**, *9*, 1–20. [CrossRef]

51. Basso, D. Deep rhodolith distribution in the Pontian Islands, Italy: A model for the paleoecology of a temperate sea. *Palaeogeogr. Palaeoclimatol. Palaeoecol.* **1998**, *137*, 173–187. [CrossRef]

52. Wentworth, C.K. A scale of grade and class terms for clastic sediments. *J. Geol.* **1922**, *30*, 377–392. [CrossRef]

53. Graham, D.J.; Midgley, N.G. Graphical representation of particle shape using triangular diagrams: And excel spreadsheet method. *Earth Surf. Proc. Land.* **2000**, *25*, 1473–1477. [CrossRef]

54. Sneed, E.D.; Folk, R.L. Pebbles in the lower Colorado River, Texas a study in particle morphogenesis. *J. Geol.* **1958**, *66*, 114–150. [CrossRef]

55. Abbott, R.T. *American Seashells; the Marine Mollusca of the Atlantic and Pacific Coasts of North America*, 2nd ed.; Van Nostrand Reinhold: New York, NY, USA, 1974; ISBN 0442202288.

56. Pettibone, M.H. *Marine Polychaete Worms of the New England Region: I. Families Aphroditidae through Trochocaetidae*; Bulletin of the United States National Museum: Washington, DC, USA, 1963; pp. 1–356.

57. Fauchald, K. *The Polychaete Worms. Definitions and Keys to the Orders, Families and Genera*; Science Series; Natural History Museum of Los Angeles County: Los Angeles, CA, USA, 1977; Volume 28, pp. 1–188.

58. Appy, T.D.; Linkletter, L.E.; Dadswell, M.J. *Annelida: Polychaeta: A Guide to the Marine Flora and Fauna of the Bay of Fundy*; Canada Fisheries and Marine Service Technical Report; Government of Canada: St. Andrews, NB, Canada, 1980; p. 920.

59. Pocklington, P. *Polychaetes of Eastern Canada: An Illustrated Key to Polychaetes of Eastern Canada including the Eastern Arctic*; Scientific Monograph Funded by Ocean Dumping Control Act Research Fund; National Museums of Canada & Department of Fisheries and Oceans: Mont-Joli, QC, Canada, 1989.

60. Gosner, K.L. *Atlantic Seashore: A Field Guide to Sponges, Jellyfish, Sea Urchins, and More*; Houghton Mifflin Harcourt: Boston, MA, USA, 2014; ISBN 0544530853.

61. Pollock, L.W. *A Practical Guide to the Marine Animals of Northeastern North America*; Rutgers University Press: New Brunswick, MA, USA, 1998; ISBN 0813523990.

62. Abbott, R.T.; Morris, P.A. *Shells of the Atlantic and Gulf Coasts and the West Indies*, 4th ed.; Houghton Mifflin Harcourt: Boston, MA, USA, 2001; ISBN 0442202288.

63. Squires, H.J. *Decapod Crustacea of the Atlantic Coast of Canada (Canadian Bulletin Fisheries Aquatic Sciences)*; Department of Fisheries and Oceans Canada: Ottawa, ON, Canada, 1990; Volume 221.

64. Anderson, M.J. A new method for non-parametric multivariate analysis of variance. *Austral Ecol.* **2001**, *26*, 32–46. [CrossRef]

65. Quinn, G.P.; Keough, M.J. *Experimental Design and Data Analysis for Biologists*; Cambridge University Press: New York, NY, USA, 2002; ISBN 0521009766.

66. Snedecor, G.; Cochran, W. *Statistical Methods*, 8th ed.; Iowa State University Press: Ames, IA, USA, 1994; ISBN 978-0-813-81561-9.

67. R Core Team. *R: A Language and Environment for Statistical Computing*; R Foundation for Statistical Computing: Vienna, Austria, 2021; Available online: https://www.r-project.org/ (accessed on 14 February 2022).

68. Venables, W.N.; Ripley, B.D. *Modern Applied Statistics with S*, 4th ed.; Springer: New York, NY, USA, 2002; ISBN 0-387-95457-0. Available online: https://www.stats.ox.ac.uk/pub/MASS4/ (accessed on 25 December 2022).

69. Oksanen, J.; Simpson, G.; Blanchet, F.G.; Kindt, R.; Legendre, P.; Minchin, P.; O'hara, R.; Solymos, P.; Stevens, M.; Szoecs, E.; et al. Vegan: Community Ecology Package, Version 2.6-4. 2022. Available online: https://CRAN.R-project.org/package=vegan (accessed on 1 January 2023).

70. Piller, W.E.; Rasser, M. Rhodolith formation induced by reef erosion in the Red Sea, Egypt. *Coral Reefs* **1996**, *15*, 191–198. [CrossRef]

71. Del Río, J.; Ramos, D.A.; Sánchez-Tocino, L.; Peñas, J.; Braga, J.C. The Punta de La Mona rhodolith bed: Shallow-water Mediterranean rhodoliths (Almuñecar, Granada, southern Spain). *Front. Earth Sci.* **2022**, *10*, 884685. [CrossRef]

72. Teichert, S.; Woelkerling, W.; Rüggeberg, A.; Wisshak, M.; Piepenburg, D.; Meyerhöfer, M.; Form, A.; Freiwald, A. Arctic rhodolith beds and their environmental controls (Spitsbergen, Norway). *Facies* **2014**, *60*, 15–37. [CrossRef]

73. Teichert, S.; Woelkerling, W.; Rüggeberg, A.; Wisshak, M.; Piepenburg, D.; Meyerhöfer, M.; Form, A.; Büdenbender, J.; Freiwald, A. Rhodolith beds (Corallinales, Rhodophyta) and their physical and biological environment at 80 ° 31′ N in Nordkappbukta (Nordaustlandet, Svalbard Archipelago, Norway). *Phycologia* **2012**, *51*, 371–390. [CrossRef]

74. Basso, D.; Nalin, R.; Nelson, C.S. Shallow-water Sporolithon rhodoliths from North Island (New Zealand). *Palaios* **2009**, *24*, 92–103. [CrossRef]

75. Bahia, R.G.; Abrantes, D.P.; Brasileiro, P.S.; Pereira Filho, G.H.; Amado Filho, G.M. Rhodolith bed structure along a depth gradient on the northern coast of Bahia state, Brazil. *Braz. J. Oceanogr.* **2010**, *58*, 323–337. [CrossRef]

76. Amado-Filho, G.M.; Maneveldt, G.; Manso, R.C.C.; Marins-Rosa, B.V.; Pacheco, M.R.; Guimarães, S. Structure of rhodolith beds from 4 to 55 meters deep along the southern coast of Espírito Santo State, Brazil. *Cienc. Mar.* **2007**, *33*, 399–410. [CrossRef]
77. Marrack, E.C. The relationship between water motion and living rhodolith beds in the southwestern Gulf of California, Mexico. *Palaios* **1999**, *14*, 159–171. [CrossRef]
78. James, D.W. Diet, movement, and covering behavior of the sea urchin *Toxopneustes roseus* in rhodolith beds in the Gulf of California, México. *Mar. Biol.* **2000**, *137*, 913–923. [CrossRef]
79. Pereira-Filho, G.H.; Veras, P.C.; Francini-Filho, R.B.; Moura, R.L.; Pinheiro, H.T.; Gibran, F.Z.; Matheus, Z.; Neves, L.M.; Amado-Filho, G.M. Effects of the sand tilefish *Malacanthus plumieri* on the structure and dynamics of a rhodolith bed in the Fernando de Noronha Archipelago, tropical West Atlantic. *Mar. Ecol. Prog. Ser.* **2015**, *541*, 65–73. [CrossRef]
80. Kamenos, N.A.; Law, A. Temperature controls on coralline algal skeletal growth. *J. Phycol.* **2010**, *46*, 331–335. [CrossRef]
81. Halfar, J.; Zack, T.; Kronz, A.; Zachos, J.C. Growth and high-resolution paleoenvironmental signals of rhodoliths (coralline red algae): A new biogenic archive. *J. Geophys. Res. Oceans* **2000**, *105*, 22107–22116. [CrossRef]
82. Steller, D.L.; Riosmena-Rodríguez, R.; Foster, M.S.; Roberts, C.A. Rhodolith bed diversity in the Gulf of California: The importance of rhodolith structure and consequences of disturbance. *Aquat. Conserv.* **2003**, *13*, S5–S20. [CrossRef]
83. Veras, P.C.; Pierozzi, I., Jr.; Lino, J.B.; Amado-Filho, G.M.; Senna, A.R.; Santos, C.S.G.; de Moura, R.L.; Passos, F.D.; Giglio, V.J.; Pereira-Filho, G.H. Drivers of biodiversity associated with rhodolith beds from euphotic and mesophotic zones: Insights for management and conservation. *Perspec. Ecol. Conserv.* **2020**, *18*, 37–43. [CrossRef]
84. Foster, M.S.; McConnico, L.M.; Lundsten, L.; Wadsworth, T.; Kimball, T.; Brooks, L.B.; Medina-López, M.; Riosmena-Rodríguez, R.; Hernández-Carmona, G.; Vásquez-Elizondo, R.M. Diversity and natural history of a *Lithothamnion muelleri-Sargassum horridum* community in the Gulf of California. *Cienc. Mar.* **2007**, *33*, 367–384. [CrossRef]
85. Himmelman, J.H.; Dumont, C.P.; Gaymer, C.F.; Vallières, C.; Drolet, D. Spawning synchrony and aggregative behaviour of cold-water echinoderms during multi-species mass spawnings. *Mar. Ecol. Prog. Ser.* **2008**, *361*, 161–168. [CrossRef]
86. Doyle, G.M.; Hamel, J.F.; Mercier, A. Small-scale spatial distribution and oogenetic synchrony in brittlestars (Echinodermata: Ophiuroidea). *Estuar. Coast. Shelf Sci.* **2014**, *136*, 172–178. [CrossRef]
87. Guo, L.; Perrie, W.; Long, Z.; Toulany, B.; Sheng, J. The impacts of climate change on the autumn North Atlantic wave climate. *Atmos. Ocean* **2015**, *53*, 491–509. [CrossRef]
88. Working Group II. *Climate Change 2022: Impacts, Adaptation and Vulnerability*; IPCC Sixth Assessment Report; IPCC: Geneva, Switzerland, 2022.

diversity

Article

Marine Invertebrate Neoextinctions: An Update and Call for Inventories of Globally Missing Species

James T. Carlton

Coastal & Ocean Studies Program, Williams College-Mystic Seaport, Mystic, CO 06355, USA; jcarlton@williams.edu

Abstract: The register of global extinctions of marine invertebrates in historical time is updated. Three gastropod and one insect species are removed from the list of extinct marine species, while two gastropods, one echinoderm, and three parasites (a nematode, an amphipod, and a louse) are added. The nine extinct marine invertebrates now recognized likely represent a minute fraction of the actual number of invertebrates that have gone extinct. Urgently needed for evaluation are inventories of globally missing marine invertebrates across a wide range of phyla. Many such species are likely known to systematists, but are either rarely flagged, or if mentioned, are not presented as potentially extinct taxa.

Keywords: extinction; habitat destruction; co-extinction; species rediscovery

1. Introduction

"The last fallen mahogany would lie perceptibly on the landscape, and the last black rhino would be obvious in its loneliness, but a marine species may disappear beneath the waves unobserved and the sea would seem to roll on the same as always."

—G. Carleton Ray

Carlton [1] introduced the concept of neoextinctions to refer to those species that have become globally extinct in historical time, as opposed to paleoextinctions over geological time. Carlton et al. [2] then summarized what was known about historical global extinctions in the sea, followed by brief updates by Carlton [3]. Additional reviews, which also included examples of regional marine extinctions ("neoextirpations," [4]) and endangered marine species, have included those of Dulvy et al. [5,6] and del Monte-Luna et al. [7].

I present here a revised and updated inventory of the current record of global marine invertebrate extinctions, as well as an appeal for the promulgation of lists of globally missing species. The threats to marine invertebrate diversity in highly vulnerable habitats that could lead to increasing numbers of extinctions in the 21st century, and the compelling rationales for understanding why extinctions matter, are not reviewed here, as these have been extensively discussed for the past two decades and more [8–13] (among many others). The burgeoning literature further flags the risks to specific threatened and endangered marine invertebrate taxa (for example, [14–23]).

2. Updated Assessment of Marine Invertebrate Global Extinctions

IUCN [24] defines a taxon as extinct "when exhaustive surveys in known and/or expected habitat, at appropriate times (diurnal, seasonal, annual), throughout its historic range have failed to record an individual." Notably, IUCN no longer suggests a specific length of time (such as 50 years [1,25])—a temporal line in the sands of the ocean—after which a species should be declared extinct, leaving consideration of what constitutes sufficiently exhaustive surveys, and thus when to "call it" for an extinction, to be somewhat subjective.

Citation: Carlton, J.T. Marine Invertebrate Neoextinctions: An Update and Call for Inventories of Globally Missing Species. *Diversity* **2023**, *15*, 782. https://doi.org/ 10.3390/d15060782

Academic Editors: Thomas J. Trott and Michael Wink

Received: 21 May 2023
Revised: 13 June 2023
Accepted: 14 June 2023
Published: 16 June 2023

Nine species—of the millions of species!—of marine invertebrates are recognized as extinct (Table 1). Six of these are here newly formally treated as extinctions. An ectoparasite and an endoparasite of the extinct Steller's sea cow—one of the most famous losses in marine biodiversity—have long been mentioned in the literature, but not previously explicitly listed as extinctions. These, along with an ectoparasitic louse from the extinct Guadalupe storm petrel (Table 1), as well as the previously listed louse from the extinct Jamaican petrel, should be considered only as examples of the loss of endo- and ectoparasites of at least 10 additional extinct marine birds and mammals [3]. If each of these extinct marine vertebrates supported only one host-specific parasite, our current number of marine invertebrate extinctions would double. It is of note that there is no requirement that a species be described for it to be declared extinct [26–28]. Indeed, "dark extinction" [29] may play a significant role in future estimates of marine invertebrate extinctions, especially of soft-bodied species in extirpated coastal habitats.

Table 1. Marine invertebrate neoextinctions [1,2].

Species	Former Geographic Range	Last Known Living	Habitat	Cause of Extinction	Comments	References
Nematoda: Chromadorea (roundworms)						
Ascaridoidea?	Alaska: Commander Islands	1766	Endoparasite of extinct Steller's sea cow, *Hydrodamalis gigas*	Co-extinction of host		[30]
Mollusca: Gastropoda (snails)						
Lottiidae: *Lottia alveus* (Conrad, 1831) (eelgrass limpet)	Labrador to New York	1929	Restricted to blades of the eelgrass *Zostera marina* in marine waters.	Marine (but not estuarine) populations of *Zostera* died out in the early 1930s due to an eelgrass disease epidemic, and the limpet never re-appeared.	References to *Lottia alveus* as being still living in the Northeast Pacific Ocean refer instead to a distinct living species, *Lottia parallela* (Dall, 1921) [31].	[32]
Potamididae: *Cerithideopsis fuscata* (Gould, 1857) (horn snail)	California: San Diego Bay	1935	Estuarine mudflats	Habitat destruction		[1]
Dialidae: *Diala exilis* (Tryon, 1866)	California: San Diego Bay and San Francisco Bay	1860s (San Diego Bay); 1860s–1870s? (San Francisco Bay)	on "salt water grass" (77, for San Diego Bay)	Habitat destruction		[33,34]
Aplysiidae: *Phyllaplysia smaragda* Clark, 1977 (sea slug)	Florida: Indian River Lagoon	1982	Restricted to blades of the manatee grass *Syringodium filiforme*	Habitat destruction		[2,35–37]
Arthropoda: Crustacea: Amphipoda (amphipods)						
Cyamidae: *Cyamus rhytinae* (Brandt, 1846) (whale louse)	Alaska: Commander Islands	1766	Ectoparasite of extinct Steller's sea cow, *Hydrodamalis gigas*	Co-extinction of host	No other cyamid amphipods have been reported from sirenians.	[2,30]
Arthropoda: Insecta: Phthiraptera (lice)						
Philapteridae: *Saemundssonia jamaicensis* Timmerman, 1962 (Jamaican petrel louse)	Jamaica	1879	Ectoparasite of extinct Jamaican petrel, *Pterodroma caribbaea*	Co-extinction of host		[3,38]
Menoponidae: *Longimenopon dominicanum* (Kellogg and Mann, 1912) (Guadalupe storm petrel louse)	Guadalupe Island, Mexico	1912	Ectoparasite of extinct Guadalupe storm petrel, *Hydrobates macrodactylus*	Co-extinction of host		[38]

Table 1. *Cont.*

Species	Former Geographic Range	Last Known Living	Habitat	Cause of Extinction	Comments	References
			Echinodermata: Asteroidea (sea stars)			
Asterinidae: *Patiriella littoralis* (Dartnall, 1970)	Tasmania	1991	Intertidal, mixed soft and hard habitat	Habitat destruction		[39]

[1] As noted in the text, Gravili et al. [40] proposed that 10 species of hydrozoans in the Mediterranean Sea had a significant chance of being extinct. Four of these species are doubtfully valid or have doubtful records [40]. Of the remaining six, three were last seen in the 1960s, but it is unclear the extent to which they have been specifically searched for, nor over what seasons or lengths of time, in their last known locations. The last three are: (1) *Eucheilota maasi* Neppi and Stiasny, 1911, described as an endemic in the Adriatic Sea [40,41], last collected in 1914, and known only from its medusa. However, Batistic and Garic [42] report medusae identified as *E. maasi* from the Adriatic Sea based on 2011–2012 collections, indicating that, if correctly identified, it is still extant. (2) *Branchiocerianthus italicus* Stechow, 1921, also described as a Mediterranean endemic last collected in 1905 when it was dredged from 300 m in the Gulf of Naples [43]. (3) *Plumularia syriaca* Billard, 1931, last collected in 1929 and only known from the Gulf of Alexandrette, Syria, at 11 m deep or greater [44]. For these latter two species, it is also similarly not clear the extent to which species-specific searches have been conducted, either by deep-sea explorations in the Gulf of Naples, or at the appropriate depths off the Syrian coast. [2] As also noted in the text, Peters et al. [19] (2013) and Cowie et al. [20] considered five species of cone shells as of questionable status, or questionably or possibly extinct. All five species again reflect the challenges that have resulted in possibly underestimating neoextinctions. *Conasprella sauros* (Garcia, 2006), with only dead shells recovered from Texas, Louisiana, and Mexico, may or may not be a fossil species [45]. The Cape Verdes *Africonus bellulus* (Rolan, 1990) has either not been reported since the 1970s [19] or has been collected sometime since the 1990s [46]; Rolan [47] only cites a 1980 work as the basis for the knowledge of this species, without collection dates or habitat data. It is provisionally accepted as a good species [47]. *Conus colmani* Rockel and Korn, 1990, noted by Cowie et al. [20] as possibly extinct, is known from many specimens, none alive, from the Queensland coast of Australia through deep-water trawling [48]. It is not discussed by Peters et al. [19]. Marshall [49] notes that it needs taxonomic re-evaluation as part of a species complex. *Conus luteus* G. B. Sowerby I, 1833, a widespread Indo-Pacific and Hawaiian species, is noted by Peters et al. [19] as having not been reported since the 1970s, but appears to have been collected alive in recent years in a number of locations, including the Marshall Islands (http://www.underwaterkwaj.com/shell/cone/Conus-luteus.htm; accessed on 1 April 2023) and Papua New Guinea [50]. It is not treated by Cowie et al. [20]. Finally, a timeline of not having been re-collected in 20 years for *Conus splendidulus* G. B. Sowerby I, 1833, from the Indian Ocean, is potentially too short to permit judgment of its status. Overall, in none of these cases does there appear to be published information on the extent of targeted searches.

Previous marine extinction treatments (noted above) have flirted with the extinct, 22 mm long, Florida sea slug *Phyllaplysia smaragda*, but failed to formally list it, despite clear statements as to its status and despite it having once existed in a site that has been thoroughly explored and re-explored. The fifth, a tiny (circa 3 mm tall) snail (*Diala exilis*), long gone from the now highly modified but well-explored bays of the California coast, was flagged in a little-known paper [33]. O'Hara et al. [39] have recently and clearly outlined the evidence that the sixth species, the small Tasmanian sea star *Patierella littoralis*, with a radius up to 22 mm, is extinct. In all three of these cases, long-term explorations in the appropriate habitats and locations have failed to detect any living individuals.

While the data are too few to suggest any biogeographic patterns, ecologically all nine species have disappeared from shallow coastal waters, where the extinction of vulnerable marine vertebrates is expected, due to either direct or indirect human-mediated forces, or where shallow water habitats can be destroyed by human activity. The exception is the apparent non-human-mediated extinction of a marine limpet (*Lottia alveus*) from the Northwest Atlantic Ocean (Table 1), unless the slime mold disease agent that caused the demise of the limpet's host plant, the eelgrass *Zostera marina*, was introduced by a human-mediated vector.

Four species are here removed from the extinct or possibly extinct list (Table 2). Two of these are marine snails that have appeared in previous treatments of global marine extinctions [2]. One, the Chinese mangrove periwinkle *Littoraria flammea*, last believed to have been collected in 1855, was rediscovered in Singapore salt marshes in 2014; it is further likely a synonym of the widespread living Indo-Pacific species *Littoraria melanostoma* (Table 2). The other, a fossil species of California limpet, *Lottia edmitchelli*, was previously thought to have survived into the Holocene, represented by a single living specimen collected in southern California in 1861. This specimen has now been re-identified as an extant species, *Lottia scabra* (Table 2). A terrestrial snail, *Omphalotropis plicosa* from Mauritius, has been misinterpreted as a marine species (Table 2), while a southern California rocky intertidal beetle, *Bembidion palosverdes*, thought gone for nearly 50 years from a mainland site, was discovered alive in 2010 on an offshore island (a refugium?) (Table 2).

Table 2. Marine invertebrates no longer considered extinct, or erroneously listed as such.

Species	Geographic Range	Habitat	Comments and Reference
Mollusca: Gastropoda (snails)			
Littorinidae: *Littoraria flammea* (Philippi, 1847) (periwinkle)	Indo-West Pacific	Mangrove and salt marsh communities	Formerly considered to have last been collected in 1855 in China, it was found living in 2014 in salt marshes near Shanghai, and may be the same as the widespread and abundant Western Pacific species *Littoraria melanostoma* (Gray, 1839) [51]
Lottiidae: *Lottia edmitchelli* (Lipps, 1966) (limpet)	Southern California	Rocky intertidal	Formerly considered to have last been collected alive in 1861 [1], the living specimen so identified is now considered to be the extant species *Lottia scabra* (Gould, 1846) [52]. *L. edmitchelli* is, further, now considered to have gone extinct by the Middle Pleistocene [52].
Assimineidae: *Omphalotropis plicosa* (Pfeiffer, 1854)	Mauritius	Tree trunks (terrestrial)	Listed as an extinct marine species by Kemp et al. [53] based on the IUCN Red List, this is a terrestrial snail, nor is it a salt marsh species (https://en.wikipedia.org/wiki/Omphalotropis_plicosa; accessed on 1 April 2023). It is not extinct [54].
Arthropoda: Insecta: Coleoptera (beetles)			
Carabidae: *Bembidion palosverdes* Kavanaugh and Erwin, 1992 (shore beetle)	California: Santa Catalina Island	Rocky intertidal	Last seen in 1964 on the Palos Verdes Peninsula, Los Angeles County, California, and thought possibly extinct [55], it was rediscovered alive in 2010 on Santa Catalina Island [56].

3. Challenges with Assessing the Global Marine Invertebrate Extinction Record

The current record of global marine invertebrate extinctions is thus extraordinarily paltry. Why is that?

I highlight here three of a number of drivers [1,2,10,57] that may have led to our current embarrassing lack of knowledge of how many, and which, species of marine invertebrates have gone extinct. These drivers are a subset of the more general challenges of accurately assessing temporal and spatial changes in historical marine biodiversity (for example, [58–65]).

3.1. Reluctance to Declare a Species Globally Missing

The marine systematics literature is richly populated with species, especially those described in the 18th and 19th centuries, that cannot be reliably recognized today, often due to apparently insufficient diagnoses or lack of the availability of the original specimens. Terms often applied to such species are **nomina dubia** (for example, [66–69]) or **incertae sedis** (for example, [70–73]). The scientific names of such species—of which there may be thousands [74]—that cannot be confidently matched today to known species are often either simply set aside without disposition, or relegated to the probable synonymy of known species. Such names form part of the "taxonomic graveyard" noted by Bouchet and Strong (2010). In more than 50 years of reading the marine taxonomic and systematic literature across all major and many smaller phyla, I have seen no suggestions that any names now considered *nomina dubia* or *incertae sedis*, based on taxa first and last described centuries ago, might refer to extinct species.

As an example, and because the Mollusca are the best known phylum of marine invertebrates, thanks in large part to centuries of seashell collectors, I analyzed the extraordinary 1258-page monograph of Coan and Valentich-Scott [75] on the marine bivalve mollusks of the Tropical Eastern Pacific (TEP), which covers a 5000 km province from Isla Cedros, Baja California, Mexico to Piura in northern Peru. Of approximately 900 species treated, I tallied nine species that have not been found since the 1860s or earlier (Table 3), along with the suggestions (from Coan and Valentich-Scott [75] or other sources) as to why these species have not been seen again. These suggestions (Table 3) include that the species in question do not actually come from the TEP ("mislabeled," "mislocalized," "extralimital", or provenance uncertain), are difficult or impossible to recognize today from their descriptions or illustrations ("nomen dubium"), or are simply a mystery ("a significant unresolved question," or "not … recognized since"). Again, however, in no case is there a suggestion that any of these species may possibly be extinct.

Table 3. Missing bivalve species in the Tropical Eastern Pacific Ocean (data from Coan and Valentich-Scott [75], unless otherwise indicated).

Family	Species	Size (mm)	Last Known Location	Last Collected	Habitat	Possible Reason for Not Being Re-Discovered (Coan and Valentich-Scott, [75], Unless Otherwise Indicated)
Chamidae	*Chama producta* Broderip, 1835 [1]	93	Mexico: Gulf of Tehuantepec	1828–1830	Sandy mud, 18 m	"Possibly a mislabeled specimen from another province."
Veneridae	*Chinopsis crenifera* (G. B. Sowerby I, 1835)	37	Ecuador: Santa Elena; Paita, Peru	<1835	---	"This species is very uncertain"; known only from Ecuador (the type locality) and Peru (the latter based on 19th century material?; see Keen [76] p. 186.)
Veneridae	*Cytherea inconspicua* G. B. Sowerby I, 1835 [2]	25	Peru: Paita, Piura	<1835	Sandy, muddy bottom	Provenance uncertain (Panamic or Peruvian?)
Veneridae	*Pitar fluctuatus* (G. B. Sowerby II, 1851)	18	Ecuador: Santa Elena, Guayas	<1851	---	"We have not found additional specimens of this distinctive species, and the type locality might be mislocalized."
Petricolidae	*Petricola amygdalina* G. B. Sowerby I, 1834	---	Ecuador: Galapagos Islands	<1834	in pteriid valves, 6–11 m	Nomen dubium or extralimital
Solenidae	*Solen oerstedii* Morch, 1860	69	Costa Rica: Puntarenas	<1860	Subtidal in mud (Huber, 2010)	"not … recognized since" (Keen [76] p. 259)
Pandoridae	*Frenamya cristata* (Carpenter, 1865)	24	Mexico: Gulf of California	<1865	---	"Only known from the type locality in the Golfo de California, Mexico"
Pandoridae	*Pandora brevifrons* G. B. Sowerby I, 1835	22	Panama: Bahia Panama	<1835	---	"In spite of intensive collecting in Panama, this species has not been found since its description in 1835, and it is possible that the types were mislocalized. However, study of specimens from adjacent and far-reaching provinces has also not yielded any material of this species."
Periplomatidae	*Periploma excurva* Carpenter, 1856	---	Mexico: Mazatlan, Sinaloa	<1856	---	"A significant unresolved question"

[1] Cardoso et al. [77] report *Chama producta* from Peru, but their material is not that species (Paul Valentich-Scott, personal communication, May 2023). Huber [78,79], in a work not online and largely inaccessible to most workers, agreed with Reeve [80], (*Chama iostoma* Conrad, 1837) that Broderip's *Chama producta* from Mexico was the same as the Indo-Pacific species *Chama limbula* Lamarck, 1819, but neither Reeve nor Huber provided evidence for this. Bouchet [81,82] treats *Chama producta* and *Chama limbula* as distinct species. [2] Size and habitat data from Huber [78], who assigns it to the genus *Pitar* without explanation.

In short, nine "missing" marine bivalves, last encountered in the mid-19th century or earlier, can be tallied in one province, and these represent only one class of one phylum.

Given that there are 62 recognized marine provinces [83], this might suggest that the number of missing species across many phyla, including short-range provincial endemics [84], could be large.

Thus, while a standard assumption in the taxonomic and systematic sciences is that historical descriptions of species that cannot be clearly interpreted today likely largely represent coarse descriptions of still-extant taxa, if they can be recognized at all, "an alternative hypothesis is that some of these early descriptions represent the only known records of species that became extinct long ago" [3].

3.2. Reluctance to Declare Missing Species as Globally Extinct

Cowie et al. [85] have recently reviewed aspects of the hesitancy to declare a species extinct, including fear of committing the "Romeo Error"—a concern of declaring a species extinct when it is not. This fear may be reinforced by the regular stream of rediscoveries of rare species, some not seen for over 100 years (for example, [2,86–91], and Table 2, herein). Further reinforcement of the Romeo Error may arise from the discovery of living individuals of species previously known only from the fossil record—most famously the coelacanth, but also with cases continuing to be reported [92].

Cowie et al. [85] remarked that, relative to the IUCN criterion noted above of a requirement for exhaustive surveys, "For a very large proportion of described species, there will never be dedicated exhaustive fieldwork, at the appropriate time and over the appropriate timeframe because they are too numerous, and knowledge is too scarce to know the time-frame and even the range to be searched." The result of setting the bar potentially unachievably high, leading authors to "not dare to declare" species extinct, suggests that extinctions will be underestimated, perhaps markedly so [85].

The specter of the Romeo Error is deeply ingrained, and further casts a shadow on especially small and poorly known species. The tiny sea slug (sacoglossan) *Stiliger vossi* Marcus and Marcus, 1960, slightly more than one millimeter long in its preserved state, was last collected in 1958 among algae in shallow water in Biscayne Bay, Florida [93]. The late Kerry Clark, a sacoglossan specialist, searched for it assiduously, but failed to find it as of 1996 [2]. It remains unreported. While we consider another Florida sea slug of larger size, *Phyllaplysia smaragda*, extinct, *S. vossi* remains indefinitely suspended between the living and dead. The "smalls" rule of invasive species science (the smaller the species, the less likely it will be categorized as non-native) works against both additions to communities [94] and deletions.

Benovic et al. [95] identified a number of hydrozoan species not seen since 1910 and known only from the Adriatic Sea, but declared none of them permanently gone. Nearly 30 years later, a change in perspective led Gravili et al. [40] to suggest that some of these species were globally extinct, as discussed further below.

3.3. When Did You Miss Me? Time Lags in Recognizing Missing or Extinct Species

Boero et al. [96] commented that "The modern-day record demonstrates that even large, once-abundant species can simply disappear without notice, suggesting that documenting the disappearance of uncommon and smaller species is a fundamental challenge." Dulvy et al. [5] have discussed the phenomenon of delayed reporting, relative to both local and global extinctions. Clear examples emerge from the limited record of marine extinctions (Table 1). The once abundant eelgrass limpet *Lottia alveus* was last found living in 1929; its disappearance was first pointed out in 1991 [32]. The once common mudflat horn snail *Cerithidea fuscata* was last collected in San Diego Bay, California, in 1935, but its disappearance was not mentioned until 1981 [97].

In more recent times, the relatively large (up to 8 cm) and colorful sea slug (nudibranch) *Felimare californiensis* (Bergh, 1879) was once common along the rocky intertidal shores of southern California: the fact that it had been last detected there in 1977 was not pointed out until 2013 [98]; it remains extant elsewhere. The large (15 cm in length) mud shrimp *Upogebia pugettensis* (Dana, 1852) began steadily disappearing from many

North American Pacific coast estuaries in the 1990s, including wholesale extirpations from some embayments, with no remarks on its absence made by marine biologists, until its widespread demise (but not global extinction) was pointed out by Chapman et al. [99].

Most taxonomic monographs do not note when a given species was last collected or seen. Species long reported by our predecessors remain on lists, and as one generation of workers follows another, it may be difficult to notice that any one species has not been seen "recently." In the monographic work noted above of 900 species of marine bivalves in the Tropical Eastern Pacific, while a small number were flagged as not having been seen since the 19th century [75], we do not know for many of the remaining hundreds of species when in fact they were last collected or seen—which additional species might have gone missing in the last 75 to 100 years, versus those whose apparent lack of recent records is "simply because no one has sought them out again" [96].

Adding to the above list, then, of those drivers that have resulted in the discovery of few marine invertebrate extinctions is the lengthy time and effort to document the details of the history of any one species, including delving into old and often obscure literature in rare journals that may not be online, recognizing the earlier names under which a species may have appeared, tracking down museum holdings, and interviewing older workers who may be, or have been, familiar with a given species. An important caveat is that, while many museum collections can now be searched online, large swaths of material of what any given museum actually holds are not yet either catalogued or if catalogued not yet downloaded, meaning that for an accurate assessment of historical collections of a species, the appropriate museum collections must be visited in person. Very few workers may find investing large amounts of time in the 18th and 19th century literature and in wading through museum collections to be worthy of their time. Finally, all museums hold large amounts of unidentified material, requiring some level of taxonomic expertise to recognize that a target species of interest is in a collection but not yet identified (that, or convincing an expert taxonomist to come along in such explorations).

Nevertheless, recording "last seen" dates across the known historical range of a species may set the stage for a broader capture of species missing (and possibly extinct) globally, a task that I suggest below be profitably pursued.

4. A Call for Inventories of Globally Missing Marine Invertebrates

The IUCN Red List Categories and Criteria of Threatened Species [24] does not define "missing" in their nine-tiered classification system of species at risk of global extinction. Martin et al. [25] have proposed, for terrestrial vertebrates, that "lost taxa"—species not yet declared globally extinct—be defined as those "that have not been reliably observed in >50 years," resurrecting a temporal metric abandoned by the IUCN for extinctions.

Despite the challenges and limitations of attempting to tilt the missing and possibly extinct species windmills, none of these impediments, including fear of the Romeo Error, should prevent promulgating inventories of missing marine invertebrate species. Such inventories would have immediate and profound value that would serve to direct targeted search efforts. Lists of missing species harvested from the literature, or by interviewing experienced systematists, could capture species characterized (1) by being relatively taxonomically robust (ideally based upon examination of original specimens) but still including those taxa suspected of being a synonym of another species, (2) by having a reasonably firm handle on the last known records within Martin et al.'s [25] 50-year window, and (3) by having occurred in habitats highly susceptible to extraordinary levels of anthropogenic disturbance if not wholesale destruction, such as in bays, estuaries, lagoons, mangroves, marshes, supralittoral shores, and many intertidal shores [1,6].

While acknowledging the many threats to deep-sea biota (for example, [100,101]) generally excluded from such lists, at least initially, would be the many hundreds if not thousands of deep-water species that may have been collected only once, and often not since the 19th century, due to the vagaries of stochastic deep-sea exploration (but see [11], relative to endemic hydrothermal vent species).

In a rare example of an attempt to detect missing species, Gravili et al. [40] assessed the status of approximately 400 species of hydroids (Phylum Cnidaria, Class Hydrozoa) in the Mediterranean Sea. Of these, 53 species have not been reported in the literature for at least 41 years, and were thus considered candidates for analysis as potential extinctions. Gravili et al. [40] argued that "The choice of 41 years as a threshold to consider a species as missing was decided based on the rather intense study of hydrozoan species in the Mediterranean in the last four decades … " (and that) "Due to intensive sampling … if a previously reported species fails to be recorded chances are good that, at least, it is more rare than before." They then evaluated these 53 species with a formula for a "confidence of extinction index," proposed originally for paleobiology by Marshall [102], and adapted by Boero et al. [96] "to analyse cases of putative extinction in recent species."

The three variables in this formula are (1) the number of years since the species was last sighted, (2) the number of years between the original description and the last sighting (first framed in Boero et al. [96] as the years between the first record (the date of first collection) and the last sighting), and (3) the number of individual years in which there is a record. The probability of extinction in this formula is thus sensitive to the choice of the demarcation year after which a species is declared missing. The formula does not capture search efforts over a given length of time or area. The rationales for the failure for admitting any of Gravili et al.'s [40] 10 statistically extinct hydroid species to the register of global marine extinctions herein are outlined in footnote 1 of Table 1.

At a family and global level, Peters et al. [19] and Cowie et al. [20] examined the worldwide conservation status of cone shells (Class Gastropoda, Family Conidae). They considered five species as of questionable conservation status or as possibly extinct. As with Gravili et al.'s [40] Mediterranean hydroids, a series of taxonomic and sampling challenges impede admitting these species at this time to the register of global extinctions (footnote 2, Table 1).

The above attempts to seek out missing species in specific taxonomic groups illuminate both the value of detecting potentially lost species and the challenges of recognizing them as extinct, in the absence of dedicated multiyear and ideally species-specific searches. As noted above, these challenges are compounded if species thought to be missing occur or occurred in deeper waters, as illustrated in the examples in Table 1, footnotes 1 and 2.

5. Epilogue

I opened this essay with the same eloquent observation of G. Carleton Ray [103] as I did 30 years ago [1]. Little has changed. While Regnier et al. [104] concluded that "marine habitats seem to have experienced few extinctions, which suggests that marine species may be less extinction prone than terrestrial or freshwater species," and while this would be welcome news if so, such a conclusion remains premature [30] in light of the striking lack of investigation of the possible or probable number of marine extinctions.

The challenges to document and verify extinctions in the sea are many, but not insurmountable [8,10,40,85] and herein. In the early decades of the 21st century, even an approximate estimate of the number of marine invertebrate species that are globally extinct eludes us. Remarkably few scientists study extinctions of marine invertebrates [2], the most speciose group of ocean animals, nor are students typically introduced to the topic as a field of study. Nevertheless, that a notable number of marine invertebrate extinctions has not been documented is not evidence that they have not occurred—or are not now occurring. The study of marine invertebrate extinctions may be rare, but extinctions may not be.

Funding: This research received no external funding.

Acknowledgments: I thank Patrick LaFollette, Paul Valentich-Scott, and Eugene Coan for advice on some molluscan taxa. I am grateful to Thomas Trott for inviting me to contribute to this collection, inspiring me to once again tackle the labyrinthine molasses of the state of our historical marine invertebrate diversity knowledge.

Conflicts of Interest: The author declares no conflict of interest.

References

1. Carlton, J.T. Neoextinctions of marine invertebrates. *Am. Zool.* **1993**, *33*, 499–509. [CrossRef]
2. Carlton, J.T.; Geller, J.B.; Reaka-Kudla, M.L.; Norse, E.A. Historical extinction in the sea. *Ann. Rev. Ecol. Syst.* **1999**, *30*, 515–538. [CrossRef]
3. Carlton, J.T. Global marine extinctions in historical time: What we know and why we don't know (a lot) more. In *Marine Extinctions—Patterns and Processes*; Brand, F., Ed.; CIESM Publisher: Monaco, 2013; pp. 23–29.
4. Raicevich, S.; Fortibuoni, T. Assessing neoextirpations in the Adriatic Sea: An historical ecology approach. In *Marine Extinctions—Patterns and Processes*; CIESM Workshop [Valencia] Monograph 45, Brand, F., Eds.; CIESM Publisher: Monaco, 2013; pp. 97–111.
5. Dulvy, N.K.; Sadovy, Y.; Reynolds, J.D. Extinction vulnerability in marine populations. *Fish Fish.* **2003**, *4*, 25–64. [CrossRef]
6. Dulvy, N.K.; Pinnegar, J.K.; Reynolds, J.D. Holocene extinctions in the sea. In *Holocene Extinctions*; Turveyed, S.T., Ed.; Oxford University Press: Oxford, UK, 2009; pp. 129–150.
7. del Monte-Luna, P.; Lluch-Belda, D.; Serviere-Zaragoza, E.; Carmona, R.; Reyes-Bonilla, H.; Aurioles-Gamboa, D.; Castro-Aguirre, J.L.; del Proo, S.A.G.; Trujillo-Millan, O.; Brook, B.W. Marine extinctions revisited. *Fish Fish.* **2007**, *8*, 107–122. [CrossRef]
8. Roberts, C.M.; Hawkins, J.P. Extinction risk in the sea. *Trends Ecol. Evol.* **1999**, *14*, 241–246. [CrossRef]
9. Harnik, P.; Lotze, H.K.; Anderson, S.C.; Finkel, Z.V.; Finnegan, S.; Lindberg, D.R.; Liow, L.H.; Lockwood, R.; McClain, C.R.; McGuire, J.L.; et al. Extinctions in ancient and modern seas. *Trends Ecol. Evol.* **2012**, *27*, 608–617. [CrossRef] [PubMed]
10. CIESM Workshop [Valencia] Monograph, 45; Brand, F. (Eds.) *Marine Extinctions—Patterns and Processes*; CIESM Publisher: Monaco, 2013; 188p.
11. Thomas, E.A.; Bohm, M.; Pollock, C.; Chen, C.; Seddon, M.; Sigwart, J.D. Assessing the extinction risk of insular, understudied marine species. *Conserv. Biol.* **2021**, *36*, e13854. [CrossRef] [PubMed]
12. Edgar, G.J.; Stuart-Smith, R.D.; Heather, F.J.; Barrett, N.S.; Turak, E.; Sweatman, H.; Emslie, M.J.; Brock, D.J.; Hicks, J.; French, B.; et al. Continent-wide declines in shallow reef life over a decade of ocean warming. *Nature* **2023**, *615*, 858–865. [CrossRef]
13. Rogers, A.D.; Miloslavich, P.; Obura, D.; Aburto-Oropreza, O. Marine Invertebrates. In *The Living Planet. The State of the World's Wildlife*; McLean, N., Ed.; Cambridge University Press: Cambridge, UK, 2023; pp. 249–269.
14. Zubillaga, A.L.; Marquez, L.M.; Croquer, A.; Bastidas, C. Ecological and genetic data indicate recovery of the endangered coral *Acropora palmata* in Los Roques, southern Caribbean. *Coral Reefs* **2008**, *27*, 63–72. [CrossRef]
15. Williams, D.E.; Miller, M.W.; Kramer, K.L. Recruitment failure in Florida Keys *Acropora palmata*, a threatened Caribbean coral. *Coral Reefs* **2008**, *27*, 697–705. [CrossRef]
16. Scheibling, R.E.; Metaxas, A. Mangroves and fringing reefs as nursery habitats for the endangered Caribbean sea star *Oreaster reticulatus*. *Bull. Mar. Sci.* **2010**, *86*, 133–148.
17. Allanson, B.R.; Msizi, S.C. Reproduction and growth of the endangered siphonariid limpet *Siphonaria compressa* (Pulmonata: Basommatophora). *Invertebr. Reprod. Dev.* **2010**, *54*, 151–161. [CrossRef]
18. Reaka, M.L.; Lombardi, S.A. Hotspots on global coral reefs. In *Biodiversity Hotspots*; Zachos, F.E., Habel, J.C., Eds.; Springer: Berlin, Germany, 2011; pp. 471–501.
19. Peters, H.; O'Leary, B.C.; Hawkins, J.P.; Carpenter, K.E.; Roberts, C.M. *Conus*: First comprehensive conservation red list assessment of a marine gastropod mollusc genus. *PLoS ONE* **2013**, *8*, e83353. [CrossRef]
20. Cowie, R.H.; Regnier, C.; Fontaine, B.; Bouchet, P. Measuring the sixth extinction: What do mollusks tell us? *Nautilus* **2017**, *131*, 3–41.
21. MacKenzie, K.; Pert, C. Evidence for the decline and possible extinction of a marine parasite species caused by intensive fishing. *Fish. Res.* **2018**, *198*, 63–65. [CrossRef]
22. Garcia-March, J.R.; Tena, J.; Henandis, S.; Vazquez-Luis, M.; Lopez, D.; Tellez, C.; Prado, P.; Navas, J.I.; Bernal, J.; Catanese, G.; et al. Can we save a marine species affected by a highly infective, highly lethal, T waterborne disease from extinction? *Biol. Conserv.* **2020**, *243*, 108498. [CrossRef]
23. Espinosa, F.; Maestre, M.; Garcia-Gomez, J.C.; Cotaina-Castro, M.I.; Pitarch-Moreno, C.; Paramio, J.M.; Maria, P.F.-S.; Garcia-Estevez, N. Joining technology and biology to solve conservation problems through translocation in the endangered limpet *Patella ferruginea*. *Front. Mar. Sci.* **2023**, *10*, 1100194. [CrossRef]
24. IUCN (International Union for the Conservation of Nature). The IUCN Red List of Threatened Species. 2023. Available online: https://www.iucnredlist.org/ (accessed on 1 April 2023).
25. Martin, T.E.; Bennett, G.C.; Fairbairn, A.; Mooers, A.O. 'Lost' taxa and their conservation implications. *Anim. Conserv.* **2022**, *28*, 14–24. [CrossRef]
26. Hawksworth, D.L.; Cowie, R.H. The discovery of historically extinct, but hitherto undescribed, species: An under-appreciated element in extinction-rate assessments. *Biodivers. Conserv.* **2013**, *22*, 2429–2432. [CrossRef]
27. Tedesco, P.A.; Bigorne, R.; Bogan, A.E.; Giam, X.; Jezequel, C.; Hugueny, B. Estimating how many undescribed species have gone extinct. *Conserv. Biol.* **2014**, *28*, 1360–1370. [CrossRef]
28. Liu, J.; Slik, F.; Zheng, S.; Lindenmayer, D.B. Undescribed species have higher extinction rates than known species. *Conserv. Lett.* **2021**, *15*, e12876.
29. Boehm, M.M.A.; Cronk, Q.C.B. Dark extinction: The problem of unknown historical extinctions. *Biol. Lett.* **2021**, *17*, 20210007. [CrossRef] [PubMed]

30. Anderson, P.K.; Domning, D.P. Steller's sea cow. In *Encyclopedia of Marine Mammals*; Perrin, W.F., Wursig, W., Thewissen, J.G.M., Eds.; Academic Press: New York, NY, USA, 2002; pp. 1178–1185.

31. Holm, G. Collecting *Lottia alveus* (Conrad, 1831) in Boundary Bay, British Columbia. *Dredgings* **2002**, *42*, 1–4.

32. Carlton, J.T.; Vermeij, G.J.; Lindberg, D.R.; Carlton, D.A.; Dudley, E.C. The first historical extinction of a marine invertebrate in an ocean basin: The demise of the eelgrass limpet *Lottia Alveus*. *Biol. Bull.* **1991**, *180*, 72–80. [CrossRef]

33. McLean, J.H. The family Dialidae (Gastropoda: Cerithioidea) in the Eastern Pacific. *Festivus* **2002**, *34*, 93–96.

34. Tryon, G.W., Jr. Description of a new species of *Rissoa*. *Am. J. Conchol.* **1866**, *2*, 12.

35. Clark, K.B. *Phyllaplysia smaragda* (Opisthobranchia: Notarchidae), a new anaspidean from Florida. *Bull. Mar. Sci.* **1977**, *27*, 651–657.

36. Clark, K.B. Rheophilic/oligotrophic lagoonal communities: Through the eyes of slugs (Mollusca: Opisthobranchia). *Bull. Mar. Sci.* **1995**, *57*, 242–251.

37. Culotta, E. Is marine biodiversity at risk? *Science* **1994**, *263*, 918–919. [CrossRef]

38. Rozsa, L.; Vas, Z. Co-extinct and critically co-endangered species of parasitic lice, and conservation-induced extinction: Should lice be reintroduced to their hosts? *Oryx* **2014**, *49*, 107–110. [CrossRef]

39. O'Hara, T.D.; Mah, M.; Hipsley, C.A.; Bribiesca-Contreras, G.; Barrett, N.S. The Derwent River seastar: Re-evaluation of a critically endangered marine invertebrate. *Zool. J. Linn. Soc.* **2018**, *186*, 483–490. [CrossRef]

40. Gravili, C.; Bevilacqua, S.; Terlizzi, A.; Boero, F. Missing species among Mediterranean non-Siphonophora Hydrozoa. *Biodivers. Conserv.* **2015**, *24*, 1329–1357. [CrossRef] [PubMed]

41. Bouillon, J.; Medel, M.D.; Pagés, F.; Gili, J.M.; Boero, F.; Gravili, C. Fauna of the Mediterranean Hydrozoa. *Sci. Mar.* **2004**, *68* (Suppl. S2), 5–438. [CrossRef]

42. Batistic, M.; Garic, R. The case of *Bougainvillia triestina* Hartlaub 1911 (Hydrozoa, Cnidaria): A 100-year long struggle for recognition. *Mar. Ecol.* **2016**, *37*, 145–154. [CrossRef]

43. Schuchert, P. The European athecate hydroids and their medusae (Hydrozoa, Cnidaria): Capitata Part 2. *Rev. Suisse Zool.* **2010**, *117*, 337–555. [CrossRef]

44. Cinar, M.E.; Yokes, M.B.; Acik, S.; Bakir, A.K. Checklist of Cnidaria and Ctenophora from the coasts of Turkey. *Turk. J. Zool.* **2014**, *38*, 677–697. [CrossRef]

45. Garcia, E.F. *Conus sauros*, a new *Conus* species (Gastropoda: Conidae) from the Gulf of Mexico. *Novapex* **2016**, *7*, 71–76.

46. Tenorio, M.J.; Abalde, S.; Pardos-Blas, J.R.; Zardoya, R. Taxonomic revision of West African cone snails (Gastropoda: Conidae) based upon mitogenomic studies: Implications for conservation. *Eur. J. Taxon.* **2020**, *663*, 1–89. [CrossRef]

47. Rolan, E. Descripcion de nuevas especies y subspecies del genero *Conus* (Mollusca, Neogastropoda) para el Archipelago de Cabo Verde. *Iberus* **1990**, *2*, 5–70.

48. Singleton, J.F. Cone News from Australia—9. Extant or extinct? *Cone Collect.* **2007**, *4*, 5–6.

49. Marshall, B. *Conus colmani* Rockel & Korn, 1990. 2022. Available online: https://www.marinespecies.org/aphia.php?p=taxdetails&id=426461 (accessed on 1 April 2023).

50. Wells, F.E. Appendix 4. List of molluscs recorded at Milne Bay, Papua New Guinea during RAP surveys. In *A Rapid Marine Biodiversity Assessment of Milne Bay Province, Papua New Guinea—Survey II (2000)*; Allen, G.R., Kinch, J.P., McKenna, S.A., Seeto, P., Eds.; RAP Bulletin of Biological Assessment 29; Conservation International: Washington, DC, USA, 2003; pp. 96–100.

51. Dong, Y.; Huang, X.; Reid, D.G. Rediscovery of one of the very few 'unequivocally extinct' species of marine molluscs: *Littoraria flammea* (Philippi, 1847) lost, found—and lost again? *J. Molluscan Stud.* **2015**, *81*, 313–321. [CrossRef]

52. Powell, C.L., II. The extinct limpet *Lottia edmitchelli* (Lipps, 1963) from the southern California bight, U.S.A. *PaleoBios* **2022**, *39*, 1–7. [CrossRef]

53. Kemp, R.; Peters, H.; Allcock, L.; Carpenter, K.; Obura, D.; Polidoro, B.; Richman, N. Chapter 3. Marine invertebrate life. In *Spineless. Status and Trends of the World's Invertebrates*; Collen, B., Bohm, M., Kemp, R., Baillie, J.E.M., Eds.; Zoological Society of London: London, UK, 2012; pp. 34–44.

54. Florens, F.B.V.; Baider, C. Relocation of *Omphalotropis plicosa* (Pfeiffer, 1852), a Mauritian endemic landsnail believed extinct. *J. Mollusc. Stud.* **2007**, *73*, 205–206. [CrossRef]

55. Kavanaugh, D.H.; Erwin, T.L. Extinct or extant? A new species of intertidal bembidiine from Palos Verdes Peninsula, California. *Coleopt. Bull.* **1992**, *46*, 311–320.

56. Caterino, M.S.; Caterino, K.J.; Maddison, D.R. Extant! Living *Bembidion palosverdes* Kavanaugh and Erwin (Coleoptera: Carabidae) found on Santa Catalina Island, California. *Coleopt. Bull.* **2015**, *69*, 410–411. [CrossRef]

57. Webb, T.J.; Mindel, B.L. Global patterns of extinction risk in marine and non-marine systems. *Curr. Biol.* **2015**, *25*, 506–511. [CrossRef]

58. Jackson, J. What was natural in the coastal oceans? *Proc. Natl. Acad. Sci. USA* **2001**, *98*, 5411–5418. [CrossRef]

59. Bartsch, I.; Tittley, I. The rocky intertidal biotopes of Helgoland: Present and past. *Helgol. Mar. Res.* **2004**, *58*, 289–302. [CrossRef]

60. Steinbeck, J.R.; Schiel, D.R.; Foster, M.S. Detecting long-term change in complex communities: A case study from the rocky intertidal zone. *Ecol. Appl.* **2005**, *15*, 1813–1832. [CrossRef]

61. Smith, T.B.; Purcell, J.; Barimo, J.F. The rocky intertidal biota of the Florida keys: Fifty-two years of change after Stephenson and Stephenson (1950). *Bull. Mar. Sci.* **2007**, *80*, 1–19.

62. Lotze, H.K. Historical reconstruction of human-induced changes in U.S. estuaries. *Ann. Rev. Ocean. Mar. Biol.* **2010**, *48*, 267–338.

63. Lotze, H.K.; McClenachan, L. Marine historical ecology: Informing the future by learning from the past. In *Marine Community Ecology and Conservation*; Bertness, M.D., Bruno, J.F., Silliman, B.R., Stachowicz, J.J., Eds.; Sinauer: Sunderland, MA, USA, 2013; pp. 165–200.

64. Gatti, G.; Bianchi, C.N.; Parravicini, V.; Rovere, A.; Peirano, A.; Montefalcone, M.; Massa, F.; Morri, C. Ecological change, sliding baselines and the importance of historical data: Lessons from combing observational and quantitative data on a temperate reef over 70 years. *PLoS ONE* **2015**, *10*, e0118581. [CrossRef] [PubMed]

65. Trott, T.J. Century-scale species incidence, rareness and turnover in a high-diversity Northwest Atlantic coastal embayment. *Mar. Biodivers.* **2015**, *46*, 33–49. [CrossRef]

66. Houbrick, R.S. Monograph of the genus *Cerithium* Bruguiere in the Indo-Pacific (Cerithiidae: Prosobranchia). *Smithson. Contrib. Zool.* **1992**, *510*, 211. [CrossRef]

67. Higo, S.; Callomon, P.; Goto, Y. *Catalogue and Bibliography of the Marine Shell-Bearing Mollusca of Japan*; Elle Scientific Publications: Yao, Japan, 1999; 938p.

68. Bastida-Zavala, J.R.; Buelina, A.S.R.; de Leon-Gonzalez, J.A.; Camacho-Cruz, K.A.; Camacho-Cruz, I.; Carmona, I. New records of sabellids and serpulids (Polychaeta: Sabellidae, Serpulidae) from the Tropical Eastern Pacific. *Zootaxa* **2016**, *4184*, 401–457. [CrossRef]

69. Mendonca, L.M.C.; Guimaraes, C.R.P.; Haddad, M.A. Taxonomy and diversity of hydroids (Cnidaria: Hydrozoa) of Sergipe, Northeast Brazil. *Zoologia* **2022**, *39*, e21032. [CrossRef]

70. Newman, W.A.; Ross, A. Revision of the balanomorph barnacles; including a catalog of the species. *San Diego Soc. Nat. Hist. Mem.* **1976**, *9*, 1–108.

71. Ng, P.K.L.; Guinot, D.; Davie, P.J.F. Systema Brachyurorum: Part I. An annotated checklist of extant brachyuran crabs of the world. *Raffles Bull. Zool.* **2008**, *17*, 1–286.

72. Bieler, R.; Petit, R.E. Catalogue of Recent and fossil "worm-snail" taxa of the families Vermetidae, Siliquariidae, and Turitellidae (Mollusca: Caenogastropoda). *Zootaxa* **2011**, *2948*, 1–103. [CrossRef]

73. Fautin, D.G. Catalog to families, genera, and species of orders Actiniaria and Corallimorpharia (Cnidaria: Anthozoa). *Zootaxa* **2016**, *4145*, 1–449. [CrossRef]

74. Bouchet, P.; Strong, E.E. Historical name-bearing types in marine molluscs: An impediment to biodiversity studies. In *Systema Naturae*; Polaszek, A., Ed.; CRC Press: London, UK, 2010; Volume 250, pp. 63–74.

75. Coan, E.V.; Valentich-Scott, P. *Bivalve Seashells of tropical West America: Marine Bivalve Mollusks from Baja California to Northern Peru*; Parts 1 and 2; Santa Barbara Museum of Natural History: Santa Barbara, CA, USA, 2012; 1258p.

76. Keen, A.M. *Seashells of Tropical West America*; Stanford University Press: Stanford, CA, USA, 1971; 1064p.

77. Cardoso, F.; Paredes, C.; Mogollon, V.; Palacios, E. La familia Chamidae (Bivalvia: Veneridae) en Peru, con la adicion de cinco nuevos registros. *Rev. Peru. Biol.* **2016**, *23*, 13–16. [CrossRef]

78. Huber, M. *Compendium of Bivalves*; Conch Books: Hackenheim, Germany, 2010; 904p.

79. Huber, M. *Compendium of Bivalves*, 2nd ed.; Conch Books: Hackenheim, Germany, 2015; 907p.

80. Reeve, L.A. Monograph of the genus *Chama*. In *Conchologia Iconica, or, Illustrations of the Shells of Molluscous Animals*; L. Reeve & Co.: London, UK, 1846; Volume 4.

81. Bouchet, P. *Chama producta* Broderip, 1835. 2023. Available online: https://www.marinespecies.org/aphia.php?p=taxdetails&id=538721 (accessed on 1 April 2023).

82. Bouchet, P. *Chama limbula* Lamarck, 1819. 2023. Available online: https://www.marinespecies.org/aphia.php?p=taxdetails&id=208495 (accessed on 1 April 2023).

83. Spalding, M.D.; Fox, H.E.; Allen, G.R.; Davidson, N.; Ferdana, Z.A.; Finlayson, M.; Halpern, B.S.; Jorge, M.A.; Lombana, A.; Lourie, S.A.; et al. Marine ecoregions of the world: A bioregionalization of coastal and shelf areas. *BioScience* **2007**, *57*, 573–583. [CrossRef]

84. Newman, W.A. Californian transition zone: Significance of short-range endemics. In *Historical Biogeography, Plate Tectonics and the Changing Environment*; Gray, J., Boucot, A.J., Eds.; Oregon State university Press: Corvallis, OR, USA, 1979; pp. 399–416.

85. Cowie, R.H.; Bouchet, P.; Fontaine, B. The sixth mass extinction: Fact, fiction or speculation? *Biol. Rev.* **2022**, *97*, 640–663. [CrossRef]

86. Anker, A.; De Grave, S. Rediscovery and range extension of the rare Caribbean alpheid shrimp, *Prionalpheus gomezi* (Crustacea: Decapoda: Alpheidae). *Mar. Biodivers. Rec.* **2012**, *5*, e107. [CrossRef]

87. Swee-Cheng, L.; Tun, K.; Goh, E. Rediscovery of the Neptune's cup sponge in Singapore: *Cliona* or *Poterion*? *Contrib. Mar. Sci.* **2012**, *2012*, 49–56.

88. Cunha, C.M.; Saad, L.O.; Lima, P.O.V. Rediscovery of Brazilian corambids (Gastropoda: Onchidorididae). *Mar. Biodivers. Rec.* **2014**, *7*, e14. [CrossRef]

89. Galea, H.R.; Di Camillo, C.G. Rediscovery and redescription of *Gymnangium sibogae* (Billard, 1913) (Cnidaria: Hydrozoa; Aglaopheniidae). *Mar. Biodivers.* **2017**, *47*, 847–857. [CrossRef]

90. Christensen, C.C. A lost species of salt marsh snail: *Blauneria gracilis* Pease, 1860 (Gastropoda: Ellobiidae) in the Hawaiian Islands. *Bishop Mus. Occas. Pap.* **2018**, *123*, 11–17.

91. Goto, R.; Ishikawa, H. An unusual habitat for bivalves: Rediscovery of the enigmatic commensal clam *Sagamiscintilla thalasemicola* (Habe, 1962) (Bivalvia: Galeommatoidea) from spoon worm's spoon. *Mar. Biodivers.* **2019**, *49*, 1553–1558. [CrossRef]

92. Valentich-Scott, P.; Goddard, J.H.R. A fossil species living off southern California, with notes on the genus *Cymatioa* (Mollusca, Bivalvia, Galeommatoidea). *ZooKeys* **2022**, *1128*, 53–62. [CrossRef] [PubMed]
93. Marcus, E.; Marcus, E. Opisthobranchs from American Atlantic waters. *Bull. Mar. Sci. Gulf Caribb.* **1960**, *10*, 129–203.
94. Carlton, J.T. Deep invasion ecology and the assembly of communities in historical time. In *Biological Invasions in Marine Ecosystems*; Rilov, G., Crooks, J.A., Eds.; Springer: Berlin, Germany, 2009; pp. 13–56.
95. Benovic, A.; Justic, D.; Bender, A. Enigmatic changes in the hydromedusan fauna of the northern Adriatic Sea. *Nature* **1987**, *326*, 597–600. [CrossRef]
96. Boero, F.; Carlton, J.T.; Briand, F.; Kiessling, W.; Chenuil, A.; Voultsiadou, E.; Twitchett, R.; Soldo, A.; Panigada, S.; Juan-Jorda, M.J.; et al. Executive Summary. In *Marine Extinctions—Patterns and Processes*; CIESM Workshop Monograph, 45, Brand, F., Eds.; CIESM Publisher: Monaco, 2013; pp. 5–19.
97. Taylor, D.W. Freshwater mollusks of California: A distributional checklist. *Calif. Fish Game* **1981**, *67*, 140–163.
98. Goddard, J.H.R.; Schaefer, M.C.; Hoover, C.; Valdes, A. Regional extinction of a conspicuous dorid nudibranch (Mollusca: Gastropoda) in California. *Mar. Biol.* **2013**, *160*, 1497–1510. [CrossRef]
99. Chapman, J.W.; Dumbauld, B.R.; Itani, G.; Markham, J.C. An introduced Asian parasite threatens northeastern Pacific estuarine ecosystems. *Biol. Invasions* **2011**, *14*, 1221–1236. [CrossRef]
100. Smith, C.R.; Glover, A.G.; Treude, T.; Higgs, N.D.; Amon, D.J. Whale-fall ecosystems: Recent insights into ecology, paleoecology, and evolution. *Ann. Rev. Mar. Sci.* **2015**, *7*, 571–596. [CrossRef] [PubMed]
101. Levin, L.A.; Alfaro-Lucas, J.M.; Colaco, A.; Cordes, E.E.; Craik, N.; Danovaro, R.; Hoving, H.-J.; Ingels, J.; Mestre, N.C.; Seabrook, S.; et al. Deep-sea impacts of climate interventions. *Science* **2023**, *379*, 978–981. [CrossRef] [PubMed]
102. Marshall, C.R. Confidence intervals on stratigraphic ranges. *Paleobiology* **1990**, *16*, 1–10. [CrossRef]
103. Ray, G.C. Ecological diversity in coastal zones and oceans. In *Biodiversity*; Wilson, E.O., Peters, F.M., Eds.; National Academy Press: Washington, DC, USA, 1988; pp. 37–50.
104. Regnier, C.; Fontaine, B.; Bouchet, P. Not knowing, not recording, not listing: Numerous unnoticed mollusk extinctions. *Conserv. Biol.* **2009**, *23*, 1214–1221. [CrossRef]

MDPI
St. Alban-Anlage 66
4052 Basel
Switzerland
www.mdpi.com

Diversity Editorial Office
E-mail: diversity@mdpi.com
www.mdpi.com/journal/diversity